Medicinal Plant Biotechnology

Edited by
Oliver Kayser and Wim J. Quax

1807–2007 Knowledge for Generations

Each generation has its unique needs and aspirations. When Charles Wiley first opened his small printing shop in lower Manhattan in 1807, it was a generation of boundless potential searching for an identity. And we were there, helping to define a new American literary tradition. Over half a century later, in the midst of the Second Industrial Revolution, it was a generation focused on building the future. Once again, we were there, supplying the critical scientific, technical, and engineering knowledge that helped frame the world. Throughout the 20th Century, and into the new millennium, nations began to reach out beyond their own borders and a new international community was born. Wiley was there, expanding its operations around the world to enable a global exchange of ideas, opinions, and know-how.

For 200 years, Wiley has been an integral part of each generation's journey, enabling the flow of information and understanding necessary to meet their needs and fulfill their aspirations. Today, bold new technologies are changing the way we live and learn. Wiley will be there, providing you the must-have knowledge you need to imagine new worlds, new possibilities, and new opportunities.

Generations come and go, but you can always count on Wiley to provide you the knowledge you need, when and where you need it!

William J. Pesce
President and Chief Executive Officer

Peter Booth Wiley
Chairman of the Board

Medicinal Plant Biotechnology

From Basic Research to Industrial Applications

Volume I

Edited by
Oliver Kayser and Wim J. Quax

WILEY-VCH Verlag GmbH & Co. KGaA

The Editors

Prof. Dr. Oliver Kayser
Rijksuniversiteit Groningen
Pharmaceutical Biology
Antonius Deusinglaan 1
9713 AV Groningen
The Netherlands

Prof. Dr. Wim J. Quax
Rijksuniversiteit Groningen
Pharmaceutical Biology
Antonius Deusinglaan 1
9713 AV Groningen
The Netherlands

1st Edition 2007
1st Reprint 2009

Library of Congress Card No.: applied for

British Library Cataloguing-in-Publication Data
A catalogue record for this book is available from the British Library.

Bibliographic information published by the Deutsche Nationalbibliothek
The Deutsche Nationalbibliothek lists this publication in the Deutsche Nationalbibliografie; detailed bibliographic data are available in the Internet at http://dnb.d-nb.de.

Composition   Detzner Fotosatz, Speyer
Printing   betz-druck GmbH, Darmstadt
Bookbinding   Litges & Dopf Buchbinderei GmbH, Heppenheim
Cover   Adam Design, Weinheim
Anniversary Logo Design   Richard Pacifico

Printed in the Federal Republic of Germany
Printed on acid-free paper

ISBN: 978-3-527-31443-0

Contents

Medicinal Plant Biotechnology. From Basic Research to Industrial Applications
Edited by Oliver Kayser and Wim J. Quax
Copyright © 2007 WILEY-VCH Verlag GmbH & Co. KGaA, Weinheim
ISBN 978-3-527-31443-0

Preface

Medicinal plant biotechnology is a discipline that should be well known and accepted as a sub field under the umbrella of pharmaceutical biotechnology. However, to our surprise we could not identify any recent contribution dealing with medicinal plants and biotechnological techniques on a high comprehensive level that the subject would deserve today. We accepted this lack as a challenge to sort out the needs of the scientific community and to identify valuable topics that must be addressed in order to provide a narrow and well-defined picture of medicinal plant biotechnology.

We identified several experts in the fields of pharmaceutical biology, biotechnology, biochemistry and genetics who were willing to spend their precious time to share their knowledge with us. Many thanks here from the editors, as without the enthusiasm, efforts, and patience of these experts this book would not have been possible.

Biotechnology in general is a fast-moving area, and this development can be recognized in our field of pharmaceutical biotechnology. In this book we have focused on medicinal plants, and have attempted to structure the latest developments into three parts. In Part I, the actual status of medicinal pharmaceutical biotechnology and the cell as a producing unit are outlined. Here, the major questions are: How do medicinal plants and biotechnology fit together, and how must we understand the plant cell as a "biofactory" that can be used in an integrative drug discovery process? It should be mentioned that endophytes as plant-related microorganisms are discussed in Part I because of the upcoming interest in this new genetic and natural product source for the future. Today, although research into endophytes is very much in its early stages, the first reports on the possibilities to use them to produce formerly plant-originated compounds make this group of organisms interesting for both academia and industry.

Whilst academic and industrial needs are different, all authors were able to provide answers for these groups, both of which are confronted with new challenges such as metabolomics, high-throughput screening, and the application of the latest recombinant DNA technologies. Whether these techniques will be transferred successfully into industrial applications is not clear, but the authors in Part I provide an outlook into this exciting future of pharmaceutical biotechnology.

Medicinal Plant Biotechnology. From Basic Research to Industrial Applications
Edited by Oliver Kayser and Wim J. Quax
Copyright © 2007 WILEY-VCH Verlag GmbH & Co. KGaA, Weinheim
ISBN 978-3-527-31443-0

In Part II, we go a step further from the well-organized cell to the nanocosmos, with special focus on genetics and molecular biology. In this part of the book, strategies are discussed to accelerate the drug discovery process based on genetic techniques such as micropropagation, combinatorial biosynthesis, and expressed sequence tag databases. Considerable attention has been paid to optimize the production of natural products of pharmaceutical relevance in cell cultures, and how production can be scaled up in bioreactors. Combinatorial biosynthesis and ways to modify physiologic traits by constructing transgenic plants will provide an idea of future techniques in natural product production. Some of these products are in the later stages of development, and we hope that we might be able to read about them in the next updated edition of this book, in Part III.

Part III deals with the future directions of medicinal plant biotechnology and examines its practical applications in industry. Authors from industry provide insight into their own production facilities, and discuss the problems and limitations. Progress here has, however, been slower than with medical and other areas of research. Because plants are genetically and physiologically more complex than single-cell organisms such as bacteria and yeasts, the necessary technologies in industrial business are developing more slowly. However, exploring ways to use genetic modification better will definitely influence this area, and therefore we must accept that our attempt to structure the latest developments was only an *attempt*, and that medicinal plant biotechnology must be considered as a complex and integrative discipline.

Our special thanks go to Steffen Pauly and others of the Wiley-VCH publishing team for their professionalism, continuous encouragement and support, which we enjoyed not for the first time in our sustainable relationship.

Special thanks also to the families of both the editors and the authors for their patience and understanding why time was spent on this project, and why they had to tolerate extended periods of negligence. We are not so naïve to believe that dedicating this book to them will compensate the missed time, but it might be a start!

We have no doubt that this book is far from complete and that some areas of interest were not touched and will have to be discussed in the future, in updated editions. However, we are convinced that we were able to provide a good "primer" to start working in medicinal biotechnology and to show how exciting the combination of medicinal plants and biotechnology can be.

Oliver Kayser   *Groningen, September 2006*
Wim J. Quax

Foreword

It is a pleasure for me to write the foreword to this unusual and excellent book on the engineering of medicinal plants! This collection of relevant articles provides a comprehensive overview of the *status quo* of gene technology and plant biotechnology with medicinal plants. The focus is twofold. First, it not only presents a comprehensive overview on the use of medicinal plants in the pharmaceutical industry and their engineering to increase yields of small molecule drugs, but also highlights the challenges to patient and consumer acceptance that surround the use of these genetically modified plants as a source of extract preparation and manufacturing.

Second, the book deals with progress made to date in engineering plants as expression hosts used to produce biopharmaceuticals – that is, protein-based drugs for human use from genetically modified plants. This topic is of utmost importance, as biopharmaceuticals are currently the mainstay products of the biotechnology market and represent the fastest-growing and, in many ways, the most exciting sector within the pharmaceutical industry.

The term "biopharmaceutical" was coined during the 1980s, when a general consensus evolved that it represented a class of therapeutics produced by means of modern biotechnologies. The recombinant DNA technology of Cohen and Boyer enabled them, in 1978, to generate the first commercial product: human insulin expressed in *E. coli*. These efforts also led to the first biotech company: on October 14th, 1980 "Genentech" went public on New York Wall Street stock exchange. Fascination about this modern biopharmaceutical and the huge potential of the new biotechnology made the stock price jump from 35 to 89 US$ in the first 20 minutes: by the evening of the same day, the market capitalization was 66 million US$!

Since then, the market for biopharmaceuticals has steadily grown, and currently almost 150 biopharmaceuticals have gained approval for general human use (EU and USA). Over this period it became clear that production capacities for biopharmaceuticals with "conventional" bioreactors would be a bottleneck, and that worldwide fermentation capacities are limited. One exciting solution to these "capacity crunches" is the use of transgenic plants to produce biopharmaceuticals.

Basically this highly innovative, but relatively new approach, is the blending of *green biotechnology* (using genetically modified plants as expression hosts) and *red biotechnology* (the biotechnological production of pharmaceuticals).

Medicinal Plant Biotechnology. From Basic Research to Industrial Applications
Edited by Oliver Kayser and Wim J. Quax
Copyright © 2007 WILEY-VCH Verlag GmbH & Co. KGaA, Weinheim
ISBN 978-3-527-31443-0

The towering German writer and natural scientist Johann Wolfgang Goethe (1749–1832) was doing research on the blending of colors as published in his scientific treatise *"Farbenlehre"* and also the English scientist and mathematician Sir Isaac Newton (1642–1727) developed a tool for color mixing, the *"color wheel"*. Light can be split up into a spectrum by sending it through a prism, and everybody admires probably the most exciting atmospheric phenomenon: rainbows caused by refraction of sunlight in raindrops. In a rainbow we see the continuous range of spectral colors and the color theory is basically a convenient tool for predicting the results of simply mixing these colors. According to that, additive color mixing of red and green is yellow, and by blocking the sunlight, the shade of the color yellow is gold – a pot of gold at the end of the biotechnology rainbow. As a homage to these brilliant scientists I would call the blending of *green* and *red biotechnology* *"Golden Biotech"*.

One important example (although not a pharmaceutical overexpressed in a plant for isolating the drug), is Golden Rice: a transgenic rice with increased content of pro-vitamin A (β-carotene) – and thus its nutritional value. The name stems from its *real* golden color, and the shiny effect is due to the high β-carotene content. The inventors Peter Beyer and Ingo Potrykus † were awarded the prestigious *nature biotech award* early 2006, "for a potential solution to eliminate the largescale vitamin A deficiency existing among the poor in countries such as India and to fight against blindness (at the beginning of the 21st century, 124 million people were estimated to be affected by vitamin A deficiency, which is responsible for 2 million deaths, and 500,000 cases of irreversible blindness annually.)"

This book nicely describes different approaches of *pink biotechnology*, its application for plant expression systems, and their advantages and limitations, and concludes by considering some of the innovations and trends likely to influence the future of engineering of medicinal plants.

Plants are by far the most abundant and cost-effective renewable resource uniquely adapted to complex biochemical synthesis. The increasing cost of energy and chemical raw materials, combined with the environmental concerns associated with conventional pharmaceutical manufacturing, will make plants even more compatible in the future. With the words of Max Planck (1858–1947) "How far advanced Man's scientific knowledge may be, when confronted with Nature's immeasurable richness and capacity for constant renewal, he will be like a marveling child and must always be prepared for new surprises," we will definitely discover more fascinating features of plant expression systems. But there is no need to wait: combining the advantages of some technologies that we have in hand by now could already lead to the ultimate plant expression system. This is what we should focus on, because, then, at the dawn of this new millennium, this would for the first time yield large-enough amounts of biopharmaceuticals to treat everybody on our planet!

An unusual feature of *Medicinal Plant Biotechnology* is that, for a book with so many facts, it is a delight to read. While easy to read, it is a guide to both, broad surveys and key papers, which are provided in convenient, but at the same time comprehensive reference lists.

I am convinced that the editors have done a great job in compiling a cutting-edge and comprehensive book on the current status of the use of medicinal plants, its genetically modifications and implications thereof. I wish this book a numerous and broad readership, and I encourage all readers to enjoy this collection of interesting contributions from scientists from academia and industry.

Jörg Knäblein

Head Microbiological Chemistry, Schering AG
Scientific Advisor and Board Member European Association of Pharma Biotech

Berlin, June 2006

List of Contributors

Friedrich Altmann
Institute of Chemistry
University of Natural Resources and
Applied Life Sciences
Muthgasse 18
1190 Vienna
Austria

Takashi Asano
Graduate School of Pharmaceutical
Sciences
Chiba University
Yayoi-cho 1-33, Inage-ku
Chiba 263-8522
Japan

Ashish Baldi
Department of Biochemical
Engineering and Biotechnology
Indian Institute of Technology – Delhi
Hauz Khas
New Delhi 110016
India

V.S. Bisaria
Department of Biochemical
Engineering and Biotechnology
Indian Institute of Technology – Delhi
Hauz Khas
New Delhi 110016
India

Donald P. Briskin
Departments of Natural Resources and
Plant Biology
University of Illinois
1101 West Peabody Drive
Urbana, IL 61801
USA

Young Hae Choi
Department of Pharmacognosy
Section Metabolomics, IBL
Leiden University
PO Box 9502
2300 RA Leiden
The Netherlands

Didier Courtois
Centre R & D Nestlé Tours
101, Avenue Gustave Eiffel
37097 Tours cedex 2
France

Birgit Dräger
Institute of Pharmacy
Martin-Luther University
Halle-Wittenberg
Hoher Weg 8
06120 Halle/Saale
Germany

Medicinal Plant Biotechnology. From Basic Research to Industrial Applications
Edited by Oliver Kayser and Wim J. Quax
Copyright © 2007 WILEY-VCH Verlag GmbH & Co. KGaA, Weinheim
ISBN 978-3-527-31443-0

Natalia Dudareva
Deptartment of Horticulture and
Landscape Architecture
Purdue University
West Lafayette, IN 47907
USA

Peter J. Facchini
Department of Biological Sciences
University of Calgary
2500 University Drive N.W.
Calgary, Alberta, T2N 1N4
Canada

Rainer Fischer
Fraunhofer Institute for Molecular
Biology and Applied Ecology
Forckenbckstr. 6
52074 Aachen
Germany

Gilbert Gorr
greenovation Biotech GmbH
Boetzinger Str. 29 b
79111 Freiburg
Germany

Jillian M. Hagel
Department of Biological Sciences
University of Calgary
2500 University Drive N.W.
Calgary, Alberta, T2N 1N4
Canada

Jerzy W. Jaroszewski
Department of Medicinal Chemistry
The Danish University of
Pharmaceutical Sciences
Universitetsparken 2
2100 Copenhagen
Denmark

Matthys K. Julsing
Pharmaceutical Biology
University of Groningen
A. Deusinglaan 1
9713 AV Groningen
The Netherlands

Oliver Kayser
Pharmaceutical Biology
University of Groningen
A. Deusinglaan 1
9713 AV Groningen
The Netherlands

Hye Kyong Kim
Department of Pharmacognosy
Section Metabolomics, IBL
Leiden University
PO Box 9502
2300 RA Leiden
The Netherlands

Wolfgang Kreis
Institute for Biology
Friedrich-Alexander University
Erlangen-Nürnberg
Staudtstr. 5
91058 Erlangen
Germany

Maja Lambert
Department of Medicinal Chemistry
The Danish University of
Pharmaceutical Sciences
Universitetsparken 2
2100 Copenhagen
Denmark

Efraim Lewinsohn
Department of Vegetable Crops
Newe Yaar Research Center
Agricultural Research Organization
30095 Ramat Yishay
Israel

Chunzhao Liu
National Key Laboratory of
Biochemical Engineering
Institute of Process Engineering
Chinese Academy of Sciences
1 Zhongguancun Bei-er-tiao
Beijing 100080
China

Erin Marasco
Department of Biochemistry
Molecular Biology and Biophysics
University of Minnesota
1479 Gortner Avenue
St. Paul, MN 55108
USA

Dinesh A. Nagegowda
Department of Horticulture and
Landscape Architecture
Purdue University
West Lafayette, IN 47907
USA

Hilde Nybom-Balsgard
Department of Crop Science
Swedish University of Agricultural
Sciences
Fjälkestadsvägen 459
291 94 Kristianstad
Sweden

Jonathan E. Page
National Research Council of Canada
Plant Biotechnology Institute
110 Gymnasium Place
Saskatoon, Saskatchewan, S7N 0W9
Canada

Friedrich Pank
Institute of Horticultural Crops
Federal Centre of Breeding Research
on Cultivated Plants
Neuer Weg 22/23
06484 Quedlinburg
Germany

Wim J. Quax
Pharmaceutical Biology
University of Groningen
A. Deusinglaan 1
9713 AV Groningen
The Netherlands

Kazuki Saito
Graduate School of Pharmaceutical
Sciences
Chiba University
Yayoi-cho 1-33, Inage-ku
Chiba 263-8522
Japan

Clare Salisbury
Gowling Lafleur Henderson LLP
Suite 2300, Four Bentall Center
1055 Dunsmuir Street
Vancouver BC V7X 1J1
Canada

Stefan Schillberg
Fraunhofer Institute for Molecular
Biology and Applied Ecology
Forckenbeckstr. 6
52074 Aachen
Germany

Claudia Schmidt-Dannert
Department of Biochemistry
Molecular Biology and Biophysics
University of Minnesota
1479 Gortner Avenue
St. Paul, MN 55108
USA

Konrad Sechley
Gowling Lafleur Henderson LLP
Suite 2300, Four Bentall Centre
1055 Dunsmuir St.
Vancouver, BC V7X 1J1
Canada

Christine Sohier
Institut Henri Beaufour –
Groupe IPSEN
c/o Centre R & D Nestlé Tours
101, Avenue Gustave Eiffel
37097 Tours cedex 2
France

A.K. Srivastava
Department of Biochemical
Engineering and Biotechnology
Indian Institute of Technology – Delhi
Hauz Khas
New Delhi 110016
India

Dan Stærk
Department of Medicinal Chemistry
The Danish University of
Pharmaceutical Sciences
Universitetsparken 2
2100 Copenhagen
Denmark

Gary Strobel
Department of Plant Sciences
Montana State University
Bozeman, MT 59717
USA

Hiroshi Sudo
Graduate School of Pharmaceutical
Sciences
Chiba University
Yayoi-cho 1-33, Inage-ku
Chiba 263-8522
Japan

Homare Tabata
Plnt Cell Culture R & D Center
Hokkaido Mitsui Chemicals, Inc.
1, Toyonuma, Sunagawa-shi
Hokkaido 073-0138
Japan

Ya'akov Tadmor
Department of Vegetable Crops
Newe Yaar Research Center
Agricultural Research Organization
PO Box 1021
Ramat Yishay 30095
Israel

Hsin-Sheng Tsay
Institute of Biotechnology
Chaoyang University of Technology,
168, Gifong E. Road
Wufong
Taichung County 41349
Taiwan

Richard M. Twyman
Department of Biology
University of York
Heslington
York, Y010 5HH
United Kingdom

Mulabagal Vanisree
Institute of Biotechnology
Chaoyang University of Technology
168, Jifong E. Road
Wufong
Taichung County 41349
Taiwan

Robert Verpoorte
Department of Pharmacognosy
Section Metabolomics, IBL
Leiden University
PO Box 9502
2300 RA Leiden
The Netherlands

Kurt Weising
Plant Molecular Systematics
Institute of Biology
University of Kassel
34109 Kassel
Germany

Michael Wink
Institute of Pharmacy and Molecular
Biotechnology (I)PMB)
University of Heidelberg
Im Neuenheimer Feld 364
69120 Heidelberg
Germany

Mami Yamazaki
Graduate School of Pharmaceutical
Sciences
Chiba University
Yayoi-cho 1-33, Inage-ku
Chiba 263-8522
Japan

Yan Zhao
National Key Laboratory of
Biochemical Engineering
Institute of Process Engineering
Chinese Academy of Sciences
1 Zhongguancun Bei-er-tiao
Beijing 100080
China

Color Plates

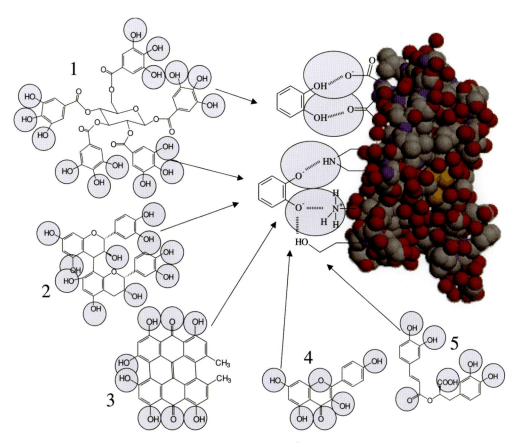

Fig. 6.6 Noncovalent modifications of proteins by secondary metabolites.
1, Pentagalloylglucose (gallotannin). 2, Dimeric procyanidin B4 (catechol
tannin). 3, Hypericine (dimeric anthraquinone). 4, Kaempherol (flavonoid).
5, Rosmarinic acid (phenylpropanoid).

Medicinal Plant Biotechnology. From Basic Research to Industrial Applications
Edited by Oliver Kayser and Wim J. Quax
Copyright © 2007 WILEY-VCH Verlag GmbH & Co. KGaA, Weinheim
ISBN 978-3-527-31443-0

Fig. 6.7 Specific interactions (ligand–receptor relationships) of secondary metabolites with proteins. 1, Canavaline (non-protein amino acid). 2, Hyoscyamine (tropane alkaloid). 3, Lupanine (quinolizidine alkaloid). 4, Physostigmine (indole alkaloid). 5, Vinblastine (dimeric monoterpene indole alkaloid). 6, Podophyllotoxin (lignan). 7, Atractyloside (diterpenes). 8, Ouabain (cardiac glycoside). 9, 12-Tetradecanoyl-phorbol-13-acetate (TPA; phorbolester). 10, Salicylic acid (phenolic acid).

Fig. 12.1 Somatic embryo-derived tubers of *Corydalis yanhusuo* formed after 6 months of culture on: (A) half-strength MS medium supplemented with 0.1 mg L^{-1} GA$_3$; and (B) half-strength MS medium supplemented with 0.51 mg L^{-1} paclobutrazol. (Scale bars: A = 9.17 mm; B = 6.31 mm.)

Fig. 12.2 Asymbiotic seed germination and *ex-vitro* establishment of plants of
Dendrobium tosaense. (A) Optimum seedling growth after 5 months on MS
basal medium + 8% banana homogenate + 1.5% sucrose. (B) Hardened
plants after 6 months in the greenhouse. (Scale bars: A = 0.82 cm; B = 6 cm.)

Fig. 12.3 (A) Induction of *Gentiana* callus from stem segments cultured
on MS basal medium with 3% sucrose, 1 mg L^{-1} α-naphthalene acetic acid
(NAA), and 0.2 mg L^{-1} kinetin for 8 weeks. (B) Established suspension
cultures from the stem-derived cells.

Fig. 12.4 Induction and multiplication of multiple shoots in the nodal explants of *Polygonum multiflorum* Thunb. Nodal explants cultured for 6 weeks on MS basal medium with 3% sucrose, 1% Difco agar, without growth regulators. (A) With 0.2 ng L^{-1} α-naphthalene acetic acid (NAA); (B) 0.2 ng L^{-1} NAA and 0.5 mg L^{-1} benzyladenine (BA); (C) 0.2 ng L^{-1} NAA and 1.0 mg L^{-1} BA; (D) 0.2 ng L^{-1} NAA and 2.0 mg L^{-1} BA; (E) 0.2 ng L^{-1} NAA and 4.0 mg L^{-1} BA; (F) 0.2 ng L^{-1} NAA and 8.0 mg L^{-1} BA; (G) 0.5 ng L^{-1} NAA and 2.0 mg L^{-1} BA; (H) 1.0 ng L^{-1} NAA and 2.0 mg L^{-1} BA; (I) 2.0 ng L^{-1} NAA and 2.0 mg L^{-1} BA; (J) 4.0 ng L^{-1} NAA and 2.0 mg L^{-1} BA; (K) 8.0 ng L^{-1} NAA and 2.0 mg L^{-1} BA; (L) *in-vitro*-propagated plantlets transferred to autoclaved soil and grown under greenhouse conditions with high humidity, after 2 weeks (M) and after 3 months (N).

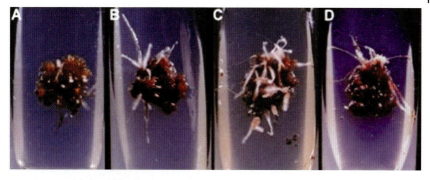

Fig. 12.5 *Salvia* callus grown on MS basal medium supplemented with 0.2 mg L^{-1} benzyladenine (BA) for periods of: (A) 8 days; (B) 16 days; (C) 24 days; and (D) 60 days.

Fig. 12.6 (A) Induction of multiple shoots from the internode explants of *Scrophularia yoshimurae*. (B) Shoot proliferation from the node explants.

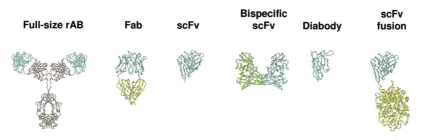

Fig. 14.1 Types of recombinant antibody expressed in plants. rAB = recombinant antibody; Fab = fragment antigen binding; scFv = single chain Fv fragment.

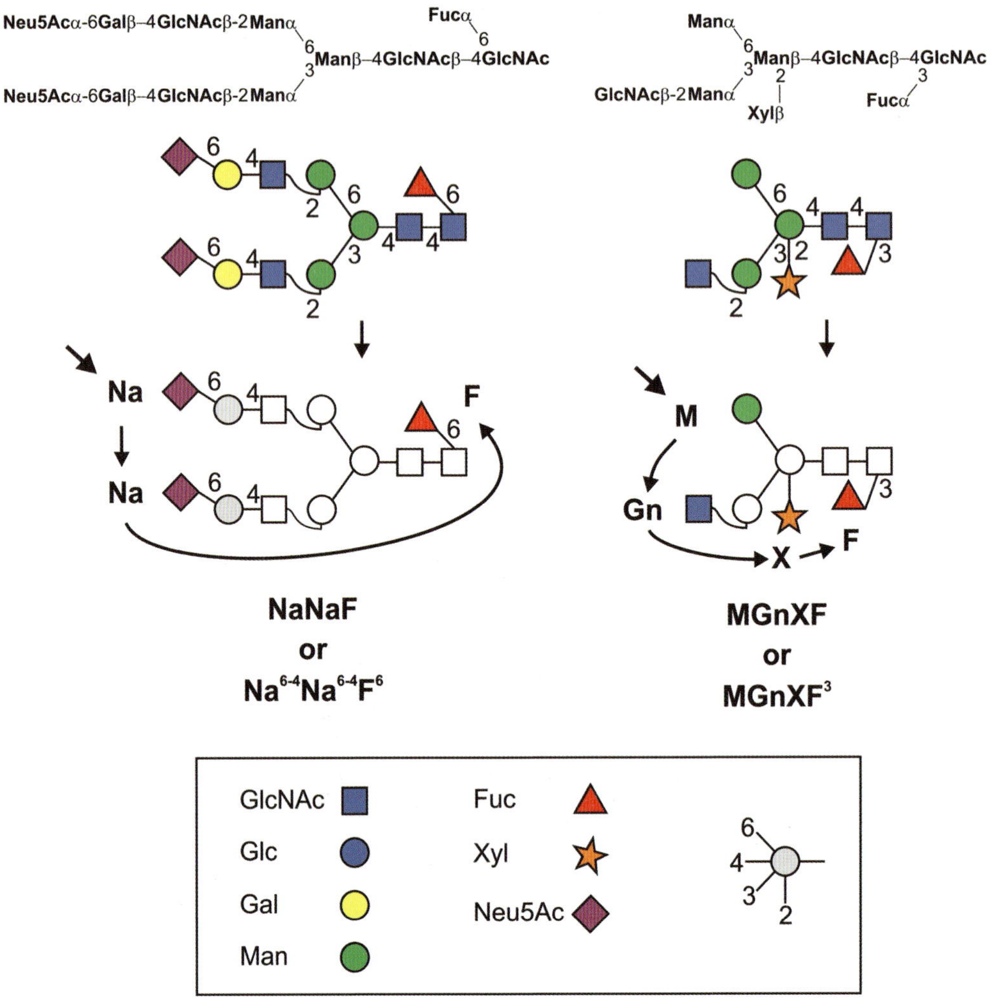

Neu5Acα-6Galβ–4GlcNAcβ-2Manα

Neu5Acα-6Galβ–4GlcNAcβ-2Manα

Manβ–4GlcNAcβ–4GlcNAc

Fucα

Manα

GlcNAcβ-2Manα

Manβ–4GlcNAcβ–4GlcNAc

Xylβ

Fucα

Na

Na

F

NaNaF
or
Na$^{6-4}$Na$^{6-4}$F^6

M

Gn

X → **F**

MGnXF
or
MGnXF^3

GlcNAc ■	Fuc ▲	
Glc ●	Xyl ★	
Gal ○	Neu5Ac ◆	
Man ●		

Fig. 15.1 The "proglycan" nomenclature for N-glycans. The upper two structures depict typical mammalian and plant complex type N-glycans, respectively. In the lower drawing the non-terminal, invariable residues have been uncolored and the invariable linkages are omitted. The arrow in the upper left corner points at the residue with which the listing of terminal sugars starts. Superscript numbers can be used to specify the linkage of sugars where alternative linkages are possible. In the case of sialylated complex type chains, the term Na$^{6-4}$ (an abbreviation of Na^6-A^4) means that Neu5Ac is linked α2,6 to a galactose which is itself in β1,4-linkage to the GlcNAc. (See also www. proglycan.com.)

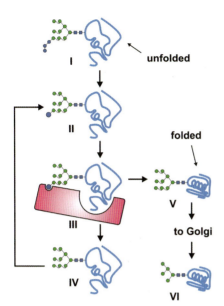

Fig. 15.2 Folding assistance by glycosylation-dependent chaperones. The unfolded protein with a glucose residue at the non-reducing end (large blue circle) is bound and chaperoned by calnexin or calreticulin (Parodi, 1999; Helenius and Aebi, 2001). Glucosidase eventually removes this glucose residue. If the protein has still not succeeded in folding properly it is re-glucosylated by a folding-sensitive glucose-transferase. Only the correctly folded glycoprotein is allowed to proceed to the Golgi body.

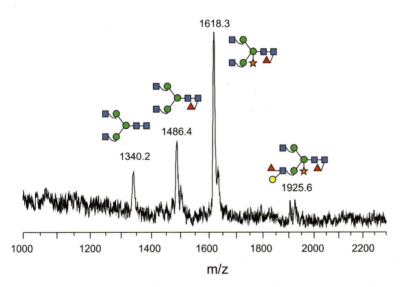

Fig. 15.4 Truly complex N-glycosylation on antibodies expressed in the moss *Physco mitrella patens*. MALDI-TOF analysis of N-glycans released from antibody expressed in a moss strain with unmodified plant-specific glycosylation. Remarkably, almost all structures are of the complex-type. While the antibody shown here still contained plant wild-type glycans with xylose and core-α1,3- fucose residues, the production of proteins lacking these two immunogenic sugar residues has been successfully accomplished with double knock-out strains of *Physco mitrella patens* (Koprivova et al., 2004; Huether et al., 2005; Jost and Gorr, 2005; M. Schuster et al., unpublished results; A. Weise et al., unpublished results).

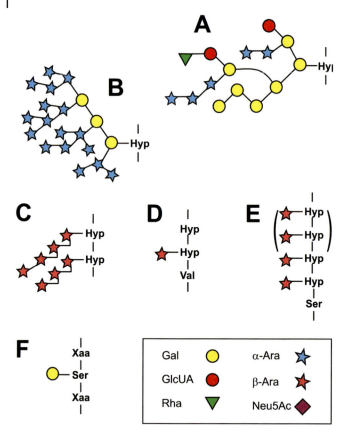

Fig. 15.5 O-linked poly-, oligo- and mono-saccharides. Due to the bewildering structural heterogeneity of plant O-glycans, and to the difficulties of separating and analyzing these glycans, the figures must be taken as arbitrary selections from a large pool of related structures. Structure A depicts a (rather small) type II arabinogalactan polysaccharide with additional sugars such as rhamnose and glucuronic acid according to Tan et al. (2004). Structure B shows a type III arabinogalactan as found on a mugwort pollen allergen (Leonard et al., 2005). Structure C represents arabinan chains on adjacent Hyp residues (Ashford et al., 1982), while D and E show the mono-arabinose moieties found in timothy grass and mugwort pollen allergens (Wicklein et al., 2004; Leonard et al., 2005). Structure F finally depicts the Gal-Ser motif thought to occur in extensins and solanaceous lectins (Lamport et al., 1973; Ashford et al., 1982). Connecting lines between sugar symbols indicate the linkage as specified in Figure 15.1.

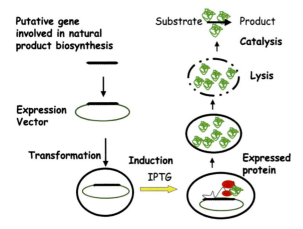

Fig. 16.1 Functional expression of a gene involved in natural product biosynthesis. The gene is putatively identified by bioinformatic means, inserted into an expression vector. Bacteria are transformed with the above construct and induced to produce high levels of active protein. The putative substrate is administered under conditions that favor catalysis, and conversion of the substrate to the expected products confirms the identity of the gene. IPTG, isopropyl-β-D-thiogalacto-pyranoside.

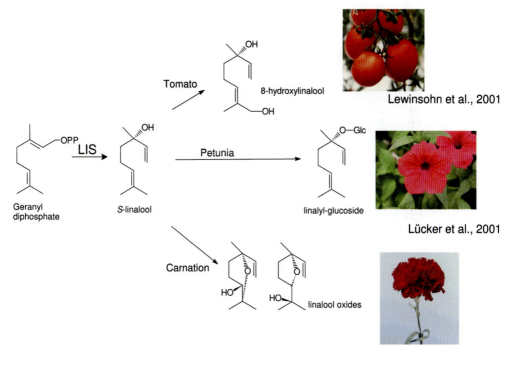

Lewinsohn et al., 2001

Lücker et al., 2001

Lavy et al., 2002

Fig. 16.2 Overexpression of the *Clarkia breweri* S-linalool synthase (LIS) in different plants and tissues results in the formation of different end products.

◄─────────── Camptothecin ───────────►

O. liukiuensis

O. pumila

O. japonica

O. kuroiwai

◄─────────── Lyalosidic acid, Harman ───────────►

Fig. 19.2 The genus *Ophiorrhiza* species distributed in Japan.

(A)

O. liukiuensis *O. kuroiwai* *O. pumila*

(B)

O. liukiuensis *O. kuroiwai* *O. pumila*

Fig. 19.3 Established tissue cultures of *Ophiorrhiza liukiuensis*, *O. kuroiwai*, and *O. pumila*. (A) Aseptic plants cultured for five weeks on half-strength MS medium containing 1% sucrose and 0.2% gellan gum in test tubes. (B) Hairy roots cultured for four weeks in B5 liquid medium containing 2% sucrose. (Reproduced from [22], with permission.)

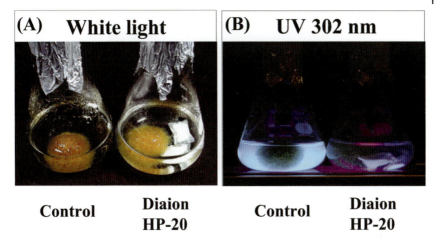

Fig. 19.6 Excretion of camptothecin into the culture medium. Hairy roots cultured for four weeks in an Erlenmeyer flask were visualized under (A) white light and (B) ultraviolet light at 302 nm. The strong fluorescence under ultraviolet irradiation is due to camptothecin having been excreted into the medium.

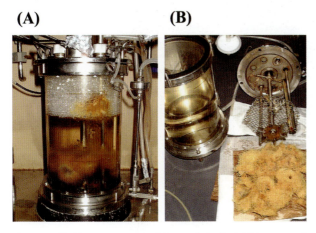

Fig. 19.7 (A) Hairy root growth of *O. pumila* in the 3-L bioreactor after eight weeks' culture. (B) Harvested hairy roots of *O. pumila* after culture.

(A)

(B)

Control **Regenerants**

Fig. 19.8 Camptothecin content of transgenic regenerated plants from hairy roots. (A) Regeneration of transgenic *O. pumila* plants from hairy roots. Regenerated shoots emerged from hairy roots after five weeks of culture under light conditions. (B) The shapes of regenerated plants cultured on half-strength MS medium containing 1% sucrose and 0.2% gellan gum in test tubes.

Fig. 21.1 *Ginkgo biloba* leaves.

Fig. 23.3 A modified airlift bioreactor for artemisinin production from
A. annua L. hairy root culture.

Fig. 23.4 A nutrient mist bioreactor for artemisinin production from
A. annua L. shoot cultures.

Part 1
Linking Plants, Genes, and Biotechnology

Medicinal Plant Biotechnology. From Basic Research to Industrial Applications
Edited by Oliver Kayser and Wim J. Quax
Copyright © 2007 WILEY-VCH Verlag GmbH & Co. KGaA, Weinheim
ISBN 978-3-527-31443-0

1
The Engineering of Medicinal Plants: Prospects and Limitations of Medicinal Plant Biotechnology

Matthys K. Julsing, Wim J. Quax, and Oliver Kayser

1.1
Introduction

The use of medicinal plants is increasing worldwide. According to the World Health Organization (WHO), approximately 80% of the world's population currently uses herbal medicines directly as teas, decocts or extracts with easily accessible liquids such as water, milk, or alcohol [1]. Although modern synthetic drugs are mostly used in developed countries, the use of herbal drugs in the western world is well accepted, and a continuously high demand for plant material and extracted natural products can be observed. The top 10 ranked plants that have received greatest interest in the USA and Europe over the past 30 years, and account for over 50% of the over-the-counter (OTC) market, are listed in Table 1.1 [2]. It should be noted that these plants are only partly of interest for biotechnology and

Table 1.1 Top ten ranked medicinal plants medicinal herbs most commonly used in the United States and Europe.

Species	Use
Hypericum perforatum	Anxiety, depression, insomnia
Echinacea purpurea	Immune stimulation
Ginkgo biloba	Dementia, Alzheimer's disease, tinnitus
Sabal serrulata	Benign prostatic hyperplasia
Tanacetum parthenium	Migraine prophylaxis
Allium sativum	Lipid-lowering, antithrombotic, fibrinolytic, anti-hypertensive, anti-atherosclerotic
Zingiber officinalis	Antiemetic
Panax ginseng	Tonic, performance enhancer, "adaptogen", mood enhancer
Valeriana officinalis	Sedative, hypnotic, anxiolytic
Ephedra distachya	Asthma, rhinitis, common cold, weight loss; enhancer of athletic performance

Medicinal Plant Biotechnology. From Basic Research to Industrial Applications
Edited by Oliver Kayser and Wim J. Quax
Copyright © 2007 WILEY-VCH Verlag GmbH & Co. KGaA, Weinheim
ISBN 978-3-527-31443-0

genetic modification, and today approaches of medicinal plant biotechnology focus more on distinct natural products and biosynthetic pathways. The main reason for this is that genetically modified plants as a source for extract preparations and for manufactured pharmaceuticals available in the pharmacy are not accepted by patients and consumers. Because of a gaining popularity for phytotherapy, and the wishful thought that products obtained from Nature are safe, patients do not consider genetically modified plants as a part of this philosophy. Hence, the approaches of medicinal plant biotechnology currently focus more on distinct natural products and biosynthetic pathways.

It is not only plants that are of great interest to the pharmaceutical industry, but also defined natural products. This situation is supported by the fact that some 25% of all drugs dispensed during the 1970s in the USA contained compounds obtained from higher plants [3]. Moreover, 11% of the 252 drugs considered to be basic and essential by the WHO are isolated and used directly from plant sources [4]. In addition, approximately 40% of pharmaceutical lead compounds for the synthetic drugs used today are derived from natural sources. It is due to this renewed interest in the use of plant sources in drug discovery that in this book we have included Chapter 6, which highlights the prospects and potential of plant-based natural products.

Today, only 10% of all medicinal plant species used are cultivated, with by far the larger majority being obtained from wild collections. Harvesting from the wild may, however, becomes problematic, as was seen in the case of *Podophyllum* and *Taxus* species, for *Piper methysticum*, for *Cimicifuga racemosa* and for *Arctostaphylos uva-ursa*, whereby both loss of genetic diversity and habitat destruction occurred. It is also not well known that conventional plant breeding methods are of major importance in medicinal plant biotechnology, due to improving agronomic and medicinal traits – a point which is discussed in Chapter 18. Thus, the domestic cultivation of medicinal plants is a well-accepted way in which to produce plant material. Moreover, such an approach also helps to overcome other problems inherent in herbal extracts, such as the standardization of extracts, variability of the plant material, minimization of toxic constituents and contaminations, increasing the content of the desired constituents, and breeding according to internationally accepted Good Agricultural Practise (GAP) guidelines. *Hypericum perforatum* and *Gingko biloba* are two examples of top-selling plants which have been cultivated for many years by European and American phytocompanies, and consequently are not greatly threatened by wild harvesting.

From this background the question arises as to whether there is a need for biotechnology and gene technology for medicinal plants. From its narrow definition, biotechnology does not focus on medicinal plants, and therefore it should be accepted in a broader sense. With regard to medicinal plants, biotechnology could be described as a method for enhancing the formation and accumulation of desirable natural products, with possible product modification in medicinal plants. Micropropagation, cell and hairy root culture as well as gene technology are all important techniques for plant propagation, but these are mostly used to improve the production and yield of desired natural products. Two well-described examples of this are artemisinin and paclitaxel, both of which are available in plants, albeit in only small quantities. As a consequence, not only are both natural products expensive, but *Taxus* species in the Himalaya region are also endangered due to unsustainable cutting and collection [5]. In an attempt to overcome these problems for both drugs, intensive research is being carried out worldwide, including combinatorial biosynthesis (see Chapter 13), or improved bioprocessing in bioreactors for both artemisinin (see Chapter 23) and paclitaxel (see Chapter 22).

The application of biotechnological techniques to medicinal plants has received considerable interest, especially when the final product is defined, purified, and natural. The manipulation of medicinal plants is well known and accepted both by scientists and consumers, if the pathways and product yield can be optimized to create precursors for semisyntheses (e.g., baccatin-III to paclitaxel), food components (e.g., vitamins), pesticide residence (e.g., *Atropa belladonna* [6,7]), and cellular storage conditions, as shown for *Mentha×piperita* with enhanced resistance against fungal attack and abiotic stress [8].

1.2
Genetic Transformation and Production of Transgenic Plants

The use of biotechnological tools in medicinal plant science is very limited as compared to other crops. However, in recent years the engineering of agronomic traits in medicinal plants led to interesting developments for genetic transformation, both *in vivo* and *in vitro*. Genetic transformation with bacterial vector systems has been widely used for several medicinal plants, including *Artemisia annua*, *Taxus* sp., *Papaver somniferum*, *Ginkgo biloba* (Chapter 21), and *Camptotheca acuminata* (Chapter 7), and species from the Solanaceae (Chapter 11), as reviewed in several chapters. Efficient gene vector systems are intensively discussed in Chapter 8, while the production of high-value phytochemicals is reviewed in Chapter 9.

The commercialization of many genetically engineered plants and plant products is currently being actively pursued by biotechnology and seed companies, and many genetically engineered plants are presently field-tested to determine their potential for commercialization. It remains an open question, however, as to which role transgenic medicinal plants will have in the future. Do we accept medicinal plants in the genomic era only as a source of chemicals, and should they acquire a new role as a genetic source or host for heterologous genes? An opportunity to discuss this question arises in Chapter 14, where the subject is plantibodies, while some answers – as well as further questions – are provided in Chapter 16.

1.3
Pathway Engineering and Combinatorial Biosynthesis

Increasing the production of pharmacologically attractive natural products represents on the main targets for the genetic manipulation of medicinal plants. An in-

1.7
Future Prospects

The commercial viability of biotechnology and gene technology in medicinal plant research is strongly influenced by the common perception of both, the plant and biotechnology. As outlined above, genetically modified medicinal plants lose their "natural" status, and are considered – erroneously – by the public as unsafe and dangerous. The crop industry learned its lesson following the rejection in Europe of genetically modified crops that were introduced into the food chain. Clearly, companies producing herbal compounds would face the same problems in obtaining permission to conduct farm-scale trails, to document the safety of the final product, and to overcome the in-principle resistance of the consumer as a strong and perhaps immovable barrier [9].

Within an open scientific environment, the discovery and development of botanical therapeutics and medicinal plant biotechnology must be accepted, as its expansion is very unlikely to cease. It is difficult to predict the future for medicinal plants, but it is likely that herbal drugs, isolated natural products and recombinant low- and high-molecular weight products will hold at least the same significance. Plants, as a renewable source with low energy consumption that can offer complex biochemical syntheses, will be even more compatible in the future. Although the real potential of plants remains unexplored, we hope that with this book we can perhaps at least "scratch the surface" slightly to provide a better understanding of medicinal plant biotechnology, as well as its possibilities and limitations.

References

1 Farnsworth, N.R. The role of ethnopharmacology in drug development. *Ciba Foundation Symposium* **1990**, *154*, 2–11; discussion 11–21.

2 Farnsworth, N. Screening plants for new medicines. In: Wilson, E. (Ed.), *Biodiversity*. National Academy Press, **1988**, pp. 83–97.

3 Farnsworth, N.R., Morris, R.W. Higher plants – the sleeping giant of drug development. *Am. J. Pharm. Sci. Support Public Health* **1976**, *148*(2), 46–52.

4 Rates, S. Plants as sources of drugs. *Toxicon* **2001**, *39*, 603–613.

5 Rikhari, H., et al. The effect of disturbance levels, forest types and associations on the regeneration of *Taxus baccata*: Lessons from the Central Himalayo. *Curr. Sci.* **2000**, 79–88.

6 Saito, K. [Molecular genetics and bio–technology in medicinal plants: studies by transgenic plants]. *Yakugaku Zasshi* **1994**, *114*(1), 1–20.

7 Saito, K. Transgenic herbicide-resistant *Atropa belladonna* using an Ri binary vector and inheritance to the transgenic trait. *Plant Cell Rep.* **1992**, *21*, 563–568.

8 Veronese, P., et al. Bioengineering mint crop improvement. *Plant Cell Tissue Organ Cult.* **2001**, *64*, 133–144.

9 Canter, P.H., Thomas, H., Ernst, E. Bringing medicinal plants into cultivation: opportunities and challenges for biotechnology. *Trends Biotechnol.* **2005**, *23*(4), 180–185.

2
Metabolomics

Young Hae Choi, Hye Kyong Kim, and Robert Verpoorte

2.1
Introduction

With the sequence of the genome of several organisms having been determined, and the speed of sequencing other genomes increasing, the emphasis of genomics is now moving towards investigations of genome function. Today, the question of the function of every single gene remains unanswered. To solve this problem it may be possible to knock out or silence every gene, but in many cases this would result in a non-viable organism. It may also be possible to examine expression patterns and to compare these with the phenotype under conditions in which the genes are expressed. In all cases, characterization of the phenotype is the key to understanding the function of the gene(s) of interest. Phenotype characterization can be achieved by means of morphological observation, or by using chemical and biochemical approaches. Biochemical characterization is achieved by proteomics, while chemical characterization is covered by metabolomics (Fiehn et al., 2000a; Schwab, 2003; Sumner et al., 2003).

Metabolomics is the youngest of the so-called "omics" methods, and ultimately concerns the analysis of all metabolites in an organism. To achieve this, a number of diverse approaches can be used (Fig. 2.1):

- *Metabolomics* aims at measuring all metabolites in an organism, both qualitatively and quantitatively. In studies of human or mammalian pathogenesis involving the analysis of materials such as urine and serum, this is referred to as *metabonomics*.
- *Metabolic profiling* aims at measuring a selected group of metabolites in an organism.
- *Metabolic fingerprinting* aims at measuring a "fingerprint" of the metabolite(s) in an organism, but without identifying all of the compounds present.

The major differences between these approaches are in the choice of whether to identify (all) compounds qualitatively and quantitatively in an organism, or to determine differences in metabolite content, followed by the identification of differ-

Medicinal Plant Biotechnology. From Basic Research to Industrial Applications
Edited by Oliver Kayser and Wim J. Quax
Copyright © 2007 WILEY-VCH Verlag GmbH & Co. KGaA, Weinheim
ISBN 978-3-527-31443-0

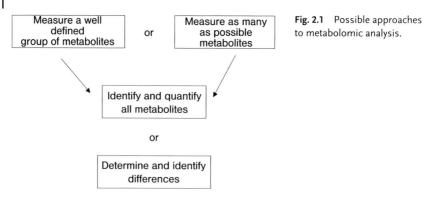

Fig. 2.1 Possible approaches to metabolomic analysis.

ent compounds. When examining a single compound or a well-defined group of compounds, this is considered to constitute a "targeted" approach.

Proteomics provides a view of the proteins present in an organism, by means of a suitable separation method. Two-dimensional (2D) gel electrophoresis is the most widely used (Jacobs et al., 2000), and from the patterns of the gel separation conclusions can be drawn on the overexpressed proteins under certain conditions. Subsequently, the proteins that seem of interest are hydrolyzed into peptides to determine their molecular weight and amino acid sequence, using mass spectrometry. When these sequences are resolved, the available large public databases can be screened for homologous peptides and/or genes and, based on homology with known peptides, a function of a protein may be proposed. Further proof is required, however, by using (bio-)chemical studies. Unfortunately, when investigating an organism in the absence of any information with regard to its genome sequence, many unknown peptides will be encountered, and this may require the encoding gene to be cloned, in order to determine its function. This will be an elaborate task, and consequently in non-sequenced organisms it might be possible to follow another strategy, by directly linking the transcriptomics (mRNA patterns) with the metabolomics data.

Hence, metabolomics is an important tool in functional genomics. In addition to this basic scientific approach, metabolomics has been applied to a wide array of other fields. For example, it is an important tool for the chemotaxonomy of botanicals, or for studying the equivalence of genetically modified and wild-type organisms, especially in the case of plants. Metabolomics also represents a major diagnostic tool, for example in the analysis of urinary metabolites using nuclear magnetic resonance (NMR) spectrometry in order to diagnose disease. Today, in drug development, metabonomics is used as a major means of detecting the toxicity of novel compounds on the kidney and liver, and more recently has attracted much attention from systems biology approaches to drug development. The different aspects of metabolomics will be discussed in detail in the following sections.

2.2
Analytical Methods

The ultimate goal of metabolomics is to measure all of the metabolites in an organism. However, the first question to be asked is how many metabolites are present in an organism? Although the answer is not known, it is possible to make an estimation, and it is commonly regarded that in any single organism there will be about the same number of metabolites as there are genes. Not all genes are involved in metabolism, but this is compensated for by a number of proteins being able to catalyze several reactions. Consequently, in a single plant approximately 30 000 low molecular-weight compounds should be present. To date, the best-studied plant is most likely tobacco, the number of identified compounds in tobacco leaves numbering approximately 3000. Within an organism, the metabolites represent a broad range of polarities and quantities, with some being highly labile. Hence, the aim of analyzing all of these compounds both qualitatively and quantitatively in a single operation is unrealistic, and cannot be achieved by any single analytical method, as each method has specific limitations.

In the case of metabolomics, the analytical tools available can be allocated to three major groups:

- Chromatographic methods: thin-layer chromatography (TLC), gas chromatography (GC), GC-mass spectrometry (MS), high-performance liquid chromatography (HPLC), HPLC-photodiode array (PDA) detector, HPLC-MS, and capillary electrophoresis (CE).
- Mass spectrometry, including MS and coupled MS-MS.
- NMR spectrometry.

2.2.1
Chromatography

Chromatographic methods have the advantage of sensitivity and specificity as they can be combined with selective detectors (MS for GC, and photodiode array detection, MS and NMR in case of HPLC) (Table 2.1).

2.2.1.1 Gas Chromatography (GC)
Although GC requires that compounds be volatile, many are neither volatile nor thermostable, and therefore require derivatization (e.g., acetylation, methylation, trimethylsilylation) prior to GC-analysis. This is an elaborate procedure which is also poorly reproducible for many compounds. The major advantage of GC is the large number of compounds that can be separated in a single analysis, and the large dynamic range (i.e., a rather broad range of quantities can be measured). Coupling GC with MS renders the method highly selective (e.g., with selective ion monitoring). Although MS detection is very sensitive, each compound has a different sensitivity, and consequently calibration curves are required for every com-

Table 2.1 Description of chromatographic methods used in metabolomics.

Methods	Target metabolites	Sample preparation	Reprodu- cibility	Resolution	Detector
TLC	General	Simple	Low	Low	UV, MS, color reagents
GC	Nonpolar (low molecular)	Elaborate derivatization	Medium	High	FID, TCD, NPD, MS
HPLC	Polar (chromophore)	Elaborate	Medium	Medium	UV, RI, MS, ELSD, fluorescence, NMR
CE	Ionic	–	Medium	High	UV, MS

ELSD, evaporative light-scattering detector; FID, flame ionization detector; NPD, nitrogen-phosphorus detector; RI, refractive index detector; TCD, thermal conductivity detector.

pound if absolute amounts of all metabolites are to be determined. GC-(MS) has proven successful in the analysis of the *Arabidopsis* metabolome, whereby over 1000 compounds were detected, about half of which were identified. GC-MS has been used for more than 40 years for targeted approaches, and consequently extensive databases of MS data in combination with GC-retention time data exist (e.g., NIST, AMDIS). These targeted approaches may relate to terpenoids (essential oils, steroids) and fatty acids (a field now also known as "lipidomics"). The main drawbacks of GC analysis is that the sample preparation methods are quite elaborate, and the average time required for a single analysis is over 1 h. Thus, high-throughput analyses are only feasible by using many GC instruments in parallel. A possible standard protocol for GC-MS analysis was developed by the Max Plank Institute of Molecular Plant Physiology (http://www.mpimp-golm.mpg.de/mms-library/details-e.html) (see Box 2.1).

Box 2.1 Derivatization method for GC-MS metabolomics
(http://www.mpimp-golm.mpg.de/mms-library/details-e.html)

- Add 80 µL methoxyamine hydrochloride in pyridine (20 mg mL^{-1}) to dried sample extract for protection of carbonyl moieties and react at 30 °C for 90 min.
- Afterwards add 80 µL N-methyl-N-trimethylsilyl-trifluoroacetamide (MSTFA) for TMS derivatization and react at 37 °C for 120 min.
- GC column; SPB 50 (polysiloxane column coated with 50% methyl and 50% phenyl groups, 30 m×0.25 mm)
- Injection temperature 230 °C
- Interface temperature 250 °C
- Ion source temperature 200 °C
- Helium flow rate 1.0 mL min^{-1}
- Oven temperature; after a 5-min solvent delay time at 70 °C, oven temperature increased at 5 °C min^{-1} to 310 °C, 1 min isocratic, cool to 70 °C followed by an additional 5-min delay.

2.2.1.2 High-Performance Liquid Chromatography (HPLC)

HPLC is the other widely used separation method for the analysis of metabolites, and can be used for both targeted and non-targeted approaches. Most often, HPLC is combined with a PDA detector, which allows recording of the ultraviolet (UV) spectra of all compounds, thus adding selectivity to the system. Unfortunately, many metabolites have no UV-absorption (e.g., many primary metabolites such as sugars, amino acids, compounds involved in tricarboxylic acid cycle, lipids, and terpenoids). However, by combining HPLC with MS these compounds can be identified, simultaneously adding further selectivity to the system. The sensitivity of HPLC-MS varies for all compounds, with sugars notably difficult to detect. For all HPLC-applications, quantitative analysis can only be achieved by creating calibration curves for each individual compound.

One important limitation of HPLC lies in the polarity window of the method. Typically, a metabolome will contain not only very nonpolar compounds that are insoluble in water and alcohols, but also highly polar compounds. Consequently, in order to cover the complete metabolome, several different systems must be employed. The number of compounds that can be separated in a gradient HPLC-system is between 100 and 200, though when combined with MS this number increases considerably, as overlapping peaks of different molecular weights can be observed as separate compounds in the MS.

As with all chromatographic methods, a major problem for HPLC is the reproducibility of separations. This occurs for several reasons, including small differences in parameters such as temperature and solvent quality, while matrix effects of the complex sample extract may influence the retention behavior of the compounds. In addition to possible small differences in batches between columns of the same manufacturer, the quality of the separation column will gradually deteriorate during prolonged use. Finally, novel columns are being introduced continuously to the market, replacing previous versions. Consequently, separations performed some years ago can no longer be reproduced, and this leads to major problems of analysis time when characterizing an organism's metabolome. Clearly, as metabolomics require methods that can be used internationally by research workers over many years, it is essential that public metabolomic databases (similar to those available for proteins and genes) are developed, and this requires long-term reproducibility of the analytical data. In this regard, several commercial alignment programs (e.g., LineUP™; Infometrix Inc., Bothell, WA, USA; www.infometrix. com) and MetAlign (Plant Research International, Wageningen, The Netherlands; www.pri.wur.nl), as well as "home-made" programs (e.g., Nielsen et al., 1998; Chen et al., 2003; Duran et al., 2003), have been designed to overcome such problems of reproducibility. However, although these programs may correct for day-to-day variations, they cannot overcome the problem of altered selectivity when column materials are produced in a different manner, or are modified ("improved"). For the analysis of a large number of compounds over a broad polarity range, a gradient program will typically take 30–60 min, though with novel equipment, using smaller-particle columns, this may be reduced to between 5 and 20 min. Hence, high-throughput analysis is not really feasible.

2.2.1.3 **Capillary Electrophoresis (CE)**

CE, which has also been applied to the analysis of metabolomes, has a very high resolution power, particularly for ionic compounds, when compared to other chromatographic methods. Moreover, its direct coupling to MS further contributes to the selectivity and provides a sensitive detection method. However, it is inherently limited to ionic compounds such as alkaloids, amino acids, or organic acids (Ishii et al., 2005).

2.2.1.4 **Thin-Layer Chromatography (TLC)**

TLC has a large potential to detect diverse group of compounds using a wide array of detection methods, including UV and coloring reagents. Although high-performance thin-layer chromatography (HPTLC) has a higher efficiency, the low resolution and reproducibility remain the limiting factors in its use for metabolomics. For metabolic fingerprinting, TLC has been used for more than 45 years in the quality control of botanicals.

2.2.2
Spectroscopy

2.2.2.1 **Mass Spectrometry (MS)**

MS detects the molecular ion of a compound after ionization and, if desired, also the fragmentation of that molecular ion. MS analysis in metabolomics uses separation of the compounds based on their molecular weight. By using high-resolution MS, compounds with different elemental formula but similar mass can be separated. Assuming that the molecular weights of the metabolites which together form the metabolome are in the range of about 50 to 2000, MS can separate about 2000 compounds. In complex mixtures, individual compounds will be detected on the basis of their molecular weight, but clearly there will be an extensive overlap of compounds with the same molecular weight. However, by using tandem MS-MS, greater selectivity can be obtained through the specific fragmentation of each compound. An example of an HPLC-MS analysis of a plant extract (*Ginkgo biloba*) is shown in Figure 2.2 (Choi et al., 2002).

MS is a very selective – and also possibly the most sensitive – method for metabolome analysis. However, the problems of reproducibility and quantitation are, as with chromatographic methods, the major constraints. Because of the different types of mass spectrometer available, the many possible variations in operating parameters, and also effects of the matrix on ionization, reproducibility is a major problem. Moreover, the efficiency of ionization of different compounds varies enormously, and may depend upon the presence of other compounds. Thus, quantitative analysis requires calibration curves for all compounds. Because MS spectrometry is a rapid process, a large number of samples can be analyzed per time unit (up to ca. 10 samples per hour). High-throughput analysis is thus feasible by means of MS. The most frequently used ionization methods of MS used in metabolomics are listed in Table 2.2.

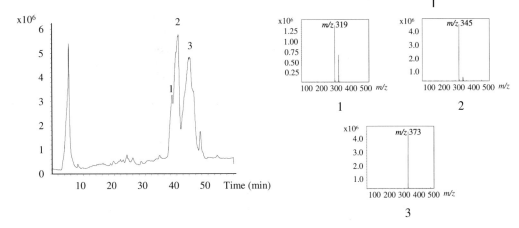

Fig. 2.2 Total ion current chromatogram of *Ginkgo biloba* extracts obtained with acetone:water (1:1) and the mass spectra of the major ginkgolic acids (1, 2, and 3) (Choi et al., 2002).

Table 2.2 Ionization methods in MS spectrometry used for metabolomics.[a]

Methods		Target metabolites [MW range]	Possible chromatographic methods connected	Advantages	Disadvantages
EI		Nonpolar (<500 MW)	GC	Identification using fragmentation	Low sensitivity of molecular weight of polar metabolites
API	ESI	Polar (<100 000 MW)	HPLC, CE	High sensitivity of molecular weight of polar metabolites	Limitation of polarity of metabolites
API	APCI	Medium polar (<1000 MW)	HPLC	Detection of metabolites soluble in nonpolar solvents	Limitation of MW of metabolites
MALDI		Polar (<100 000 MW)	HPLC	Detection of peptides and proteins	Lower sensitivity of low molecular weight metabolites

[a] Resolution varies depending on separators (e.g., quadrupole, ion trap, time of flight).

APCI, Atmospheric chemical ionization; API, Atmospheric ionization;
EI, Electron impact; ESI, Electrospray ionization;
MALDI, Matrix-assisted laser desorption.

2.2.2.2 **Nuclear Magnetic Resonance (NMR) Spectrometry**

NMR spectrometry is a physical measurement of the resonance of magnetic nuclei such as ^1H and ^{13}C in a strong magnetic field. Because of small local differences in the magnetic field in a molecule, each proton and carbon will show a different resonance, resulting in a highly specific spectrum for each compound. Such a spectrum is highly reproducible, and in ^1H NMR the intensity of a signal of a proton is directly correlated with the molar concentration – that is, all compounds in a mixture can be compared, without the need for a calibration curve. One single internal standard can be used to make the quantitative analysis of all compounds in a mixture. Thus, both in terms of reproducibility and quantitation, ^1H NMR spectrometry has a major advantage over chromatographic methods and MS. Many previous reports (see also Chapter 3) have shown that the specific characteristics of NMR make it an excellent tool for a macroscopic, non-targeted analysis of the total metabolome (Ward et al., 2003; Choi et al., 2004a,b,c, 2005, 2006; Hendrawati et al., 2006; Khatib et al., 2006; Yang et al., 2006).

The main disadvantage of NMR is that it is less sensitive than the other methods mentioned. Typically, the amount of material needed for an NMR analysis is about 50 mg biomass (dry weight), though in practice this is similar to what is required for the other methods. For NMR the complete extract is needed for non-destructive analysis, after which the sample can be kept for further analysis. In contrast, in other methods only part of the extract is used for the analysis, though this is destructive and requires back-up material for future use. The sensitivity of NMR can be improved by increasing the measuring time, and also by improving the spectrometer (e.g., cryo- or coldprobe), and this has been achieved in recent years. In NMR a higher field strength of the magnet improves sensitivity, and also influences the spectrum of a compound; the stronger the field the better the separation of the signals, and second-order spectra become first-order. This may hamper direct comparison of spectra recorded at different field strengths, but the problem can be resolved by using two-dimensional (2D)-NMR spectrometry. A 2D-J-resolved spectrum shows in the second dimension the coupling constants of each proton signal as a set of signals at the chemical shift of the proton. By projection of these signals onto the chemical shift axis, all proton signals become singlets (Hendrawati et al., 2006; Khatib et al., 2006; Yang et al., 2006). An example of an application of projected J-resolved spectra to the tobacco metabolome analysis (Choi et al., 2006) is shown in Figure 2.3. In this way the resolution is improved, but more importantly the projected spectrum is independent of the field strength, which makes this a constant characteristic suitable for long-term inclusion into a public database. The disadvantage of this approach is that signals can no longer be compared quantitatively, although relative comparison for one compound in different mixtures is still possible.

An advantage of NMR spectroscopy is that the time required to record a ^1H NMR spectrum is quite short (depending on the equipment and concentration of the sample, a ^1H NMR spectrum can be obtained within 10 min), which makes high-throughput analysis an option. As NMR spectra are recorded in solution, there is a limitation to the polarity range that can be covered using this technique. At least two different solvents are needed to cover the nonpolar and polar compounds. In

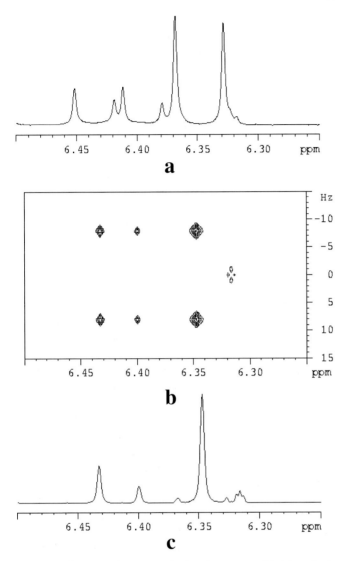

Fig. 2.3 (a) ^{1}H NMR, (b) 2-D-J-resolved, and (c) projected 1-D-J-resolved spectrum of healthy *Nicotiana tabacum* leaves in the range of δ 6.25–δ 6.50, showing the signals of isomeric chlorogenic acids (Choi et al., 2006).

order to cover the diverse polarity of metabolites, a two-phase extraction method using a mixture of chloroform, methanol and water has been applied to plant metabolomics (Choi et al., 2004a,b,c, 2005). In order to reduce preparation time, a simple direct extraction method using deuterated NMR solvents has been developed for sample preparation (Choi et al., 2006; Hendrawati et al., 2006; Khatib et al., 2006; Yang et al., 2006).

NMR results obtained using correct sample preparation allow a direct comparison of all compounds present. The data obtained also have all the desired characteristics needed for inclusion into a public database on metabolomic data.

2.2.3
Identification of Metabolites

One problem in metabolomic studies of new organisms is the identification of all metabolites. Many can easily be recognized by comparison with databases of well-known primary metabolites occurring in most living cells. However, particularly in chemically advanced organisms such as plants, secondary metabolism is very well developed and produces a wide variety of species-specific compounds. Retention behavior in chromatography, UV spectra and mass spectra are useful to identify well-known compounds, but in the case of rare or novel compounds further structure elucidation is necessary.

NMR-spectrometry is by far the most powerful tool for structure elucidation. For ^{1}H NMR spectra, only small amounts of material are needed, and direct coupling with liquid chromatography (LC) provides good spectra that can help to determine the structure. For more complex compounds, however, ^{13}C NMR spectra and various 2D-NMR spectra are needed to establish the carbon skeleton of the molecule and determine the positions of the protons. This requires larger amounts of material, which can be obtained by preparative chromatography, or repeated analytical chromatography and collection of the peak(s) of interest, for example on a solid-phase extraction column(s) (LC-SPE-NMR). In crude extracts using NMR-metabolomics, much information can be acquired by applying various 2D-NMR spectrometric methods such as correlated spectroscopy (COSY, to reveal proton–proton couplings), 2D-J-resolved spectra (to show the coupling constants of various proton signals), heteronuclear multiple quantum coherence (HMQC, to determine direct proton–carbon couplings), and heteronuclear multiple bond coherence (HMBC, to determine long-range proton–carbon couplings). These methods can in fact also be used as a metabolomics tool as they allow more compounds to be quantified. An example of the identification of metabolites in *Brassica rapa* leaves using 2D-NMR spectrum is shown in Figure 2.4 (Liang et al., 2006).

2.2.4
Sample Treatment

A key step in the analysis of a metabolome is the sample preparation. This includes all steps from harvesting, extracting and sample purification prior to the actual analysis. Each of these steps must be optimized and validated for a reliable reproducible analysis.

The first step is to collect the material which, in order to avoid any (bio)chemical change in the material, should be frozen immediately after harvesting. In plants, for example, immediately after harvesting a defense reaction will occur in the cells, which includes the hydrolysis of glycosides and oxidation of phenolics that subsequently bind to cell walls.

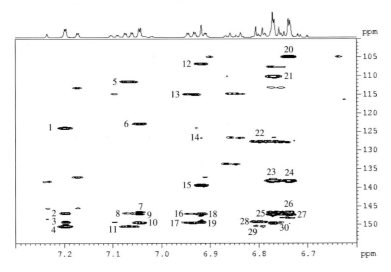

R$_1$ = OCH$_3$, R$_2$ = OH,   5-hydroxyferuloyl malate (**1**)
R$_1$ = OH, R$_2$ = H,   caffeoyl malate (**2**)
R$_1$ = H, R$_2$ = H,   coumaroyl malate (**3**)
R$_1$ = OCH$_3$, R$_2$ = H,   feruloyl malate (**4**)
R$_1$ = OCH$_3$, R$_2$ = OCH$_3$, sinapoyl malate (**5**)

Fig. 2.4 Heteronuclear multiple bond coherence (HMBC) spectrum of aromatic moiety of phenylpropanoids of *Brassica rapa* leaves in the range of δ 6.60–δ 7.30 of ^{1}H and δ 100–δ 150 of ^{13}C, showing the long-range C–H couplings. 1: H-2′/C-6′ of compound 4. 2: H-2′/C-7′ of compound 4. 3: H-2′/C-3′ of compound 4. 4: H-2′/C-4′ of compound 4. 5: H-6′/C-2′ of compound 4. 6: H-2′/C-6′ of compound 2. 7: H-2′/C-3′ of compound 2. 8: H-6′/C-7′ of compound 4. 9: H-2′/C-7′ of compound 2. 10: H-2′/C-4′ of compound 2. 11: H-6′/C-4′ of compound 4. 12: H-2′ and 6′/C-2′ and 6′ of compound 5. 13: H-6′/C-2′ of compound 2. 14: H-2′ and 6′/C-1′ of compound 5. 15: H-2′ and 6′/C-4′ of compound 5. 16: H-6′/C-7′ of compound 2. 17: H-6′/C-4′ of compound 2. 18: H-2′ and 6′/C-7′ of compound 5. 19: H-2′ and 6′/C-3′ and 5′ of compound 5. 20: H-6′/C-2′ of compound 1. 21: H-2′/C-6′ of compound 1. 22: H-5′/C-1′ of compound 4. 23: H-2′/C-4′ of compound 1. 24: H-6′/C-4′ of compound 1. 25: H-2′/C-7′ of compound 1. 26: H-6′/C-5′ of compound 1. 27: H-6′/C-7′ of compound 1. 28: H-5′/C-3′ of compound 4. 29: H-5′/C-4′ of compound 4. 30: H-2′/C-3′ of compound 1 (Liang et al., 2006).

The next step is to grind the material to liberate the metabolites from the cells. Cells of microorganisms can be destroyed by using a French press, but plant material containing lignified cells requires mechanical grinding. During the extraction, (bio)chemical reactions may occur in the material, resulting in changes in the metabolome. This can be avoided by drying the material before extraction (either by heat or freeze-drying), keeping the material at low temperature, or grinding in the presence of a solvent that denatures enzymes which could convert the metabolites. Denaturation can also be achieved by a brief microwave treatment.

To extract the metabolites, a broad spectrum of solvents can be used. Nonpolar solvents (e.g., hydrocarbons), medium-polar solvents (e.g., ethyl acetate, diethyl ether, chloroform, dichloromethane), and polar solvents (e.g., methanol, ethanol, water), and any combination of these solvents can be used. Each group will extract a different type of metabolite(s); some metabolites will completely dissolve in a certain solvent, but others only partly. pH is also a factor that affects the profile of metabolites extracted; for example, alkaloids are soluble in nonpolar solvents at high pH, and in aqueous solvents at low pH. A two-phase solvent system consisting of chloroform:methanol:water (2 : 1 : 1) has been used to extract both polar and nonpolar compounds in a single extraction (Choi et al., 2004a,b,c, 2005) (see Box 2.2).

This method was also applied to metabolic analysis for plant cell culture materials after washing the materials with deionized water (Suhartono et al., 2005), while a modified method with addition of NH_4OH was used for benzylamine alkaloids (Kim et al., 2005).

This two-phase extraction method was found to be problematic, however, when handling large numbers of samples because of the long processing time. Hence, a simple direct extraction method was developed for sample preparation, using deuterated NMR solvents (Choi et al., 2006; Hendrawati et al., 2006; Khatib et al., 2006; Yang et al., 2006) (see Box 2.3).

For the metabolic profiling of *Strychnos* species (Frédérich et al., 2004), methanol-d_4 containing trifluoroacetic acid was used, with the analysis focusing on the indole alkaloids.

For chromatographic analytical methods, the sample must be dissolved in a solvent that is suitable for injection into the GC or HPLC system. For HPLC, the solvent must be closely related to the eluent of the HPLC-separation system. For GC analysis, derivatization is required, and several different methods have been re-

---

Box 2.2   Two-phase extraction method for NMR analysis (Choi et al., 2004a,b,c, 2005).

- Add 2 mL chloroform, 1 mL methanol, and 1 mL water to 50–100 mg plant material.
- Ultrasonicate for 5 min, then centrifuge at 3000 rpm (1300 g) for 20 min at 4 °C.
- Take upper phase of methanol-water fraction and lower phase of chloroform-methanol fraction.
- Evaporate both fractions using rotary evaporator and add 1 mL chloroform-*d* to chloroform-methanol fraction and 1 mL of methanol-d_4 and H_2O-d_2 (KH_2PO_4 buffer, pH 6.0) to methanol-water fraction for NMR measurements.

Box 2.3 Single-solvent extraction method for NMR metabolomics
(Choi et al., 2006; Hendrawati et al., 2006; Khatib et al., 2006;
Yang et al., 2006).

- Add 0.75 mL methanol-d_4 and 0.75 mL H_2O-d_2 (KH_2PO_4 buffer, pH 6.0)
  to 50–100 mg plant material.
- Vortex for 1 min.
- Ultrasonicate for 10 min.
- Centrifuge at 13 000 rpm (11 000 g) for 10 min at room temperature.
- Transfer 800 µL of supernatant to NMR tube.

---

ported for this purpose (Fiehn et al, 2000b; Roessner et al., 2000). In the case of NMR, direct measurement in aqueous systems is possible (e.g., direct analysis of urine by NMR spectrometry), though better results are obtained using deuterated solvents. These solvents can be used for the primary extraction of the material, which is a simple and rapid method suited to high-throughput metabolome analysis. For MS analysis, different approaches can be used, for example injection of the dissolved extract or applying the extract on a sample probe introduced into the mass spectrometer.

When working with solvents it is essential to be aware of the risks of artifact formation (e.g., transesterifications, reactions with solvents, or reactions with contaminations in solvents). One should also be aware of the presence of different additives (e.g., 1% ethanol in chloroform, antioxidants in ethers) (Svendsen and Verpoorte, 1983) or contaminants (e.g., phthalates from plastics, detergents from dishwashers in glassware) (Middleditch, 1989).

As no single solvent is capable of extracting all metabolites, different extraction methods must be used to analyze all of the metabolites produced by an organism. In order to analyze the minor compounds in a metabolome, further fractionation steps might be included, either in the form of liquid-liquid extractions, solid-phase extraction, or perhaps fractionation by preparative column chromatography or centrifugal partitioning chromatography.

Whichever method is chosen, a thorough validation is always required, from harvesting to final analysis, to include intra-day and day-to-day variability of the procedure. This should result in a robust, reproducible method preferably suited to high-throughput metabolome analysis.

2.3
Data Handling

The next stage in the process is to handle the large number of data sets obtained from the many samples that have been analyzed under different conditions. The analytical output must be identified and quantified, and chemometric methods such as multivariate data analysis have been shown to be excellent tools for this

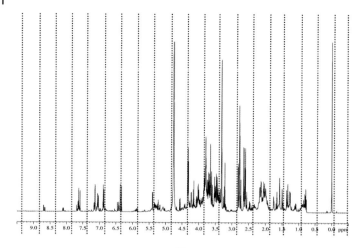

a

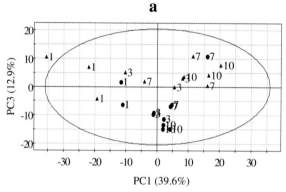

b

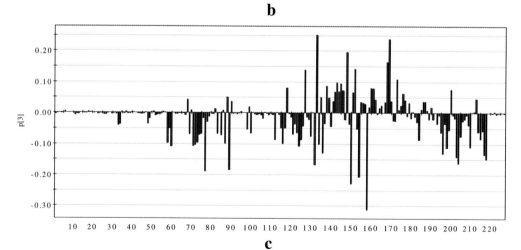

c

Fig. 2.6 (a) Data bucketing of NMR spectra; (b) score plot; and (c) loading plot of principal component analysis of healthy and infected *Nicotiana tabacum* leaves by tobacco mosaic virus (TMV). ●; lower leaves of healthy plants, ▲; local-infected leaves of TMV-infected plants, The ellipse represents the Hotelling T2 with 95% confidence in score plots. Numberings on the plot are the dates after infection (Choi et al., 2006).

Clearly, in order to make metabolomics a globally used tool, standardization of the methods is required, and a generally accepted format for the data is required. For transcriptomics data such a format has been defined by the Microarray Gene Expressing Data (MGED) Society (http://www.mged.org). Standards for Minimum Information About a Microarray Experiment (http://www.mged.org/Workgroups/MIAME/miame.html; see also Brazma et al., 2001) have been defined through an international initiative. The research community has embraced this standard, and many major journals now require compliance with MIAME for any new submission (Anonymous, 2002). Whilst for metabolomics data the discussion is ongoing, it might be difficult to achieve a standard that will be accepted worldwide. For example, all efforts during the past 40 years to arrive at a standardized database for mass spectra have been unsuccessful. Here, a major problem is – and will continue to be – the variability in the data, due to the different MS methods applied. It is likely that NMR-spectra will in time become the simplest data for a globally accepted public database, particularly as the original dataset of the spectra can be stored and later translated into any desirable format. Whilst chromatograms with unknown peaks will be more difficult to store, treated chromatographic data resulting in tables of identified compounds and relative intensities can be produced; however, full quantitative data for all compounds are much more difficult to produce.

2.4
The General Set-Up of a Metabolomics Project

The first step in a metabolomics project is to define the problem to be studies. Based on that, a choice must be made as to which analytical methods are most appropriate to answer the question. It may be necessary even to combine different methods, especially when both major and minor compounds are of importance. This includes deciding whether to use a targeted approach only on certain groups of metabolites, or a non-targeted approach in which as many metabolites as possible can be analyzed. For both approaches, the next decision is whether to identify and quantify all of the compounds observed, or to focus only on those compounds that clearly show different levels under the different experimental conditions.

To facilitate this choice, preliminary experiments are required to identify the degree of variability in the planned experiments. When the methods have been selected, the next step is qualitative, to analyze a series of representative samples, to identify as many compounds as possible by comparison with known databases, and to determine the experimental (e.g., sample preparation, extraction method, analytical method) and biological variability for one genotype (e.g., diurnal variations, developmental stage, external conditions). Once this dataset has been established, the next step is to measure the organism under different experimental conditions, to determine variations by statistical means (e.g., multivariate analysis), and to combine these data with other data such as transcriptomics. The strength of multivariate analysis lies in the fact that any type of data can be used for the analysis, including physiological (e.g., blood pressure) and morphological data, as the original data are dimensionless.

2.5
Applications

Besides applications in functional genomics, the metabolomic approach has become very useful in studies of genetically modified organisms, notably to identify possible differences from the wild-type organism. The use of metabolic profiles has also been used for the quality control of food and medicinal plants. Likewise, when identifying new leads for drug development from natural sources, the metabolomics approach would be of great value for the rapid identification of known active compounds. Finally, the use of metabolomics might also be envisaged for studying the activity of medicinal plants. The measurement of pharmacological activity in a living organism (be it a cell culture, an animal, or a patient) of extracts with different compositions may allow the identification of a compound – or indeed a combination of compounds – that correlate with such activity. In this way, pro-drugs or synergism might also be recognized, in addition to new modes of drug action and drug targets.

Clearly, this systems biology approach will represent a major challenge over the next few years for the study of medicinal plants. Moreover, metabolomics represents a totally new approach to drug discovery that should replace the single compound–single target concept that for many years has been the paradigm of drug development, but which is producing fewer novel medicines each year.

2.6
Conclusions

The analysis of the full metabolome of an organism requires a combination of analytical methods, since no single technique can cover the vast range of spectra of the compounds present. Moreover, these methods must be robust and very reproducible, since large quantities of data will be needed if metabolomics is to be applied to areas such as functional genomics or systems biology. The techniques employed are mature chromatographic methods such as GC-MS and HPLC, or physical approaches such as mass spectrometry and NMR, all of which have long histories in the analysis of natural products. Ultimately, the data produced will require powerful biostatistical methods for their analysis, and multivariate approaches and neural networks are being developed for this purpose.

Today, despite no international conventions existing for data storage and presentation relating to the metabolomics approach, a variety of international databases for MS and NMR contain the spectra of very many pure compounds. Whilst public databases with metabolome profiles do not yet exist, there is clear need for their future development in the advancement of metabolomics.

References

Anonymous. Microarray standards at last. *Nature* **2002**, *419*, 323.

Brazma, A., Hingamp, P., Quackenbush, J., Sherlock, G., Spellman, P., Stoeckert, C., Aach, J., Ansorge, W., Ball, C.A., Causton, H.C., Gaasterland, T., Glenisson, P., Holstege, F.C., Kim, I.F., Markowitz, V., Matese, J.C., Parkinson, H., Robinson, A., Sarkans, U., Schulze-Kremer, S., Stewart, J., Taylor, R., Vilo, J., Vingron, M. Minimum information about a microarray experiment (MIAME)-toward standards for microarray data. *Nat. Genet.* **2001**, *29*, 365–371.

Chen, F., Duran, A.L., Blount, J.W., Sumner, L.W., Dixon, R.A. Profiling phenolic metabolites in transgenic alfalfa modified in lignin biosynthesis. *Phytochemistry* **2003**, *64*, 1013–1021.

Choi, Y.H., Yoo, K.-P. Kim, J. Supercritical fluid extraction of bilobalide and ginkgolides from *Ginkgo biloba* leaves using a mixture of CO_2, methanol, and water, *Chromatographia* **2002**, *56*, 753–757.

Choi, H.-K., Choi, Y.H., Verberne, M., Lefeber, A.W.M., Erkelens, C., Verpoorte, R. Metabolic fingerprinting of wild type and transgenic tobacco plants by ^{1}H NMR and multivariate analysis technique. *Phytochemistry* **2004a**, *65*, 857–864.

Choi, Y.H., Kim, H.K., Hazekamp, A., Erkelens, C., Lefeber, A.W.M., Verpoorte, R. Metabolomic differentiation of *Cannabis sativa* cultivars using ^{1}H NMR spectroscopy and principal component analysis. *J. Nat. Prod.* **2004b**, *67*, 953–957.

Choi, Y.H., Tapias, E.C., Kim, H.K., Lefeber, A.W.M., Erkelens, C., Verhoeven, J.T.J., Brzin, J., Zel, J., Verpoorte, R. Metabolic discrimination of *Catharanthus roseus* leaves infected by phytoplasma using ^{1}H-NMR spectroscopy and multivariate data analysis. *Plant Physiol.* **2004c**, *135*, 2398–2410.

Choi, Y.H., Sertic, S., Kim, H.K., Wilson, E.G., Michopoulosa, F., Lefeber, A.W.M., Erkelens, C., Kricun, S.D.P., Verpoorte, R. Classification of *Ilex* species based on metabolomic fingerprinting using NMR and multivariate data analysis. *J. Agric. Food Chem.* **2005**, *53*, 1237–1245.

Choi, Y.H., Kim, H.K., Linthorst, H.J.M., Hollander, J.G., Lefeber, A.W.M., Erkelens, C., Nuzillard, J.-M., Verpoorte, R. NMR metabolomics to revisit the tobacco mosaic virus infection in *Nicotiana tabacum* leaves. *J. Nat. Prod.* **2006**, *69*, 742–748.

Duran, A.L., Yang, J., Wang L., Sumner, L.W. Metabolomics spectral formatting, alignment and conversion tools (MSFACTs). *Bioinformatics* **2003**, *19*, 2283–2293.

Eriksson, L., Johansson, E., Kettaneh-Wold, N., Wold, S. *Multi-and megavariate data analysis.* Umetrics, Umeå, Sweden, **2001**.

Fiehn, O., Kopka, J., Dormann, P., Altmann, T., Trethewey, R.N., Willmitzer, L. Metabolite profiling for plant functional genomics. *Nat. Biotechnol.* **2000a**, *18*, 1157–1161.

Fiehn, O., Kopka, J., Trethewey, R.N., Willmitzer, L. Identification of uncommon plant metabolites based on calculation of elemental compositions using gas chromatography and quadrupole mass spectrometry. *Anal. Chem.* **2000b**, *72*, 3573–3580.

Frédérich, M., Choi, Y.H., Angenot, L., Harnischfeger, G., Lefeber, A.W.M., Verpoorte, R. Metabolomic analysis of *Strychnos nux-vomica*, *icaja* and *ignatii* extracts by ^{1}H nuclear magnetic resonance spectrometry and multivariate analysis techniques, *Phytochemistry* **2004**, *65*, 1993–2001.

Hendrawati, O., Yao, Q., Kim, H.K., Linthorst, H.J.M., Erkelens, C., Lefeber, A.W.M., Choi, Y.H., Verpoorte, R. Metabolic differentiation of *Arabidopsis* treated with methyl jasmonate using nuclear magnetic resonance spectroscopy. *Plant Sci.* **2006** (in press).

Ishii, N., Soga T., Nishioka T., Tomita M. Metabolome analysis and metabolic simulation. *Metabolomics* **2005**, *1*, 29–37.

Jacobs, D.I., Van der Heijden, R., Verpoorte, R. Proteomics in plant biotechnology and secondary metabolism research. *Phytochem. Anal.* **2000**, *11*, 277–287.

Khatib, A., Wilson, E.G., Kim, H.K., Lefeber, A.W.M., Erkelens, C., Choi, Y.H., Verpoorte, R. Application of two dimensional J-resolved nuclear magnetic resonance spectrum to differentiation of beer. *Anal. Chim. Acta* **2006**, *559*, 264–270.

Kim, H.K., Choi, Y.H., Erkelens, C., Lefeber, A.W.M., Verpoorte, R. Metabolic fingerprinting of *Ephedra* species using ¹H-NMR spectroscopy and principal component analysis. *Chem. Pharm. Bull.* **2005**, *53*, 105–109.

Liang, Y.-S., Kim, H.K., Lefeber, A.W.M., Erkelens, C., Choi, Y.H., Verpoorte, R. Identification of phenylpropanoids in *Brassica rapa* leaves treated with methyl jasmonate using two-dimensional nuclear magnetic resonance spectroscopy. *J. Chromatogr. A* **2006**, *1112*, 148–155.

Massart, D.L., Vandeginste, B.G.M., Deming, S.N., Michotte, Y., Kauffman, L. *Chemometrics: A Textbook*. Elsevier, New York, **1988**.

Middleditch, B.S. *Analytical artifacts*. Elsevier, Amsterdam, The Netherlands, **1989**.

Nielsen, N.-P.V., Carstensen, J.M., Smedsgaard, J. Aligning of single and multiple wavelength chromatographic profiles for chemometric data analysis using correlation optimised warping. *J. Chromatogr. A* **1998**, *805*, 17–35.

Roessner, U., Wagner, C., Kopka, J., Trethewey, R.N., Willmitzer, L. Simultaneous analysis of metabolites in potato tuber by gas chromatography-mass spectrometry. *Plant J.* **2000**, *23*, 131–142.

Schwab, W. Metabolome diversity: too few genes, too many metabolites? *Phytochemistry* **2003**, *62*, 837–850.

Suhartono, L., Van Iren, F., De Winter, W., Roytrakul, S., Choi, Y.H., Verpoorte, R. Metabolic comparison of cryopreserved and normal cells from *Tabernaemontana divaricata* suspension cultures. *Plant Cell Tiss. Org. Cult.* **2005**, *83*, 59–66.

Svendsen, A.B., Verpoorte, R. *Chromatography of alkaloids part A: thin-layer chromatography*. Elsevier, Amsterdam, The Netherlands, **1983**, pp. 51–50.

Sumner, L.W., Mendes, P., Dixon, R.A. Plant metabolomics: large-scale phytochemistry in the functional genomics era. *Phytochemistry* **2003**, *62*, 817–836.

Ward, J., Harris C., Lewis, J., Beale, M.H. Assessment of ¹H NMR spectroscopy and multivariate analysis as a technique for metabolite fingerprinting of *Arabidopsis thaliana*. *Phytochemistry* **2003**, *62*, 949–958.

Yang, S.Y., Kim, H.K., Lefeber, A.W.M., Erkelens, C., Angelova, N., Choi, Y.H., Verpoorte, R. Application of two-dimensional nuclear magnetic resonance spectroscopy to quality control of ginseng commercial products. *Planta Med.* **2006**, *72*, 364–369.

3
HPLC-NMR Techniques for Plant Extract Analysis

Dan Stærk, Maja Lambert, and Jerzy W. Jaroszewski

3.1
Introduction

Mankind has relied on medicinal plants since time immemorial. Written records about the use of medicinal plants include the Chinese Pen Tsao, or The Great Herbal, compiled by Shen Nung who lived around 3000 BC, the Ebers Papyrus, the earliest record of Egyptian medicine from 1600 BC, the Hindu holy books, the Vedas, dated from before 1000 BC (some of them much older), and *De Materia Medica* by Dioscorides from 78 AD. Despite recent advances in biotechnology, synthetic combinatorial chemistry, and high-throughput screening, natural products from plants continue to serve as a major source of new chemical entities for pharmaceutical research [1, 2]. Although chemists and pharmacologists have explored many thousands of plants, most investigations have been restricted to the structural characterization of a few major constituents or to the assessment of a single biological activity *in vitro*. Therefore, the potential of even relatively thoroughly investigated plants must ironically be regarded as still largely unexplored. In addition, although the total number of plant species on the Earth and their extinction rate remain the matter of debate [3], it is clear that chemical and pharmacological knowledge exists about only a small percentage of the total number of species. The disappearance of many biotopes rich in endemism means that many of as-yet unexplored plants will never be investigated. An acceleration of the process of mapping the chemical diversity of plants is therefore of utmost importance. This can only be achieved if new methods for the rapid dereplication of natural products from plant extracts – preferably methods that are more sensitive than those available today – are developed.

Most studies on secondary metabolites from plants are related to a biological activity that is based on traditional use, or has been observed through screening programs using *in-vitro* bioassays [4–7]. The traditional process of isolating and characterizing bioactive natural products is a very time-consuming process. In addition, only a single or a few biological activities can be monitored during a bioactivity-guided isolation process, leading to the risk of overlooking potentially impor-

Medicinal Plant Biotechnology. From Basic Research to Industrial Applications
Edited by Oliver Kayser and Wim J. Quax
Copyright © 2007 WILEY-VCH Verlag GmbH & Co. KGaA, Weinheim
ISBN 978-3-527-31443-0

tant activities. The absence of structural information about the active constituents until the isolation procedure has been completed leads to the risk of isolating already known or otherwise trivial compounds. For these reasons, there is a need for tools providing full or partial information about the structure of extract constituents as early as possible in the fractionation process. The targeted isolation of natural products, guided by the novelty of their chemical structures and followed by screening of pure compound libraries (which can, in principle, be done for a large variety of bioactivities), is thus an attractive alternative to the traditional bioactivity-guided isolation approach [8, 9].

Finally, modern biotechnology creates the possibility of accessing the chemical diversity of natural products and of translating genetic diversity of organisms into an expanded chemical space of small molecules [10, 11] through other approaches than screening and isolation. Thus, transgenic plants, studies on regulation of metabolic networks, engineering of biosynthetic pathways, and metabolomics/metabonomics [12–15] create the need for high-throughput phytochemistry and research tools providing system-wide chemical data. Given the unsurpassed structure–identification capabilities of NMR spectroscopy and the current ability of this spectroscopic technique to operate at the microgram and submicrogram levels, hyphenation of separation methods with NMR is of particular interest for modern natural products research.

3.2
Hyphenation of Separation Techniques and Spectroscopic Methods

The coupling of two or more analytical techniques into one integrated technique is usually indicated by use of a hyphen, as in gas chromatography-mass spectrometry (GC-MS). The hyphenation of high-performance liquid chromatography (HPLC) with different detection techniques that provide some structural information has contributed considerably to the progress in natural products research over the past few decades. The HPLC detectors can be categorized into those giving a single quantitative response, such as single-wavelength ultraviolet detectors (HPLC-UV), fluorescence detectors (HPLC-fluorescence), and evaporative light-scattering detectors (HPLC-ELS), and those giving a series of spectra at small time-intervals throughout the chromatographic run. While the former category only gives peak-information about the eluting compounds, the latter can provide valuable structural information. The last category includes hyphenation of HPLC with photodiode array detection (HPLC-PDA or HPLC-DAD) [16], with mass spectrometry (HPLC-MS and HPLC-MS^n) [17], with circular dichroism (HPLC-CD) [18], and with infrared spectroscopy (HPLC-IR) [19]. However, none of these spectroscopic methods allows rigorous structure elucidation. Many previously characterized natural products may be identified using spectral databases, but overall the spectral information obtained from these techniques is insufficient. This contrasts the detailed spectral information obtained from nuclear magnetic resonance (NMR). NMR spectroscopy has the advantage of being a nonselective and nondestructive technique that will detect all hydrogen-containing compounds (for ^{1}H NMR spectroscopy) present above the detection limit. In addition, full structural assignment including relative stereochemistry is available through two-dimensional (2D) experiments, and carbon data are available through heteronuclear [1]H-^{13}C correlation experiments.

The initial hyphenation of HPLC and NMR was carried out with iron magnets during the late 1970s and early 1980s [20–26]. The more recent development of superconducting shielded magnets with higher field strengths has improved the sensitivity and allowed positioning of the HPLC systems close to the magnet. A modern HPLC-NMR system consists of a shielded superconducting magnet (at present typically 400–600 MHz proton frequencies) and a supplementary mass spectrometer coupled directly to the HPLC system, as shown schematically in Figure 3.1. The computer-controlled HPLC system consists typically of a pump capable of delivering a nonpulsating binary or less frequent ternary solvent gradient, an automated injection system, and a column oven to improve reproducibility of the chromatographic run. The eluate is split usually in a 1:19 ratio, with the higher proportion directed to a PDA detector and then the NMR flow-cell through small inner-diameter capillaries. The lower proportion is directed to the more sensitive but also destructive MS detector. The UV- and MS signals are detected by the hyphenation software, and these signals trigger further events according to the preprogrammed system operation mode.

In contrast to traditional NMR operation, where NMR tubes are situated in a rotating spinner to increase effective field homogeneity, the flow-cell is designed as a nonrotating vertical glass tube with radiofrequency (RF) coils attached directly to the middle of the glass tube, and both ends connected to HPLC capillaries [27]. Most applications are performed with "inverse configuration" flow-probes, where the RF coil tuned and matched to the ^{1}H NMR frequency is fixed directly to the glass tube, and the second coil, tuned and matched typically to the ^{13}C NMR frequency, is placed outside the latter. However, normal configuration flow-probes –

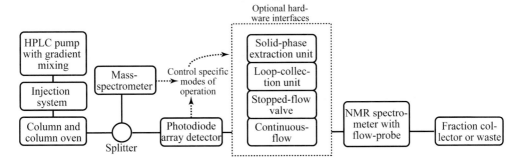

Fig. 3.1 Schematic representation of a typical, hyphenated NMR system. The HPLC column eluate is split in a 1:19 ratio and analyzed in parallel by the mass spectrometer and by the photodiode array detector, and the NMR spectrometer, respectively. Various optional hardware interfaces are used for different HPLC-NMR modes, as illustrated by the dashed box.

that is, with the ^{13}C coil placed inside and closer to the sample for increased sensitivity of ^{13}C detection – are also available. It should also be pointed out, that the active volume of NMR flow-cells (30, 60, 120, and 200 μL in current use) account for only one-half of the total cell volume. Thus, because of free diffusion of the analyte within the cell, only one-half of the analyte present in the flow-cell will actually contribute to the NMR signal.

3.3
Direct HPLC-NMR Methods

There are two fundamentally different ways of hyphenating HPLC and NMR spectroscopy: direct hyphenation, which involves the acquisition of NMR data with the analyte dissolved in the HPLC mobile phase (i.e., with the column effluent), and indirect hyphenation, where the NMR spectra are acquired after the analyte has been isolated from the mobile phase [28, 29].

The direct HPLC-NMR methods use NMR spectrometers similarly to the use of other HPLC detectors, but with adaptations necessary in order to accommodate the lower sensitivity of NMR. A major problem with direct HPLC-NMR methods is related to a limited dynamic range of the analogue-to-digital converter of NMR spectrometers. HPLC is preferably performed with standard nondeuterated solvents, which means that signals from compounds present in low concentrations (milli- or submillimolar) should be detected in the presence of much stronger signals from the solvent (e.g., pure water corresponds to 55 M H_2O). The problem can be partly solved by using deuterated solvents for HPLC, but for most solvents other than D_2O the price is too high for routine applications. In addition, the use of D_2O can complicate mass spectra of compounds containing exchangeable hydrogens, and can also possibly affect the separation via isotope effects. For these reasons it is important to apply techniques that can suppress multiple solvent signals, including suppression of solvent resonances that change position under gradient conditions. For 1D spectra, this can be achieved by a 1D NOESY pulse sequence with pre-saturation of solvent signals during relaxation delay and mixing time [30]. Another possibility is the WET pulse sequence (water suppression enhanced through T_1 effects) [31] that applies a series of variable tip-angle, solvent-selective RF pulses, followed by dephasing field-gradient pulses. The WET pulse sequence is easily combined with standard 2D NMR experiments to give excellent solvent suppression.

3.3.1
Continuous-Flow HPLC-NMR

The continuous-flow HPLC-NMR mode is the simplest mode of operation, and it does not require any hardware interfaces other than a capillary connection to the NMR flow-probe (Fig. 3.1). ^{1}H NMR spectra are acquired continuously during the entire HPLC separation, and the data are saved as a pseudo 2D NMR dataset with the ^{1}H NMR spectral window along one axis and the separation time along the other axis. Because of the limited residence time of the analytes in the NMR flow-cell, it is necessary to make a compromise between the signal-to-noise (S/N) ratio of the NMR data (the number of scans for each 1D dataset), and the resolution along the HPLC separation axis (the total number of spectra). These numbers are inherently inversely proportional. Early studies on the relation between the flow rate, flow-cell volume, and acquisition time have been performed [32–34] in order to optimize continuous-flow HPLC-NMR experiments. Today, most applications of continuous-flow HPLC-NMR to natural products are performed on reversed-phase C_{18} columns ranging between 4 and 8 mm i.d., at flow rates of 0.3 to 1.2 mL min^{-1}, adding 16 to 64 scans per spectrum, and using flow-cells with detection volumes of 60 to 120 µL [18, 35–46]. This results in typical analysis times ranging from 20 min to a few hours, although experiments with a reduced flow rate of 0.15 mL min^{-1} and 256 scans per spectrum, resulting in analysis times exceeding 12 h, have been reported [47].

The main advantage of continuous-flow HPLC-NMR is its simplicity, although the technique is less sensitive than other HPLC-NMR modes discussed below. This is nmostly because of the limited residence time of the analytes in the flow-cell, but also because the volume of most HPLC peaks exceeds the volume of the NMR flow-cell, thereby placing much of the analyte outside the cell during data acquisition. The detection limit of the secoiridoid swertiamarin (MW 374 Da) was shown to be 20 µg in the continuous-flow mode after separation at a flow rate of 1 mL min^{-1}, using a 60-µL flow-probe, a spectrometer operating at 500 MHz, and acquiring 16 scans for each spectrum [48]. A detection limit of 24 µg was found for a minor constituent (MW 568 Da) of an artificial mixture of carotenoids eluted with a flow rate of 0.3 mL min^{-1}, using a 120-µL flow-probe operating at 600 MHz, and acquiring 128 scans for each spectrum [46].

For the analysis of crude extracts by continuous-flow HPLC-NMR, the amount injected onto the column is typically well above 1 mg (depending on the complexity of the mixture and the loading capacity of the column), but for partly purified fractions good spectra can be acquired with smaller amounts. One example is a study employing combined use of HPLC-MS and continuous-flow HPLC-NMR to a partially purified extract of *Vernonia fastigiata* [49]. Separation of 700 µg of the extract was performed on a C_{18} column eluted with a water-acetonitrile gradient or with a water-methanol gradient. This provided two sets of spectra with no overlap of regions affected by solvent suppression, and allowed rapid identification of nine sesquiterpene lactones. Because the continuous-flow HPLC-NMR mode is independent of other peak-selection methods, it is normally used in initial studies of the major components of plant extracts, followed by HPLC-NMR investigations under static conditions.

3.3.2
Stopped-Flow HPLC-NMR

HPLC-NMR systems can be operated in the stopped-flow mode, which gives the time needed to acquire 1D spectra with increased S/N ratio compared to those obtained in continuous-flow mode or, if necessary, to perform 2D NMR experiments needed for full structure elucidation of extract constituents. As the name implies, the flow is stopped when the analyte of interest enters the flow-cell. The compounds of interest for stopped-flow acquisition are usually detected using an upstream PDA-detector or MS, and either threshold limits in UV- and/or mass TIC chromatograms, selective ion chromatograms, or daughter ions from MS/MS, are used to trigger the stopped-flow valve for exact positioning of the analyte in the flow-cell (see Fig. 3.1). All operations are carefully timed and automated by hyphenation software that controls both the HPLC system, including the PDA detector and stopped-flow valve, the mass spectrometer, and the NMR data acquisition. The stopped-flow mode allows careful optimization of shims, and especially the solvent suppression benefits from improved field homogeneity. Tuning, matching, and pulse calibration can also be more carefully optimized under static conditions. This, combined with prolonged data accumulation times, provides the possibility of acquiring spectra with reasonable S/N ratio for compounds other than the major constituents. Prolonged periods with stopped flow may cause diffusion-mediated band broadening on the HPLC column, and therefore repeated stops and restarts of the flow may lead to unacceptable loss of sensitivity. This problem can be solved by using separate chromatographic runs for each analyte of interest, but for most applications related to natural products, where extracts contain many different compounds and chromatographic runs exceed 30 min, this is not an attractive solution.

The stopped-flow HPLC-NMR mode has been used for investigations of constituents of *Hypericum perforatum* extract [50], where UV and MS data were used to trigger stopped-flow measurements. A single pre-purification step by solid-phase extraction allowed acquisition of good quality ¹H NMR spectra of minor constituents at the level of 0.1%.

Another study [51] used the continuous-flow mode for initial investigation of dichloromethane extract of *Swertia calycina*, followed by stopped-flow ¹H NMR and 2D WETCOSY experiments for more reliable structural characterization of one of the extract constituents, sweroside. A comparison of the ¹H NMR spectrum obtained from continuous-flow and from stopped-flow HPLC-NMR is shown in Figure 3.2. Several other applications of the stopped-flow HPLC-NMR mode, including use of COSY, TOCSY, NOESY, ROESY, and HSQC experiments, have been reported [37, 46, 52].

Another important application of stopped-flow HPLC-NMR is the determination of absolute configuration of secondary alcohols by the Mosher method [53–55]. The traditional way of analyzing Mosher's esters involves the formation of several milligrams of each diastereomer and subsequent preparative-scale isolation for NMR data acquisition in tubes. In the HPLC-NMR approach, only a few micrograms of the secondary alcohol are necessary. Formation of the esters is performed in HPLC vials, and the reaction mixture is analyzed by stopped-flow HPLC-NMR immediately after ester formation is complete. Thus, the reduced volume and increased sensitivity of the NMR flow-probe makes it possible to establish the absolute stereochemistry of natural products at a very low concentration level.

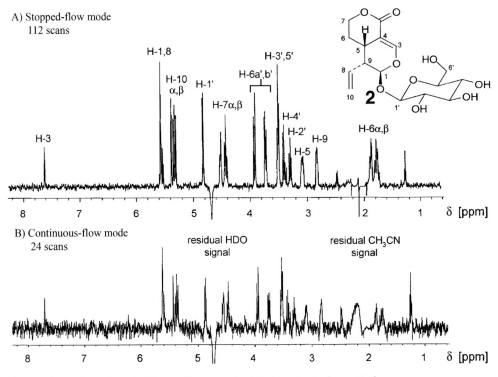

A) Stopped-flow mode
112 scans

H-3

H-1,8

H-10 α,β

H-1'

H-7α,β

H-6a',b'

H-3',5'

H-4'

H-2'

H-5

H-9

H-6α,β

8   7   6   5   4   3   2   1   δ [ppm]

B) Continuous-flow mode
24 scans

residual HDO
signal

residual CH₃CN
signal

8   7   6   5   4   3   2   1   δ [ppm]

Fig. 3.2 Comparison of stopped-flow mode (A) and continuous-flow mode (B) HPLC-NMR spectra of sweroside from the crude extract of *Swertia calycina*. Crude extract (1 mg) was injected onto a 150×3.9 mm i.d. C₁₈ column eluted with a linear gradient of acetonitrile in water at a flow-rate of 1 mL min⁻¹. Spectra were acquired at 500 MHz using a 60 μL inverse detection flow-probe. (Reproduced with permission from [51]; Copyright John Wiley & Sons, Ltd.)

Partly overlapping peaks can be investigated with the time-slice mode of operation. In this mode, the flow is stopped at short intervals during the peak elution, acquiring a series of ¹H NMR spectra under stopped-flow conditions. The method is essentially stopped-flow performed at regular intervals, and as such it requires the stopped-flow valve. Upstream UV or MS detectors can be used to control time-slicing of selected HPLC peaks, or time-slicing can be applied to the entire chromatogram, in order not to overlook compounds that do not respond to these detectors. One example of the latter method is analysis of beer [56].

For compounds with MW around 500 Da, the detection limit in stopped-flow experiments at the ¹H observation frequency of 600 MHz is about 100 ng [57]. The acquisition of 2D ¹H-¹H data with 1 μg can be performed in a few hours, whereas overnight acquisition of 2D ¹H-¹³C correlation data requires approximately 20 μg of material [48].

3.3.3
Loop-Collection HPLC-NMR

The loop-collection mode makes use of a so-called peak sampling unit (see Fig. 3.1). The system contains 12 or 36 capillary loops, each of which can be made a part of the HPLC eluent transfer path through the use of automated valves, triggered by a detector signal threshold. This makes it possible to collect 12 or 36 compounds in individual loops without stopping the chromatographic run, thereby removing diffusion problems associated with stopped-flow HPLC-NMR. Once chromatography is completed, the loop content is transferred to the NMR flow-cell through connecting capillaries. At this post-chromatographic stage, the loop-collection unit can be regarded as an autosampler for the NMR system, and both short- and long-term NMR experiments can be conducted independently from the chromatographic separation. Another advantage of the loop-collection mode compared to the stopped-flow mode is that an additional washing step of the NMR flow-cell can be introduced between each transfer of the analyte, thereby avoiding any sample carryover.

Loop-collection HPLC-NMR has proven to be a valuable method in the field of natural products research [58, 59]. One example is a study employing HPLC-NMR-MS/MS for identification of quercetin and phloretin glycosides in apple peel extract [58]. Several closely eluting peaks, as determined by UV- and TIC MS-traces, were observed in the chromatogram. Threshold levels in UV spectra as well as detection of masses corresponding to quercetin (m/z 302) and phloretin (m/z 274) in MS/MS experiments, were used to trigger loop collection. This study successfully identified nine quercetin and phloretin glycosides in varying amounts using 1D TOCSY spectra. The 1D ^{1}H NMR spectra for the least-abundant component, rutin, which was present in an amount of approximately 4 μg, was obtained in 1.5 h.

The loop-collection method suffers from the same restrictions, with respect to sensitivity and solvent suppression, as the stopped-flow method. In principle, 2D ^{1}H-^{1}H and inverse ^{1}H-^{13}C correlation spectra can be obtained, but this requires a high analyte concentration in the chromatographic elution band, a requirement inherent to the use of direct-type HPLC-NMR methods. The limit of detection is thus, in principle, the same as for stopped-flow HPLC-NMR.

3.4
Indirect HPLC-NMR Methods

The only indirect HPLC-NMR method developed to date is the HPLC-SPE-NMR technique. Here, separation of the analyte from the HPLC mobile phase is achieved by means of solid-phase extraction [29]. Indirect NMR hyphenation is therefore, in fact, an automated, on-line, microscale version of traditional procedures involving "preparative" isolation of mixture components by chromatography, followed by their examination using NMR tubes.

One major disadvantage of direct HPLC-NMR methods is the relative incompatibility between standard HPLC solvents and NMR spectroscopy. This problem is only partly solved with modern solvent peak-suppression techniques, because analyte resonances under or close to the suppressed peak are also affected. The limited detectability due to limited loading capacity of the columns is another disadvantage, which is further amplified by the fact that the separated analytes are often distributed in peak elution volumes that are much larger than the active volume of the NMR flow-cell. In addition, the HPLC solvent gradients typically used in continuous-flow experiments create problems with changing magnetic susceptibility and field homogeneity as well as shifts of analyte resonances, thereby complicating identification based on reference chemical shift values found in literature or in spectral databases. The newly developed hyphenation of HPLC and NMR intervened by an automated solid-phase extraction unit (HPLC-SPE-NMR) circumvents many of the problems listed above, while still working as a fully integrated and automated technique.

3.4.1
HPLC-SPE-NMR

Solid-phase extraction has previously been used for on-line concentration of samples before chromatographic separation and NMR analysis – that is, as the SPE-HPLC-NMR technique [60]. However, the loading capacity of the HPLC column is still a factor limiting NMR performance of this method. Solid-phase extraction has also been used for off-line analyte isolation from HPLC fractions prior to traditional NMR analysis [61]. Another more recent approach used off-line preparative-scale HPLC followed by on-line SPE-NMR [62,63]. The first reported on-line HPLC-SPE-NMR coupling [64] employed post-column dilution of effluent from of a reversed-phase column with D_2O to facilitate analyte trapping on a C_{18} guard column. The guard column was subsequently back-flushed with CD_3CN in order to desorb the analyte and transfer it into the NMR flow-cell in a narrow elution band. This study showed an increment in the S/N ratio of 1.5 to 2.3 compared to stopped-flow HPLC-NMR. Modern, commercially available HPLC-SPE-NMR systems include an on-line and fully automated solid-phase extraction unit (see Fig. 3.1). In this setup, a PDA detector and/or MS are used to trigger the SPE trapping of analytes similarly to triggering containment of the analyte band in the loop-storage mode.

The principle of operation of the HPLC-SPE-NMR system is shown schematically in Figure 3.3. In the trapping stage, the HPLC eluate is diluted with water to decrease the eluting power. The analytes of interest are then individually adsorbed (trapped) on small cartridges, each containing a suitable SPE material equilibrated with water, via automated valve switching. In the drying stage, pressurized nitrogen gas is used to dry the cartridges, thereby avoiding the appearance of signals from nondeuterated HPLC and post-column dilution solvents in the NMR spectra. SPE cartridges are stored in a nitrogen atmosphere to avoid contamination and to reduce the risk of analyte oxidation. Trapped analytes are transferred to the NMR flow-cell in the elution stage, where dried cartridges are flushed with a deuterated organic solvent.

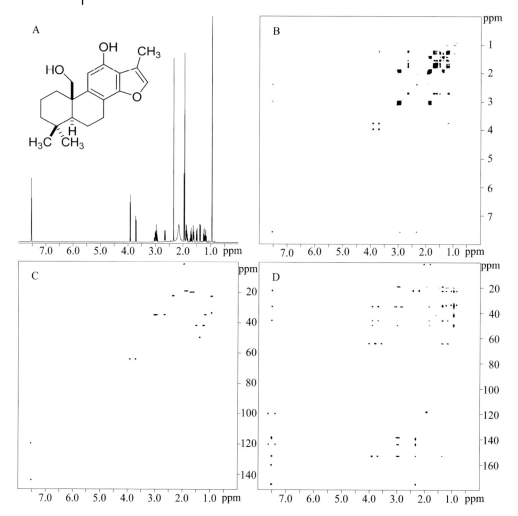

Fig. 3.4 (A) 600 MHz 1D ^{1}H NMR, (B) COSY, (C) HSQC, and (D) HMBC spectra acquired using the HPLC-SPE-NMR mode. The spectra allow full structural assignment including ^{13}C resonances of a diterpene from the crude root extract of a lamiaceous plant. Crude extract (1.5 mg) was injected onto a 150×4.6 mm i.d. C$_{18}$ column eluted with a linear gradient of acetonitrile in water at a flow rate of 0.8 mL min^{-1}. Post-column dilution with water at 0.8 mL min^{-1} and six repeated trappings on a 10×2 mm i.d. SPE cartridge was followed by 30-min drying with pressurized N$_2$-gas and elution with acetonitrile-d_3. The spectra were acquired using a 30 µL inverse-detection flow-probe. All spectra were acquired without solvent suppression (unpublished results).

allowed linear accumulation of an isoflavonoid in up to seven repeated trappings [70]. One study reported a correlation between trapping efficiency and polarity in a series of compounds in an artificial mixture [74], whereas another demonstrated complex trapping relationships for a series of iridoid glycosides [67]. Therefore, the choice of SPE material and the post-column dilution ratio as well as the solvent used for cartridge elution should be considered on a case-to-case basis.

3.5
Cryogenically Cooled NMR Probes

Most of the noise observed in NMR spectra is thermal electronic noise created by the resistance of the receiver coil [75], and thus significant reduction of the noise level can be achieved by cryogenic cooling. In theory, the noise level upon cooling of the transceiver from room temperature to boiling liquid helium temperature will drop by a factor of 22 [76]. However, current technical implementations offer an improvement in S/N ratio by a factor of about 4 (for nonresistive samples) in comparison to spectra acquired with the same samples using room-temperature probes. This is mainly due to the loss of the filling factor (ratio between sample volume and coil detection volume), since the sample must be maintained at room temperature and appropriate insulation space must be left between the sample and the cold transceiver. Modern commercial cryogenic probes use recirculating, cold helium gas rather than liquid helium to cool the transceiver coil to about 20 K. Preamplifiers are enclosed in the same Dewar system and are cooled to about 80 K [77]. A significant advance is the recent development of convertible cryogenic probes that allow change between conventional NMR tubes and flow-cell inserts, which enables use of a single cryogenic probe for various applications and saves probe cool-down and warm-up time.

The first, and still the most popular, area of application of cryogenic flow-probes is in pharmaceutical analysis and drug metabolism [77]. As mentioned earlier, the combined advantage of the cryogenic probe and HPLC-SPE-NMR was demonstrated in the study of Greek oregano extract [65], and further proliferation of the cryogenic flow-probe technology in the field of natural products research is expected.

3.6
Miniaturization

There are at least three main reasons for the development of miniaturized NMR probes and their hyphenation with separation techniques. First, many samples for which NMR analyses are required are limited in size; this applies especially to samples of biological origin. Second, the mass-sensitivity (sensitivity per mass unit) increases with decreased dimensions of the NMR transceiver coils [75,78]. Third, miniaturized separation techniques usually work with higher concentrations of the analyte in the chromatographic bands, as the analyte concentration is typically inversely proportional to the square of the column diameter. This results in an increased analyte-to-solvent ratio for a given amount of injected material, making solvent suppression easier. Moreover, low solvent consumption in capillary LC separations allows the use of fully deuterated eluents.

Traditional NMR flow-cells are constructed with vertical cavities (tubes) placed along the main magnetic field of a superconducting magnet with a vertical bore. Since the NMR experiments involve the creation and detection of transverse magnetization, the magnet geometry necessitates the use of saddle-type (Helmholtz) transceiver coils. While the above-mentioned miniaturization benefits apply for the saddle-type coils [79], miniaturization enables use of solenoidal design with cell axes perpendicular to the magnet bore. For a given coil volume, solenoidal coils perform about three times better than the saddle-type coils [75]. For commercial capillary NMR probes, which have total cell volumes of 5 or 10 µL, the mass sensitivity is five- to six-fold higher than for conventional probes. This is roughly equivalent to a cryogenically cooled probe with a 3-mm sample tube.

To date, most published reports employing capillary probes with solenoidal microcoils use direct analyte injection [80–83]. However, several studies combining capillary NMR probes with capillary HPLC separation of natural products have been published [84, 85]. An example is shown in Figure 3.5.

Capillary NMR probes with solenoidal microcoils are capable of detecting low-nanogram quantities of analytes (¹H detection), and the methodology is adaptable for high-throughput applications, both for pressure-driven and electro-driven separations [86–91]. The small size of the microcoils makes construction of probes with multiple cells feasible [92–94]. Moreover, since these probes contain a single transceiver coil, there is no sensitivity loss for direct ¹³C detection related to filling-factor differences that are typical for inverse-detection dual-coil probes. However, hundreds of micrograms are still required to obtain ¹³C NMR spectra in a reasonably short time [81, 87, 95], and these amounts are not easily achievable in the capillary HPLC mode.

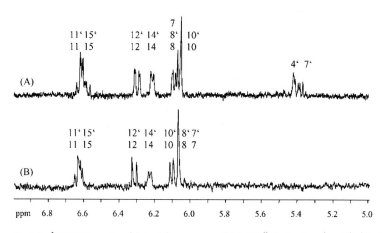

Fig. 3.5 ¹H NMR spectra obtained from a mixture of carotenoid standards using a stopped-flow capillary HPLC-NMR system (15 cm×250 µm C$_{30}$ column) with acetone-*d$_6$*: D$_2$O gradient separation (5 µL min⁻¹, 5 µL capillary NMR probe with detection volume of 1.5 µL, 500-nL column injection volume). (A) 50 ng of *E*-lutein; (B) 50 ng of *E*-zeaxanthin. (Reproduced with permission from [85]; Copyright Elsevier.)

3.7
Conclusions

Of the two novel and maturing techniques in the area of NMR hyphenation, HPLC-SPE-NMR and stopped-flow capillary HPLC-NMR offer the highest sensitivity improvements, and are the most promising solutions. Both techniques have advantages and disadvantages. While the latter method exhibits the highest mass-sensitivity and is well suited for mass-limited samples and possibly for high-throughput applications, the higher absolute analyte amounts available through HPLC-SPE-NMR are advantageous when no severe limitations to the sample size exist. Moreover, at least in the drug discovery context, the HPLC-NMR experiment will often be an introductory experiment prior to compound isolation, with the aim of performing biological tests. Thus, standard-bore HPLC columns used for HPLC-NMR can be immediately used to isolate sufficient amounts of material for bioassays, whereas capillary HPLC-NMR must be regarded as a purely analytical tool. Moreover, the use of cryogenic probes in the case of HPLC-SPE-NMR matches the mass-sensitivity advantage of microcoils, which have yet to be combined with the cryogenic technology. The combination of HPLC-SPE-NMR with microcoil probe design and with cryogenic technology will provide further sensitivity advancement, at least for compounds compatible with solid-phase extraction, a property which is fulfilled by the majority of secondary plant metabolites. Thus, it can be concluded that NMR hyphenation offers flexibility, robustness, and sensitivity that are appropriate for a wide range of applications related to medicinal plants.

References

References marked * are special of interest; those marked ** are general papers of considerable interest

1 **Cragg, G.M., Newmann, D.J., Snader, K.M. Natural products in drug discovery and development. *J. Nat. Prod.* **1997**, *60*, 52–60. [Useful review of the role of natural products in drug discovery.]

2 **Newmann, D.J., Cragg, G.M., Snader, K.M. Natural products as sources of new drugs over the period 1981–2002. *J. Nat. Prod.* **2003**, *66*, 1022–1037. [Useful review of the role of natural products in drug discovery.]

3 Pitman, N.C.A., Jørgensen, P.M. Estimating the size of the world's threatened flora. *Science* **2002**, *298*, 989.

4 Chadwick, D.J., Marsh, J. *Ethnobotany and the Search for New Drugs.* John Wiley & Sons, Chichester, **1994**.

5 Iwu, M.M., Wootton, J.C. *Ethnomedicine and Drug Discovery.* Elsevier, Amsterdam, **2002**.

6 Polya, G. *Biochemical Targets of Plant Bioactive Compounds.* Taylor and Francis, London, **2003**.

7 Zhang, L., Demain, A.L. *Natural Products – Drug Discovery and Therapeutic Medicine.* Humana Press, Totowa, **2005**.

8 ** Bindseil, K.U., Jakupovic, J., Wolf, S.D., Lavayre, J., Leboul, J., van der Pyl, D. Pure compound libraries: a new perspective for natural product based drug discovery. *Drug Discov. Today* **2001**, *4*, 206–220. [An important review of pure compound libraries based on natural products.]

9 ** Koehn, F.E., Carter, G.T. The evolving role of natural products in drug discovery. *Nature Rev. Drug Discov.* **2005**, *4*, 206–220. [An important review of the renewed interest of natural products in drug discovery.]

10 Larsson, J., Gottfries, J., Bohlin, L., Backlund, A. Expanding the ChemGPS chemical space with natural products. *J. Nat. Prod.* **2005**, *68*, 985–991.

11 Feher, M., Jonathan, M.. Schmidt, J.M. Property distributions: differences between drugs, natural products, and molecules from combinatorial chemistry. *J. Chem. Inf. Comput. Sci.* **2003**, *43*, 218–227.

12 Romeo, J.T., Dixon, R.A. (Eds.), *Phytochemistry in the Genomics and Post-Genomics Eras.* Pergamon, Amsterdam, **2002**.

13 Sumner, L.W., Mendes, P., Dixon, R.A. Plant metabolomics: large-scale phytochemistry in the functional genomics era. *Phytochemistry* **2003**, *62*, 817–836.

14 Zhang, W., Franco, C., Curtin, C., Conn, S. To stretch the boundary of secondary metabolite production in plant cell-based bioprocessing: anthocyanin as a case study. *J. Biomed. Biotechnol.* **2004**, *5*, 264–271.

15 Sweetlove, L.J., Fernie, A.R. Regulation of metabolic networks: understanding metabolic complexity in the systems biology era. *New Phytologist* **2005**, *168*, 9–24.

16 Huber, L., George, S.A. *Diode Array Detection in HPLC* . Marcel Dekker, New York, **1993**.

17 Niessen, W.M.A. *Liquid Chromatography-Mass Spectrometry*, 2nd edn. Marcel Dekker, New York, **1999**.

18 Bringmann, G., Messer, K., Wohlfarth, M., Kraus, J., Dumbuya, K., Rückert, M. HPLC-CD on-line coupling in combination with HPLC-NMR and HPLC-MS/MS for the determination of the full absolute stereostructure of new metabolites in plant extracts. *Anal. Chem.* **1999**, *71*, 2678–2686.

19 Somsen, G.W., Gooijer, C., Brinkmann, U.A.Th. Liquid chromatography-Fourier-transform infrared spectrometry. *J. Chromatogr. A* **1999**, *856*, 213–243.

20 Watanabe, N., Niki, E. Direct-coupling of FT-NMR to high-performance liquid chromatography. *Proc. Jpn. Acad. Ser. B* **1978**, *54*, 194–199.

21 Bayer, E., Albert, K., Nieder, M., Grom, E., Keller, T. On-line coupling of high-performance liquid chromatography and nuclear magnetic resonance. *J. Chromatogr.* **1979**, *186*, 497–507.

22 Haw, J.F., Glass, T.E., Hausler, D.W., Dorn, H.C. Direct coupling of a liquid chromatograph to a continuous flow hydrogen nuclear magnetic resonance detector for analysis of petroleum and synthetic fuels. *Anal. Chem.* **1980**, *52*, 1135–1140.

23 Bayer, E., Albert, K., Nieder, M., Grom, E., An, Z. Examples of application of direct HPLC-NMR-coupling with high resolution. *Fresenius Z. Anal. Chem.* **1980**, *304*, 111–116.

24 Buddrus, J., Herzog, H. Coupling of HPLC and NMR. 1. Analysis of flowing liquid-chromatographic fractions by proton magnetic resonance. *Org. Magn. Res.* **1980**, *13*, 153–155.

25 Haw, J.F., Glass, T.E., Dorn, H.C. Continuous flow high field nuclear magnetic resonance detector for liquid chromatographic analysis of fuel samples. *Anal. Chem.* **1981**, *53*, 2327–2332.

26 Bayer, E., Albert, K., Nieder, M., Grom, E., Wolff, G., Rindlisbacher, M. On-line coupling of liquid chromatography and high-field nuclear magnetic resonance spectrometry. *Anal. Chem.* **1982**, *54*, 1747–1750.

27 Albert, K. On-line use of NMR detection in separation chemistry. *J. Chromatogr. A* **1995**, *703*, 123–147.

28 ** Jaroszewski, J.W. Hyphenated NMR methods in natural products research, part 1: direct hyphenation. *Planta Med.* **2005**, *71*, 691–700. [A comprehensive review of direct HPLC-NMR methods in natural products research.]

29 ** Jaroszewski, J.W. Hyphenated NMR methods in natural products research, part 2: HPLC-SPE-NMR and other new trends in NMR hyphenation. *Planta Med.* **2005**, *71*, 795–802. [A comprehensive review of HPLC-SPE-NMR methods in natural products research.]

30 Nicholson, J.K., Foxal, P.J.D., Spraul, M., Farrant, R.D., Lindon, J.C. 750 MHz ^{1}H and ^{1}H-^{13}C NMR spectroscopy of human blood plasma. *Anal. Chem.* **1995**, *67*, 793–811.

31 Smallcombe, S.H., Patt, S.L., Keifer, P.A. WET solvent suppression and its applications to LC NMR and high-resolution

NMR spectroscopy. *J. Magn. Reson. Ser. A* **1995**, *117*, 295–303.

32 Albert, K., Nieder, M., Bayer, E., Spraul, M. Continuous-flow nuclear magnetic resonance. *J. Chromatogr.* **1985**, *346*, 17–24.

33 Griffiths, L. Optimization of NMR and HPLC conditions for LC-NMR. *Anal. Chem.* **1995**, *67*, 4091–4095.

34 Griffiths, L. Optimization of liquid chromatography-NMR spectroscopy. *Magn. Reson. Chem.* **1997**, *35*, 257–261.

35 Bringmann, G., Rückert, M., Messer, K., Schupp, O., Louis, A.M. Acetogenic isoquinoline alkaloids CXXI. Use of on-line high-performance liquid chromatography-nuclear magnetic resonance spectrometry coupling in phytochemical screening studies: rapid identification of metabolites in *Dioncophyllum thollonii*. *J. Chromatogr. A* **1999**, *837*, 267–272.

36 Bringmann, G., Günthe, C., Schlauer, J., Rückert, M. HPLC-NMR on-line coupling including the ROESY technique: direct characterization of naphthylisoquinoline alkaloids in crude plant extracts. *Anal. Chem.* **1998**, *70*, 2805–2811.

37 * Garo, E., Wolfender, J.-L., Hostettmann, K. Prenylated flavanones from *Monotes engleri*: on-line structure elucidation by LC/UV/NMR. *Helv. Chim. Acta* **1998**, *81*, 754–763. [An illustrative example of stopped-flow HPLC-NMR.]

38 Cavin, A., Potterat, O., Wolfender, J.-L., Hostettmann, K., Dyatmyko, W. Use of on-flow LC/^{1}H NMR for the study of an antioxidant fraction from *Orophea enneandra* and isolation of a polyacetylene, lignane, and a tocopherol derivative. *J. Nat. Prod.* **1998**, *61*, 1497–1501.

39 Ioset, J.-R., Wolfender, J.-L., Marston, A., Gupta, M.P., Hostettmann, K. Identification of two isomeric meroterpenoid naphthoquinones from *Cordia linnaei* by liquid chromatography-mass spectrometry and liquid chromatography-nuclear magnetic resonance spectroscopy. *Phytochem. Anal.* **1999**, *10*, 137–142.

40 Wolfender, J.-L., Verotta, L., Belvisi, L., Fuzzati, N., Hostettmann, K. Structural investigation of isomeric oxidised forms of hyperforin by HPLC-NMR and HPLC-MS^n. *Phytochem. Anal.* **2003**, *14*, 290–297.

41 Waridel, P., Wolfender, J.-L., Lachavanne, J.-B., Hostettmann, K. Identification of the polar constituents of *Potamogeton* species by HPLC-UV with post-column derivatization, HPLC-MS^n and HPLC-NMR, and isolation of a new *ent*-labdane diglycoside. *Phytochemistry* **2004**, *65*, 2401–2410.

42 Gavidia, I., Seitz, H.U., Pérez-Bermúdez, P., Vogler, B. LC-NMR applied to the characterisation of cardiac glycosides from three micropropagated *Isoplexis* species. *Phytochem. Anal.* **2002**, *13*, 266–271.

43 Setzer, W.N., Vogler, B., Bates, R.B., Schmidt, J.M., Dicus, C.W., Nakkiew, P., Haber, W.A. HPLC-NMR/HPLC-MS analysis of the bark extract of *Stauranthus perforatus*. *Phytochem. Anal.* **2003**, *14*, 54–59.

44 Renukappa, T., Roos, G., Klaiber, I., Vogler, B., Kraus, W. Application of high-performance liquid chromatography coupled to nuclear magnetic resonance spectrometry, mass spectrometry, and bioassay for the determination of active saponins from *Bacopa monniera* Wettst. *J. Chromatogr. A* **1999**, *847*, 109–116.

45 Pusecker, K., Albert, K., Bayer, E. Investigation of hop bitter acids by coupling of high-performance liquid chromatography to nuclear magnetic resonance spectroscopy. *J. Chrom. A* **1999**, *836*, 245–252.

46 * Dachtler, M., Glaser, T., Kohler, K., Albert, K. Combined HPLC-MS and HPLC-NMR on-line coupling for the separation and determination of lutein and zeaxanthin stereoisomers in spinach and in retina. *Anal. Chem.* **2001**, *73*, 667–674. [An illustrative example of stopped-flow HPLC-NMR.]

47 Waridel, P., Wolfender, J.-L., Lachavanne, J.-B., Hostettmann, K. *Ent*-labdane glycosides from the aquatic plant *Potamogeton lucens* and analytical evaluation of the lipophilic extract constituent of various *Potamogeton* species. *Phytochemistry* **2004**, *65*, 945–954.

48 Wolfender, J.-L., Ndjoko, K., Hostettmann, K. The potential of LC-NMR in phytochemical analysis. *Phytochem. Anal.* **2001**, *12*, 2–22.

49 Vogler, B., Klaiber, I., Roos, G., Walter, C.U., Holler, W., Sandor, P., Kraus, W. Combination of LC-MS and LC-NMR as a tool for the structure determination of natural products. *J. Nat. Prod.* **1998**, *61*, 175–178.

50 * Hansen, S.H., Jensen, A.G., Cornett, C., Bjørnsdottir, I., Taylor, S., Wright, B., Wilson, I.D. High-performance liquid chromatography on-line coupled to high-field NMR and mass spectrometry for structure elucidation of constituents of *Hypericum perforatum* L. *Anal. Chem.* **1999**, *71*, 5235–5241. [An illustrative example of stopped-flow HPLC-NMR.]

51 * Wolfender, J.-L., Rodriguez, S., Hostettmann, K., Hiller, W. Liquid chromatography/ultra violet/mass spectrometric and liquid chromatography/nuclear magnetic resonance spectroscopic analysis of crude extracts of gentianaceae species. *Phytochem. Anal.* **1997**, *8*, 97–104. [An illustrative example of stopped-flow HPLC-NMR.]

52 * Bringmann, G., Wohlfarth, M., Rischer, H., Schlauer, J., Brun, R. Extract screening by HPLC coupled to MS-MS, NMR, and CD: a dimeric and three monomeric naphthylisoquinoline alkaloids from *Ancistrocladus griffithii*. *Phytochemistry* **2002**, *61*, 195–204. [An illustrative example of stopped-flow HPLC-NMR.]

53 * Queiroz, E.F., Wolfender, J.-L., Raelison, G., Hostettmann, K. Determination of the absolute configuration of 6-alkylated α-pyrones from *Ravensara crassifolia* by LC-NMR. *Phytochem. Anal.* **2003**, *14*, 34–39. [An illustrative example of stopped-flow HPLC-NMR.]

54 * Guilet, D., Guntern, A., Ioset, J.-R., Queiroz, E.F., Ndjoko, K., Foggin, C.M., Hostettmann, K. Absolute configuration of a tetrahydrophenanthrene from *Heliotropium ovalifolium* by LC-NMR of its Mosher esters. *J. Nat. Prod.* **2003**, *66*, 17–20. [An illustrative example of stopped-flow HPLC-NMR.]

55 * Seco, J.M., Tseng, L.-H., Godejohann, M., Quinoa, E., Riguera, R. Simultaneous enantioresolution and assignment of absolute configuration of secondary alcohols by directly coupled HPLC-NMR of 9-AMA esters. *Tetrahedron Asymm.* **2002**, *13*, 2149–2153. [An illustrative example of stopped-flow HPLC-NMR.]

56 * Gil, A.M., Duarte, I.F., Godejohann, M., Braumann, U., Maraschin, M., Spraul, M. Characterization of the aromatic composition of some liquid foods by nuclear magnetic resonance and mass spectrometric detection. *Anal. Chim. Acta* **2003**, *488*, 35–51. [An illustrative example of time-sliced HPLC-NMR.]

57 Corcoran, O., Spraul, M. LC-NMR-MS in drug discovery. *Drug Discov. Today* **2003**, *8*, 624–631.

58 * Lommen, A., Godejohann, M., Venema, D.P., Hollman, P.C.H., Spraul, M. Application of directly coupled HPLC-NMR-MS to the identification and confirmation of quercetin glycosides and phloretin glycosides in apple peel. *Anal. Chem.* **2000**, *72*, 1793–1797. [An illustrative example of loop-collection HPLC-NMR.]

59 Tseng, L.H., Braumann, U., Godejohann, M., Lee, S.S., Albert, K. Structure identification of aporphine alkaloids by on-line coupling of HPLC-NMR with loop storage. *J. Chin. Chem. Soc.* **2000**, *47*, 1231–1236.

60 deKoning, J.A., Hogenboom, A.C., Lacker, T., Albert, K., Brinkmann, U.A.Th. On-line trace enrichment in hyphenated liquid chromatography-nuclear magnetic resonance spectroscopy. *J. Chromatogr. A* **1998**, *813*, 55–61.

61 Wilcox, C.D., Phelan, R.M. The use of solid-phase extraction columns to effect simple offline LC/MS, LC/NMR, and LC/FTIR. *J. Chromatogr. Sci.* **1986**, *24*, 130–133.

62 Nyberg, N.T., Baumann, H., Kenne, L. Application of solid-phase extraction coupled to an NMR flow-probe in the analysis of HPLC fractions. *Magn. Reson. Chem.* **2001**, *39*, 236–240.

63 Nyberg, N.T., Baumann, H., Kenne, L. Solid-phase extraction NMR studies of chromatographic fractions of saponins from *Quillaja saponaria*. *Anal. Chem.* **2003**, *75*, 268–274.

64 ** Griffiths, L., Horton, R. Optimization of LC-NMR III – Increased signal-to-noise ratio through column trapping. *Magn. Reson. Chem.* **1998**, *36*, 104–109. [The work presented represents a milestone in development of the HPLC-SPE-NMR method.]

65 * Exarchou, V., Goodejohann, M., van Beek, T.A., Gerothanassis, I.P., Vervoort, J. LC-UV-solid-phase extraction-NMR-MS combined with a cryogenic flow probe and its application to the identification of compounds present in Greek oregano. *Anal. Chem.* **2003**, *75*, 6288–6294. [An illustrative example of application of the HPLC-SPE-NMR method.]

66 * Wang, C.-Y., Lee, S.-S. Analysis and identification of lignans in *Phyllanthus urinaria* by HPLC-SPE-NMR. *Phytochem. Anal.* **2005**, *16*, 120–126. [An illustrative example of application of the HPLC-SPE-NMR method.]

67 * Seger, C., Godejohann, M., Tseng, L.H., Spraul, M., Girtler, A., Sturm, S., Stuppner, H. LC-DAD-MS/SPE-NMR hyphenation. A tool for the analysis of pharmaceutically used plant extracts: Identification of isobaric iridoid glycoside regiomers from *Harpagophytum procumbens. Anal. Chem.* **2005**, *77*, 878–885. [An illustrative example of application of the HPLC-SPE-NMR method.]

68 * Clarkson, C., Stærk, D., Hansen, S.H., Jaroszewski, J.W. Hyphenation of solid-phase extraction with liquid chromatography and nuclear magnetic resonance: application of HPLC-DAD-SPE-NMR to identification of constituents of *Kanahia laniflora. Anal. Chem.* **2005**, *77*, 3547–3553. [An illustrative example of application of the HPLC-SPE-NMR method.]

69 * Christophoridou, S., Dais, P., Tseng, L.-H., Spraul, M. Separation and identification of phenolic compounds in olive oil by coupling high-performance liquid chromatography with postcolumn solid-phase extraction to nuclear magnetic resonance spectroscopy (LC-SPE-NMR). *J. Agric. Food Chem.* **2005**, *53*, 4667–4679. [An illustrative example of application of the HPLC-SPE-NMR method.]

70 * Lambert, M., Stærk, D., Hansen, S.H., Sairafianpour, M., Jaroszewski, J.W. Rapid extract dereplication using HPLC–SPE–NMR: Analysis of isoflavanoids from *Smirnowia iranica. J. Nat. Prod.* **2005**, *68*, 1500–1509. [An illustrative example of application of the HPLC-SPE-NMR method.]

71 * Miliauskas, G., van Beek, T.A., de Ward, P., Venskutonis, R.P., Sudhölter, E.J.R. Identification of radical scavenging compounds in *Rhaponticum carthamoides* by means of LC-DAD-SPE-NMR. *J. Nat. Prod.* **2005**, *68*, 168–172. [An illustrative example of application of the HPLC-SPE-NMR method.]

72 * Pukalskas, A., van Beek, T.A., de Ward, P. Development of a triple hyphenated HPLC-radical scavenging detection-DAD-SPE-NMR system for the rapid identification of antioxidants in complex plant extracts. *J. Chrom. A* **2005**, *1074*, 81–88. [An illustrative example of application of the HPLC-SPE-NMR method.]

73 Lambert, M., Stærk, D., Hansen, S.H., Jaroszewski, J.W. HPLC–SPE–NMR hyphenation in natural products research: optimization of analysis of *Croton membranaceus* extract. *Magn. Reson. Chem.* **2005**, *43*, 771–775.

74 * Sandvoss, M., Bardsley, B., Beck, T.L., Lee-Smith, E., North, S.E., Moore, P.J., Edwards, A.J., Smith, R.J. HPLC-SPE-NMR in pharmaceutical development: capabilities and applications. *Magn. Reson. Chem.* **2005**, *43*, 762–770. [An illustrative example of application of the HPLC-SPE-NMR method.]

75 ** Hoult, D.I., Richards, R.E. Signal-to-noise ratio of nuclear magnetic resonance experiment. *J Magn. Reson.* **1976**, *24*, 71–85. [Seminal account of factors affecting sensitivity of NMR detection.]

76 Styles, P., Soffe, N.F., Scott, C.A., Cragg, D.A., Row, F., White, D.J., White, P.C.J. A high-resolution NMR probe in which the coil and preamplifier are cooled with liquid helium. *J. Magn. Reson.* **1984**, *60*, 397–404.

77 ** Kovacs, H., Moskau, D., Spraul, M. Cryogenically cooled probes – a leap in NMR technology. *Progr. Nucl. Magn. Reson. Spectr.* **2005**, *46*, 131–155. [A review of theory and applications of cryogenic probe technology.]

78 Peck, T.L., Magin, R.L., Lauterbur, P.C. Design and analysis of microcoils for NMR spectroscopy. *J. Magn. Reson. B* **1995**, *108*, 114–124.

79 Schlotterbeck, G., Ross, A., Hochstrasser, R., Senn, H., Kühn, T., Marek, D., Schett, O. High-resolution capillary tube NMR.

A miniaturized 5-μL high-sensitivity TXI probe for mass-limited samples, off-line LC NMR, and HT NMR. *Anal. Chem.* **2002**, *74*, 4464–4471.

80 Hu, J. F., Yoo, H.-D., Williams, C.T., Garo, E., Cremin, P.A., Zeng, L., Vervoort, H.C., Lee, C.M., Hart, S.M., Goering, M.G., O'Neil-Johnson, M., Eldridge, G.R. Miniaturization of the structure elucidation of novel natural products – two trace antibacterial acylated caprylic alcohol glycosides from *Arctostaphylos pumila. Planta Med.* **2005**, *71*, 176–180.

81 Hu, J.-F., Garo, E., Yoo, H.-D., Cremin, P.A., Zeng, L., Goering, M.G., O'Neil-Johnson, M., Eldridge, G.R. Application of capillary-scale NMR for the structure determination of phytochemicals. *Phytochem. Anal.* **2005**, *16*, 127–133.

82 Yoo, H.-D., Cremin, P.A., Zeng, L., Garo, E., Williams, C.T., Lee, C.M., Goering, M.G., O'Neil-Johnson, M., Eldridge, G.R, Hu, J.-F. Suaveolindole, a new mass-limited antibacterial indolosesquiterpene from *Greenwayodendron suaveolens* obtained via high-throughput natural products chemistry methods. *J. Nat. Prod.* **2005**, *68*, 122–124.

83 Hu, J.-F., Garo, E., Yoo, H.-D., Cremin, P.A., Goering, M.G., O'Neil-Johnson, M., Eldridge, G.R. Cyclolignans from *Scyphocephalium ochocoa* via high-throughput natural product chemistry methods. *Phytochemistry* **2005**, *66*, 1077–1082.

84 * Xiao, H.B., Krucker, M., Putzbach, K., Albert, K. Capillary liquid chromatography-microcoil ^{1}H nuclear magnetic resonance spectroscopy and liquid chromatography-ion trap mass spectrometry for on-line structure elucidation of isoflavones in *Radix astragali. J. Chromatogr. A* **2005**, *1067*, 135–143. [An illustrative example of application of capillary HPLC-NMR.]

85 * Putzbach, K., Krucker, M., Grynbaum, M.D., Hentschel, P., Webb, A.G., Albert, K. Hyphenation of capillary high-performance liquid chromatography to microcoil magnetic resonance spectroscopy – determination of various carotenoides in a small-sized spinach samples. *J. Pharm. Biomed. Anal.* **2005**, *38*, 910–917. [An illustrative example of application of capillary HPLC-NMR.]

86 Olson, D.L., Peck, T.L., Webb, A.G., Magin, R.L., Sweedler, J.V. High-resolution microcoil ^{1}H NMR for mass-limited, nanoliter-volume samples. *Science* **1995**, *270*, 1967–1970.

87 Olson, D.L., Norcross, J.A., O'Neil-Johnson, M., Molitor, P.F., Detlefsen, D.J., Wilson, A.G., Peck, T.L. Microflow NMR: concepts and capabilities. *Anal. Chem.* **2004**, *76*, 2966–2974.

88 Bailey, N.J.C., Marshall, I.R. Development of ultrahigh-throughput NMR spectroscopic analysis utilizing capillary flow NMR technology. *Anal. Chem.* **2005**, *77*, 3947–3953.

89 Jansma, A., Chuan, T., Albrecht, R.W., Olson, D.L., Peck, T.L., Geierstanger, B.H. Automated microflow NMR: routine analysis of five-microliter samples. *Anal. Chem.* **2005**, *77*, 6509–6515.

90 Webb, A.G. Nuclear magnetic resonance coupled microseparations. *Magn. Reson. Chem.* **2005**, *43*, 688–696.

91 ** Webb, A.G. Microcoil nuclear magnetic resonance spectroscopy. *J. Pharm. Biomed. Anal.* **2005**, *38*, 892–903. [A review of capillary probes with solenoidal microcoils.]

92 Wang, H., Ciobanu, L., Webb, A. Reduced data acquisition time in multi-dimensional NMR spectroscopy using multiple-coil probes. *J. Magn. Reson.* **2005**, *173*, 134–139.

93 Wang, H., Ciobanu, L., Edison, A.S., Webb, A.G. An eight-coil high-frequency probehead design for high-throughput nuclear magnetic resonance spectroscopy. *J. Magn. Reson.* **2004**, *170*, 206–212.

94 Jayawickrama, D.A., Sweedler, J.V. Dual microcoil NMR probe coupled to cyclic ce for continuous separation and analyte isolation. *Anal. Chem.* **2004**, *76*, 4894–4900.

95 Kaerner, A., Smittka, T., Unpublished results.

4
Plant-Associated Microorganisms (Endophytes) as a New Source of Bioactive Natural Products

Gary Strobel

Abstract

Endophytic microorganisms are to be found in virtually every higher plant. These organisms reside in the living tissues of the host plant, and do so in a variety of relationships ranging from symbiotic to pathogenic. Endophytes may contribute to their host plant by producing a plethora of substances that provide protection and survival value to the plant. Ultimately, these compounds, once isolated and characterized, may also have potential for use in modern medicine. Novel antibiotics, antimycotics, immunosuppressants, and anticancer compounds are only a few examples of what has been found after the isolation and culturing of individual endophytes, followed by purification and characterization of some of their natural products. The potential for the discovery of new drugs that may be effective candidates for treating newly developing diseases in humans is vast.

4.1
Introduction

The need for new and useful compounds to provide assistance and relief in all aspects of the human condition is ever-growing. Drug resistance in bacteria, the appearance of new life-threatening viruses, the recurrent problems of diseases in persons with organ transplants, and the tremendous increase in the incidence of fungal infections in the world's population all underscore our inadequacy to cope with these medical problems. Environmental degradation, loss of biodiversity, and spoilage of land and water also add to problems facing mankind, and each of these in turn can have health-related consequences.

Endophytes – microorganisms that reside in the tissues of living plants – are relatively unstudied as potential sources of novel natural products for use in modern medicine. However, some of the most extensive and comprehensive investigations on natural products produced by endophytes have been carried out on the *Neotyphodium* sp. found on grasses [1]. Alkaloids synthesized by this fungus in its grass

Medicinal Plant Biotechnology. From Basic Research to Industrial Applications
Edited by Oliver Kayser and Wim J. Quax
Copyright © 2007 WILEY-VCH Verlag GmbH & Co. KGaA, Weinheim
ISBN 978-3-527-31443-0

hosts have been implicated in fescue toxicosis in rangeland animals [1]. The chemistry and biology of this and other grass endophytes have been reviewed elsewhere [2]. Unfortunately, the fact that these studies are so comprehensive may lead to the conclusion that endophytes only produce toxic compounds in their respective hosts, and that they hold no promise for any medicinal applications whatsoever [2]. This is simply not the case, and as endophytes have been increasingly examined from a plethora of sources an overwhelming number have been found to produce natural products with promising potential for medicinal applications. This is especially true when the higher plants that live in the world's rainforests are examined for endophytes, and their secondary products are tested against bacterial and fungal pathogens of mankind.

Among the approximately 300 000 higher plant species that exist on the Earth, each individual plant of the billions that exist is most likely a host to one or more endophytes (G. Strobel, unpublished results). Only a handful of these plants (grass species) have ever been completely studied relative to their endophytic biology [2]. Consequently, the opportunity to identify new and interesting endophytic microorganisms among the myriads of plants in different settings and ecosystems is vast. The aim of this chapter is to provide some insight into the occurrence of these microorganisms in Nature, to identify the products that they make, and how some of these organisms are beginning to show some potential for human use. The majority of the chapter discusses the rationale for study, the methods used, and details examples of a number of endophytes isolated and studied in the author's laboratory over the course of many years. However, the chapter also includes some specific examples illustrating the work of others in this emerging field of bioprospecting the microbes of the world's rainforests.

4.2
Why Are There Needs for New Medicines?

Today, there is a general call for new antibiotics, and chemotherapeutic agents that are highly effective and possess low toxicity. This search is driven by the development of resistance in infectious microorganisms (e.g., *Staphylococcus*, *Mycobacterium*, *Streptococcus*) to existing drugs, and by the menacing presence of naturally resistant organisms. The ingress to the human population of new disease-causing agents such as AIDS, Ebola virus and SARS requires the discovery and development of new drugs to combat them. Not only do diseases such as AIDS require drugs that target them specifically, but new therapies are also needed for the treatment of ancillary infections that arise a consequence of a weakened immune system. Furthermore, others who are immunocompromised (e.g., cancer and organ transplant patients) are at risk of infection by opportunistic pathogens, such as *Aspergillus*, *Cryptococcus*, and *Candida*, that normally are not major problems in the human population. In addition, more drugs are needed to efficiently treat parasitic protozoan and nematodal infections such as malaria, leishmaniasis, trypanomiasis, and filariasis. Perhaps surprisingly, malaria alone claims more lives each year than any other single infectious agent, with the exception of AIDS and tuberculosis (TB) [3]. However, it is the enteric diseases that claim more lives each year than any other disease complex, and unfortunately the large majority of those dying are children [3].

Novel natural products, and the organisms that create them, offer major opportunities for innovation in drug discovery. Exciting possibilities exist for those who are willing to venture into the wild and unexplored territories of the world to experience the excitement and thrill of engaging in the discovery of endophytes, their biology, and potential usefulness.

4.3
Natural Products in Medicine

Even with untold centuries of human experience behind us and a movement into a modern era of chemistry and automation, it remains evident that natural product-based compounds have had an immense impact on modern medicine. For instance, about 40% of currently available prescription drugs are based such compounds, while between 1981 and 2002 well over 50% of the new chemical products registered by the Food and Drug Administration as anticancer, antimigraine and antihypertensive agents were either natural products or derivatives thereof [4]. Furthermore, between 1989 and 1995, excluding biologics, some 60% of all drug approvals and pre-new drug applications were made for products of natural origin [5]. Likewise, between 1983 and 1994 over 60% of all approved and pre-NDA stage cancer drugs were of natural origin, as were 78% of all newly approved antibacterial agents [6]. Many other examples abound that illustrate the value and importance of natural products from plants and microorganisms in modern civilizations. One modern example of a natural product that has made an enormous impact on medicine is that of taxol [7–9].

More recently, however, natural product research efforts have lost favor with many major drug companies, who now prefer to employ a combinatorial chemistry approach that involves the automated synthesis of structurally related small molecules. In addition, many drug companies have developed interests in making products that have a larger potential profit base than anti-infectious drugs. These synthetic compounds may provide social benefits, relieve allergenic responses, reduce the pain of arthritis, or even sooth the stomach. Strangely, it appears that this loss of interest in natural products is not only an economically driven decision but can also be attributed to the enormous effort and expense required to identify a biological source, to isolate an active natural product, to decipher its structure, and then to begin the long road to product development. It is now also apparent that combinatorial chemistry and other synthetic chemistry methodology, which revolves around certain basic chemical structures, serves as a never-ending supply of products to feed the screening robots of the drug industry. Today, within many large pharmaceutical companies, the progress of professionals is based primarily upon the numbers of compounds that can be produced and sent to the screening machines. However, this approach tends to work against the numerous steps needed even to find one compound in natural product discovery. It is important to emphasize at this point that the primary purpose of combinatorial chemistry should be to complement and assist the efforts of natural product drug discovery and development, and not to supersede it [5]. For this reason, a few larger companies have retained an interest in natural products chemistry. In fact, the natural product often serves as a lead molecule, the activity of which can be enhanced by manipulation through combinatorial and synthetic chemistry. Traditionally, natural products have been the "pathfinder" compounds, with an untold diversity of chemical structures unparalleled by even the largest combinatorial libraries.

4.4
Endophytic Microbes

It may also be true that a reduction in interest for natural products used in drug development has occurred as a result of people growing weary of dealing with the traditional sources of bioactive compounds, including plants from temperate zones and microbes from a plethora of soil samples gathered in different parts of the world by armies of collectors. In other words, why continue to do the same thing when robots, combinatorial chemistry, and molecular biology have arrived on the scene? Furthermore, the logic and rationale for the time and effort spent on drug discovery using a target- and site-directed approach has been overwhelming.

While combinatorial synthesis produces compounds at random, secondary metabolites – defined as low molecular-weight compounds not required for growth in pure culture – are produced as an adaptation for specific functions in nature [10]. Shutz points out that certain microbial metabolites seem to be characteristic of certain biotopes, both on an environmental as well as organismal level [11]. Accordingly, it appears that the search for novel secondary metabolites should center on organisms that inhabit unique biotopes. Thus, it behooves the investigator to carefully study and select the biological source before proceeding, rather than to take a totally random approach in selecting the source material. Careful study also indicates that organisms and their biotopes that are subjected to constant metabolic and environmental interactions should produce even more secondary metabolites [11]. Endophytes are microbes that inhabit such biotopes, namely higher plants, which is why they are currently considered as a wellspring of novel secondary metabolites offering the potential for exploitation of their medical benefits [12].

In addition, it also is extremely helpful for the investigator interested in exploiting endophytes to have access to – or to have some expertise in – microbial taxonomy, and this includes modern molecular techniques involving sequence analyses of 16S and 18S rDNA. Currently, endophytes are viewed as an outstanding source of bioactive natural products because there are so many of them occupying literally millions of unique biological niches (higher plants), and growing in so many unusual environments. Thus, it would appear that a myriad of biotypical factors associated with plants can be important in the selection of a plant for study. It may be the case that these factors govern which microbes are present in the plant, as well as the biological activity of the products associated with these organisms.

What are endophytes? Bacon et al. [2] provided an inclusive and widely accepted definition of endophytes as: "*Microbes that colonize living, internal tissues of plants without causing any immediate, overt negative effects*". While the symptomless nature of endophyte occupation in plant tissue has prompted focus on symbiotic or mutualistic relationships between endophytes and their hosts, the observed biodiversity of endophytes suggests they can also be aggressive saprophytes or opportunistic pathogens. Both fungi and bacteria are the most common microbes existing as endophytes. It would seem that other microbial forms most certainly exist in plants as endophytes such as mycoplasmas, rickettsia and archebacteria, but as yet evidence for their existence has been presented. In fact, it may be the case that the majority of microbes existing in plants are not culturable with common laboratory techniques, which makes their presence and role in plants even more intriguing. Although the most frequently isolated endophytes are the fungi [13], at the outset it is important to realize that the vast majority of plants have not been studied for any endophytic association. Thus, enormous opportunities exist for the recovery of novel fungal forms, including genera, biotypes, as well as species in the myriad of plants yet to be studied. Hawksworth and Rossman [14] have estimated there may be as many as one million different fungal species, yet only about 100 000 have been described to date. As more evidence accumulates, however, estimates keep rising as to the actual number of fungal species. For instance, Dreyfuss and Chapela [15] estimate there may be at least one million species of endophytic fungi alone. It seems clear that endophytes are a rich and reliable source of genetic diversity, and may represent many previously undescribed species. Finally, in our experience, novel microbes (as defined at the morphological and or molecular levels) often have associated with them, novel natural products. This fact alone helps to eliminate the problems of dereplication in compound discovery

4.5
Rationale for Plant Selection

It is important to understand that the methods and rationale used seem to provide the best opportunities to isolate novel endophytic microorganisms at the genus, species, or biotype levels. Since the number of plant species in the world is so great, creative and imaginative strategies must be used to quickly narrow the search for endophytes displaying bioactivity [16].

A specific rationale for the collection of each plant for endophyte isolation and natural product discovery is used. Several hypotheses govern this plant selection strategy, as follows:

1. Plants from unique environmental settings, especially those with an unusual biology, and possessing novel strategies for survival, are seriously considered for study.

2. Plants that have an ethnobotanical history (used by indigenous peoples) and are related to the specific uses or applications of interest are selected for study. These plants are chosen either by direct contact with local peoples, or via local literature. Ultimately, it may be learned that the healing powers of the botanical source may in fact not be related to the natural products of the plant, but rather to the endophyte inhabiting the plant.

3. Plants that are endemic, having an unusual longevity, or have occupied a certain ancient land mass, such as Gondwanaland, are also more likely to lodge endophytes with active natural products than other plants.

4. Plants growing in areas of great biodiversity, it follows, also have the prospect of housing endophytes with great biodiversity.

Just as plants from a distinct environmental setting are considered to be a promising source of novel endophytes and their compounds, so too are plants with an unconventional biology. For example, an aquatic plant, *Rhyncholacis penicillata*, was collected from a river system in southwest Venezuela where the harsh aquatic environment subjected the plant to constant beating by virtue of rushing waters, debris, and tumbling rocks and pebbles [17]. These environmental insults created many portals through which common phytopathogenic oomycetes could enter the plant. Still, the plant population appeared to be healthy, possibly due to protection from an endophytic product. This was the environmental biological clue used to pick this plant for a comprehensive study of its endophytes. Eventually, an unusual and potent antifungal strain of *Serratia marcescens*, living both as an epiphyte and an endophyte, was recovered from *R. penicillata*. This bacterium produces oocydin A, a novel antioomycetous compound, having the properties of a chlorinated macrocyclic lactone (Fig. 4.1) [17]. It is conceivable that the production of oocydin A by *S. marcescens* is directly related to the endophyte's relationship with its higher plant host. Currently, oocydin A is being considered for agriculture use to control the ever-threatening presence of oomycetous fungi such as *Pythium* spp. and *Phytophthora* spp. Oocydin A also has activity against a number of rapidly dividing cancer cell lines [17].

Fig. 4.1 Oocydin A, a chlorinated macrocyclic lactone isolated and characterized from a strain of *Serratia marcescens*, obtained from *Rhyncholacis penicillata* (stereochemistry unknown).

Plants with an ethnobotanical history, as mentioned above, also are likely candidates for study since the medical uses for which the plant was selected may relate more to its population of endophytes than to the plant biochemistry itself. For example, a sample of the snakevine, *Kennedia nigriscans*, from the Northern Territory of Australia, was selected for study since its sap has traditionally been used as bush medicine. In fact, this area was selected for plant sampling since it has been home to the world's longest standing human civilization, the Australian Aborigines. The snakevine is harvested, crushed and heated in an aqueous brew by local Aborigines in southwest Arnhemland to treat cuts, wounds and infections. As it turned out, the plant contains an entire suite of streptomycetes [18]. One in particular has unique partial 16S rDNA sequences when compared to those in GenBank. The organism was designated *Streptomyces* NRRL 30562 [19]. It produces a series of actinomycins including actinomycins D, Xo_β, and X_2 among others (U.F. Castillo and G.A. Strobel, unpublished results). Furthermore, it produces broad-spectrum novel peptide antibiotics termed "munumbicins", that are discussed below. It seems likely that some of the healing properties in plants, as discovered by indigenous peoples, might be facilitated by compounds produced by one or more specific plant-associated endophytes as well as the plant products themselves. This appears to be an excellent example illustrating this point.

In addition, it is worth noting that some plants generating bioactive natural products have associated endophytes that produce the same natural products. Such is the case with taxol, a highly functionalized diterpenoid and famed anticancer agent that is found in *Taxus brevifolia* and other yew species (*Taxus* spp.) [8]. In 1993, a novel taxol-producing fungus, *Taxomyces andreanae*, was isolated and characterized from the yew, *Taxus brevifolia* [20].

4.6
Endophytes and Biodiversity

Among the myriad of ecosystems on Earth, those having the greatest general biodiversity of life seem to be those that also have the greatest number and most diverse endophytes. Tropical and temperate rainforests are the most biologically diverse terrestrial ecosystems. The most threatened of these spots cover only 1.44% of the land's surface, yet they harbor over 60% of the world's terrestrial biodiversity [16]. In addition, each of the 20 to 25 areas identified as supporting the world's greatest biodiversity also support unusually high levels of plant endemism [16]. As such, one would expect, with high plant endemism, that there also should exist specific endophytes which may have evolved with the endemic plant species. Biological diversity implies chemical diversity because of the constant chemical innovation required to survive in ecosystems where the evolutionary race to survive is most active. Tropical rainforests are a remarkable example of this type of environment, because competition is great, resources are limited, and selection pressure is at its peak. This gives rise to a high probability that rainforests are a source of novel molecular structures and biologically active compounds [21].

Bills et al. described a metabolic distinction between tropical and temperate endophytes through statistical data which compared the number of bioactive natural products isolated from endophytes of tropical regions to the number of those isolated from endophytes of temperate origin [22]. Not only did these authors find that tropical endophytes provided more active natural products than temperate endophytes, but they also noted that a significantly higher number of tropical endophytes produced a larger number of active secondary metabolites than did fungi from other substrata. This observation suggests the importance of the host plant as well as the ecosystem in influencing the general metabolism of endophytic microbes.

4.7
Endophytes and Natural Products

Tan and Zou believe the reason why some endophytes produce certain phytochemicals, originally characteristic of the host, might be related to a genetic recombination of the endophyte with the host that occurred in evolutionary time [12]. This is a concept that was originally proposed as a mechanism to explain why *T. andreanae* might be producing taxol [23]. Thus, if endophytes can produce the same rare and important bioactive compounds as their host plants, this would not only reduce the need to harvest slow-growing and possibly rare plants, but also help to preserve the world's ever-diminishing biodiversity. Furthermore, it is recognized that a microbial source of a high-value product may be easier and more economic to produce effectively, thereby reducing its market price.

All aspects of the biology and interrelatedness of endophytes with their respective hosts is a vastly underinvestigated and exciting field [24–27]. Thus, more background information on a given plant species and its microorganismal biology would be exceedingly helpful in directing the search for bioactive products. Presently, no one is quite certain of the role of endophytes in nature and what appears to be their relationship to various host plant species. While some endophytic fungi appear to be ubiquitous (e.g., *Fusarium* spp., *Pestalotiopsis* spp., and *Xylaria* spp.), one cannot state definitively that endophytes are truly host-specific or even systemic within plants, any more than assume that their associations are chance encounters. Frequently, many endophytes of the same species are isolated from the same plant, and only one or a few biotypes of a given fungus will produce a highly biologically active compound in culture [28]. A great deal of uncertainty also exists between what an endophyte produces in culture and what it may produce in nature. It does seem possible that the production of certain bioactive compounds by the endophyte *in situ* may facilitate the domination of its biological niche within the plant, or even provide protection to the plant from harmful invading pathogens. Furthermore, little information exists relative to the biochemistry and physiology of the interactions of the endophyte with its host plant. It would seem that many factors changing in the host as related to the season, and other factors including age, environment, and location may influence the biology of the endophyte. Indeed, further research at the molecular level must be conducted in the field to study endophyte interactions and ecology. All of these interactions are probably chemically mediated for some purpose in Nature. An ecological aware-ness of the role these organisms play in nature will provide the best clues for tar-geting particular types of endophytic bioactivity with the greatest potential for bio-prospecting.

4.7.1
Isolation, Preservation and Storage of Endophytic Cultures for Product Isolation

Detailed techniques for the isolation of microbial endophytes are outlined in a number of reviews and technical articles [17–20, 24–27]. If endophytes are being obtained from plants growing in polar regions, the dry tropics, or some temperate areas of the world, one can expect to acquire from none to only one or two endo-phytic cultures per plant sample (0.5–10.0) centimeter limb piece. However, from the wet tropics this number may rise to 20 to 30 or even more microbes per plant piece.

Given limited fermentation capabilities, it becomes necessary to preserve fresh-ly isolated microbes for work in the future. Generally, preservation in an aqueous 15% glycerol solution at −70 °C is an exceedingly good procedure for saving cul-tures until work on them can proceed at a later date [24–27]. It is also critical to la-bel and store cultures for patent and publication purposes.

4.7.2
Some Examples of Bioactive Natural Products from Endophytes

The following section illustrates some examples of natural products obtained from endophytic microbes, and their potential in the pharmaceutical and agrochemical arenas. Although many of the examples are taken from our studies, this review is by no means inclusive of all natural product investigations in endophytes.

4.7.2.1   Endophytic Fungal Products as Antibiotics
Fungi are the most commonly isolated endophytic microbes, usually appearing as fine filaments growing from the plant material on the agar surface. Generally, the most commonly isolated fungi are in the group *Fungi Imperfecti* or *Deuteromycetes*. Basically, they produce asexual spores in or on various fruiting structures. Also, it is quite common to isolate endophytes that are producing no fruiting structures whatsoever, such as *Mycelia Sterilia*. Quite commonly, endophytes do produce sec-ondary metabolites when placed in culture. However, the temperature, the compo-sition of the medium and the degree of aeration will affect the amount and types of compounds produced by an endophytic fungus. On occasion, endophytic fungi produce antibiotics. Natural products from endophytic fungi have been observed to inhibit or kill a wide variety of harmful microorganisms including, but not lim-ited to phytopathogens, as well as bacteria, fungi, viruses, and protozoans that af-

fect humans and animals. Some examples of bioactive products from endophytic fungi are described in the following sections.

Cryptosporiopsis cf. *quercina* is the imperfect stage of *Pezicula cinnamomea*, a fungus commonly associated with hardwood species in Europe. It was isolated as an endophyte from *Tripterigeum wilfordii*, a medicinal plant native to Eurasia [29]. On Petri plates, *C. quercina* demonstrates excellent antifungal activity against some important human fungal pathogens, including *Candida albicans* and *Trichophyton* spp. A unique peptide antimycotic, termed cryptocandin, was isolated and characterized from *C. quercina* [29]; this compound contains a number of peculiar hydroxylated amino acids and a novel amino acid, 3-hydroxy-4-hydroxy methyl proline (Fig. 4.2). The bioactive compound is related to the known antimycotics, the echinocandins, and the pneumocandins [30]. As is generally true, not one but several bioactive and related compounds are produced by an endophytic microbe. Thus, other antifungal agents related to cryptocandin are also produced by *C.* cf. *quercina*. Cryptocandin is also active against a number of plant pathogenic fungi, including *Sclerotinia sclerotiorum* and *Botrytis cinerea*. Cryptocandin and its related

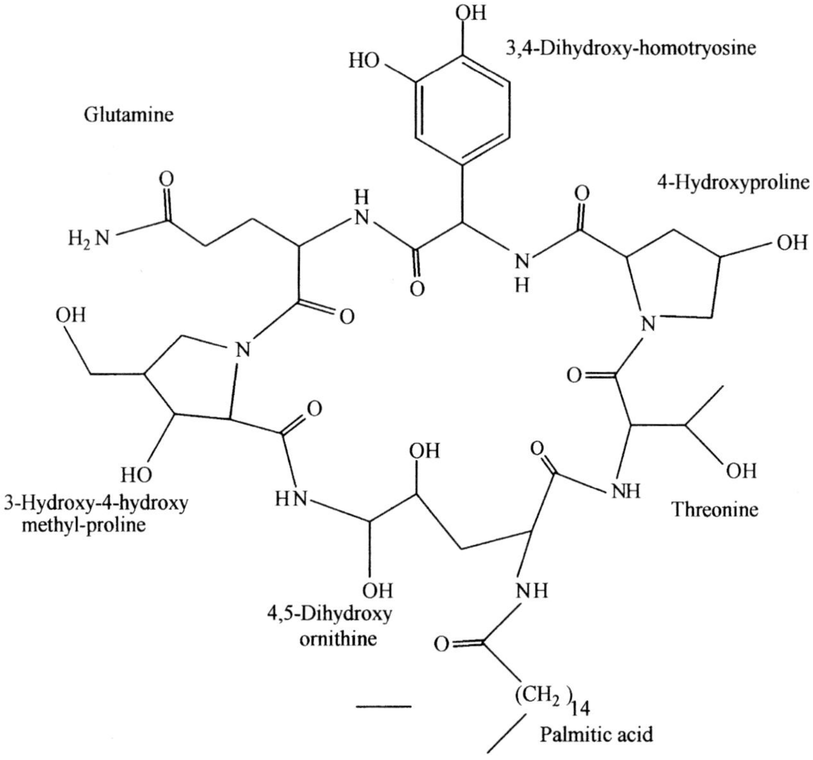

Fig. 4.2 Cryptocandin A, an antifungal lipopeptide obtained from the endophytic fungus, *Cryptosporiopsis* cf. *quercina*. (No stereochemistry is intended.)

compounds are currently being considered for use against a number of fungi causing diseases of the skin and nails.

Cryptocin, a unique tetramic acid, is also produced by *C. quercina* (see above) (Fig. 4.3) [31]. This unusual compound possesses potent activity against *Pyricularia oryzae*, the causal organism of rice blast, one of the worst plant diseases in the world, as well as a number of other plant pathogenic fungi [31]. The compound was generally ineffective against a general array of human pathogenic fungi. Nevertheless, with minimum inhibitory concentrations (MICs) against *P. oryzae* at 0.39 μg mL^{-1}, this compound is being examined as a natural chemical control agent for rice blast and is being used as a model to synthesize other antifungal compounds.

As mentioned earlier, *P. microspora* is a common rainforest endophyte [24–26]. It transpires that enormous biochemical diversity does exist in this endophytic fungus, and many secondary metabolites are produced by various strains of this widely dispersed organism. One such secondary metabolite is ambuic acid, an antifungal agent, which has been recently described from several isolates of *P. microspora* found as representative isolates in many of the world's rainforests (Fig. 4.4) [32]. This compound, and another endophyte product, terrein, have been used as models to develop new solid-state NMR tensor methods to assist in the characterization of the molecular stereochemistry of organic molecules. The rationale and methods used and developed for this novel method of chemical characterization are discussed elsewhere [33,34].

A strain of *P. microspora* was also isolated from the endangered tree *Torreya taxifolia*, and produces several compounds having antifungal activity; these include pestaloside, an aromatic β-glucoside, and two pyrones – pestalopyrone and hydroxypestalopyrone [35]. These products also possess phytotoxic properties. Other newly isolated secondary products obtained from *P. microspora* (endophytic on *Taxus brevifolia*) include two new caryophyllene sesquiterpenes – pestalotiopsins A and B [36]. Additional new sesquiterpenes produced by this fungus are 2-α-hydroxydimeninol and a highly functionalized humulane [37, 38]. Variation in the amount

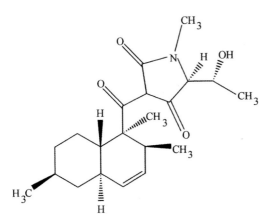

Fig. 4.3 Cryptocin, a tetramic acid antifungal compound found in *Cryptosporiopsis* cf. *quercina*

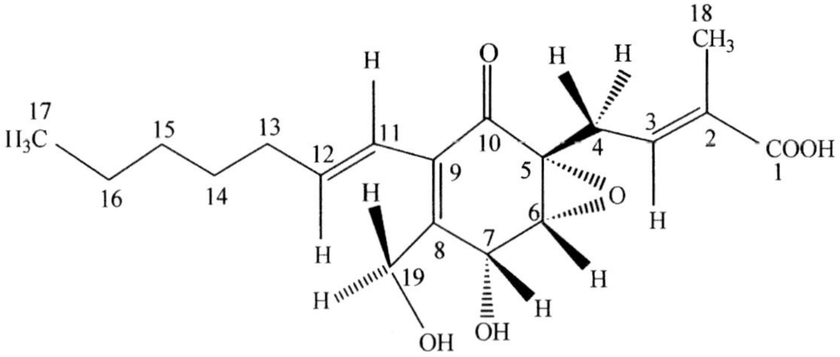

Fig. 4.4 Ambuic acid, a highly functionalized cyclohexenone produced by a number of isolates of *Pestalotiopsis microspora* found in rainforests around the world. This compound possesses antifungal activity and has been used as a model compound for the development of solid-state NMR methods for the structural determination of natural products.

and types of products found in this fungus depends on both the cultural conditions as well as the original plant source from which it was isolated.

Pestalotiopsis jesteri is a newly described endophytic fungal species from the Sepik river area of Papua New Guinea. This fungus produces jesterone and hydroxy-jesterone, both of which exhibit antifungal activity against a variety of plant pathogenic fungi [39]. These compounds are highly functionalized cyclohexenone epoxides. Jesterone, subsequently, has been prepared by organic synthesis with complete retention of biological activity (Fig. 4.5) [40]. Jesterone is one of only a few products from endophytic microbes in which total synthesis of a bioactive product has been successfully accomplished.

Phomopsichalasin, a metabolite from an endophytic *Phomopsis* sp., represents the first cytochalasin-type compound with a three-ring system replacing the cytochalasin macrolide ring. This metabolite exhibits antibacterial activity in disk diffusion assays (at a concentration of 4 µg per disk) against *Bacillus subtilis*, *Salmonella gallinarum*, and *Staphylococcus aureus*. It also displays a moderate activity against the yeast *Candida tropicalis* [41].

An endophytic *Fusarium* sp. from the plant, *Selaginella pallescens*, collected in the Guanacaste Conservation Area of Costa Rica, was screened for antifungal activity. A new pentaketide antifungal agent, CR377, was isolated from the culture broth of

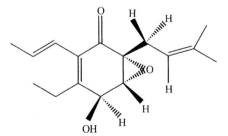

Fig. 4.5 Jesterone, a cyclohexenone epoxide from *Pestaliotiopsis jesteri* has antioomycete activity.

the fungus and showed potent activity against *Candida albicans* in agar diffusion assays [42].

Colletotric acid, a metabolite of *Colletotrichum gloeosporioides*, an endophytic fungus isolated from *Artemisia mongolica*, displays antibacterial activity against bacteria as well as against the fungus, *Helminthsporium sativum* [43]. Another *Colletotrichum* sp., isolated from *Artemisia annua*, produces bioactive metabolites that also showed antimicrobial activity. *A. annua* is a traditional Chinese herb that is well recognized for its synthesis of artemisinin (an antimalarial drug) and its ability to inhabit many geographically different areas. The *Colletotrichum* sp. found in *A. annua* not only produced metabolites with activity against human pathogenic fungi and bacteria, but also metabolites that were fungistatic to plant pathogenic fungi [44].

A novel antibacterial agent, guignardic acid, was isolated from the endophytic fungus *Guignardia* sp. Interestingly, the organism was obtained from the medicinal plant, *Spondias mombin* of the tropical plant family Anacardiaceae found in Brazil. The compound was isolated by UV-guided fractionation of the fermentation products of this fungus. The compound is the first member of a novel class of natural compounds containing a dioxolanone moiety formed by the fusion of 2-oxo-3-phenylpropanoic acid and 3-methyl-2-oxobutanoic acid, which are products of the oxidative deamination of phenylalanine and valine, respectively (Fig. 4.6). The structure was determined by classical spectroscopic methods and confirmed by organic synthesis [45].

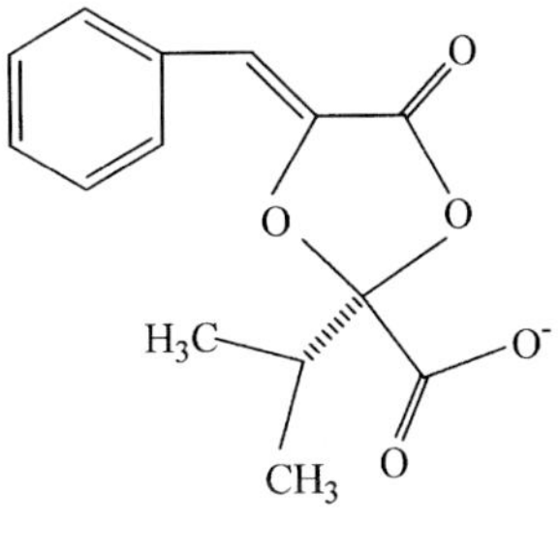

Fig. 4.6 Guignardic acid from *Guignardia* sp. obtained from *Spondias mombin*, an Anacardiaceaeous plant in Brazil.

4.7.2.2 Endophytic Bacterial Products as Antibiotics

Only a limited number of bacterial species are known to be associated with plants, with one of the most common genus encountered being *Pseudomonas* spp. *Pseudomonas* spp. have representative biotypes and species that are epiphytic, endophytic, and pathogenic. They have been reported from every continent, including the Antarctic. Some of these species produce phytotoxic compounds as well as antibiotics. The ecomycins are produced by *Pseudomonas viridiflava* [46]; this bacterium is generally associated with the leaves of many grass species, and is located on and within the tissues [46]. The ecomycins represent a family of novel lipopeptides, and have masses of 1153 Da and 1181 Da, respectively. Besides common amino acids such as alanine, serine, threonine, and glycine, some unusual amino acids are incorporated into the structure of the ecomycins, including homoserine and β-hy-

mulated into an artificial mixture. This mixture not only mimics the antibiotic effects of the volatile compounds produced by the fungus, but also was used to confirm the identity of the majority of the volatiles emitted by this organism [56]. Each of the five classes of volatile compounds produced by the fungus had some microbial effects against the test fungi and bacteria, but none was lethal. Collectively, however, they acted synergistically to cause death in a broad range of plant and human pathogenic fungi and bacteria. The most effective class of inhibitory compounds was the esters, of which isoamyl acetate was the most biologically active, though in order to be lethal it must be combined with other volatiles [56]. The composition of the medium on which *M. albus* grows dramatically influences the type of volatile compounds produced [57]. The ecological implications and potential practical benefits of the "mycofumigation" effects of *M. albus* are very promising, given the fact that soil fumigation using methyl bromide will soon be illegal in the United States. Methyl bromide is not only a hazard to human health, but it has been implicated in causing destruction of the ozone layer. The potential use of mycofumigation to treat soil, seeds, and plants will soon be a reality. The artificial mixture of volatile compounds may also have value in treating seeds, fruits and plant parts in storage and while being transported. In addition, *M. albus* is already in a limited market for the treatment of human wastes. Its gases have both inhibitory and lethal effects on fecal-inhabiting organisms such as *Escherichia coli* and *Vibrio cholera*.

Using *M. albus* as a screening tool, it has now been possible to isolate other endophytic fungi producing volatile antibiotics. The newly described *M. roseus* was twice obtained from tree species found in the Northern Territory of Australia. This fungus is equally effective as *M. albus* in causing inhibition and death of test microbes in the laboratory [58]. Other interesting *M. albus* isolates have been obtained from several plant species growing in the Northern Territory and the jungles of the Tesso Nilo area of Sumatra, Indonesia [59,60].

In addition, for the first time, a nonmuscodor species (*Gliocladium* sp.) was discovered as a volatile antibiotic producer. The volatile components of this organism are totally different than those of either *M. albus* or *M. roseus*. In fact, the most abundant volatile inhibitor is [8]-annulene, formerly used as a rocket fuel and discovered for the first time as a natural product. However, the bioactivity of the volatiles of this *Gliocladium* sp. is not as good or comprehensive as that of the *Muscodor* spp. [61].

4.7.2.5 Antiviral Compounds from Endophytes

Another fascinating use of products from endophytic fungi is the inhibition of viruses. Two novel human cytomegalovirus (hCMV) protease inhibitors, cytonic acids A and B, have been isolated from solid-state fermentation of the endophytic fungus *Cytonaema* sp. Their structures as were elucidated as *p*-tridepsides isomers by MS and NMR methods [62]. It is apparent that the potential for the discovery of compounds having antiviral activity from endophytes is in its infancy. The fact, however, that some compounds have been found already is promising. The main limitation to compound discovery to date is probably related to an absence of common antiviral screening systems in most compound discovery programs.

4.7.2.6 Endophytic Fungal Products as Anticancer Agents

Taxol and some of its derivatives represent the first major group of anticancer agents to be produced by endophytes (Fig. 4.7). Taxol, a highly functionalized diterpenoid, is found in each of the world's yew (*Taxus*) species, but was originally isolated from *Taxus brevifolia* [8]. The original target diseases for this compound were ovarian and breast cancers, but now it also is used to treat a number of other human tissue-proliferating diseases. The presence of taxol in yew species prompted the study of their endophytes. By the early 1990s, however, no endophytic fungi had been isolated from any of the world's representative yew species. After several years of effort, a novel taxol-producing endophytic fungus, *Taxomyces andreanae*, was discovered in *Taxus brevifolia* [20]. The most critical line of evidence for the presence of taxol in the culture fluids of this fungus was the electrospray mass spectrum of the putative taxol isolated from *T. andreanae*. In electrospray mass spectroscopy, taxol usually gives two peaks, one at mass 854 which is M+H$^+$, and the other at 876 which is M+Na$^+$. Fungal taxol had an identical mass spectrum to authentic taxol [23]. Subsequent ^{14}C-labeling studies showed the presence of fungal-derived taxol in the culture medium [23]. These early studies set the stage for a more comprehensive examination of the ability of other *Taxus* species and many other plants to yield endophytes producing taxol.

Some of the most commonly found endophytes of the world's yews and many other plants are *Pestalotiopsis* spp. [24–27]. One of the most frequently isolated en-

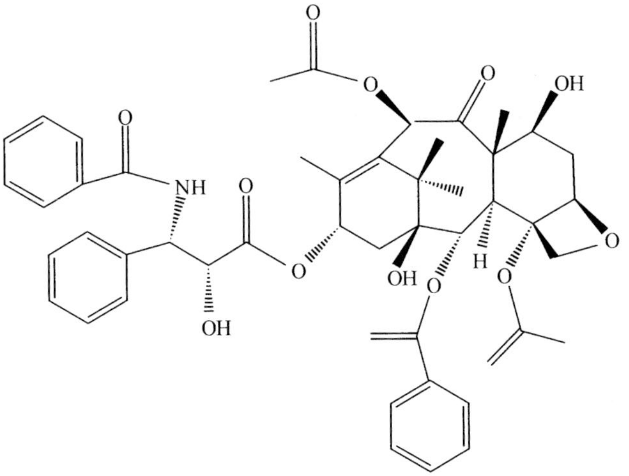

Fig. 4.7 Taxol, the world's first billion-dollar anticancer drug is produced by many endophytic fungi. It too, possesses outstanding antioomycete activity.

dophytic species is *Pestalotiopsis microspora* [24]. An examination of the endophytes of *Taxus wallichiana* yielded *P. microspora*, and a preliminary monoclonal antibody test indicated that it might produce taxol (30). After preparative thin-layer chromatography (TLC), a compound was isolated and shown by spectroscopic techniques to be taxol. Labeled (^{14}C) taxol was produced by this organism when it was administered several ^{14}C precursors [63]. Furthermore, several other *P. microspora* isolates were obtained from a bald cypress tree in South Carolina and also shown to produce taxol [28]. This was the first indication that endophytes residing in plants other than *Taxus* spp. were producing taxol. Therefore, a specific search was conducted for taxol-producing endophytes on continents not known for any indigenous *Taxus* spp. This included investigating the prospects that taxol-producing endophytes exist in South America and Australia. From the extremely rare (and previously thought extinct) Wollemi Pine (*Wollemia nobilis*), *Pestalotiopsis guepini* was isolated which was shown to produce taxol [64]. Also, quite surprisingly, a rubiaceous plant, *Maguireothamnus speciosus*, yielded a novel fungus, *Seimatoantlerium tepuiense*, that produces taxol. This endemic plant grows on the top of the tepuis in the Venzuelan-Guyana border in southwest Venezuela [65]. Fungal taxol production was also noted in *Periconia* sp. and *Seimatoantlerium nepalense*, another novel endophytic fungal species [66, 67]. Simply, it appears that the distribution of those fungi producing taxol is worldwide, and not confined to the endophytes of yews. The ecological and physiological explanation for the wide distribution of taxol-producing fungi seems to be related to the fact that taxol is a fungicide, and that the most sensitive organisms to it are plant pathogens such as *Pythium* spp. and *Phytophthora* spp. [68]. These pythiaceous organisms are some of the world's most important plant pathogens, and are strong competitors with endophytic fungi for niches within plants. In fact, their sensitivity to taxol is based on their interaction with tubulin in an identical manner as in rapidly dividing human cancer cells [9, 68]. Thus, bona fide endophytes may be producing taxol and related taxanes to protect their respective host plant from degradation and disease caused by these pathogens.

Other investigators have also made observations on taxol production by endophytes, including the discovery of taxol production by *Tubercularia* sp. isolated from the Chinese yew (*Taxus mairei*) in the Fujian province of southeastern mainland China [69]. At least three endophytes of *Taxus wallichiana* produce taxol, including *Sporormia minima* and *Trichothecium* sp. [70]. Using HPLC and ESIMS, taxol has been discovered in *Corylus avellana* cv. Gasaway [71]. Several fungal endophytes of this plant (filbert) produce taxol in culture [71]. It is important to note, however, that taxol production by all endophytes in culture is in the range of submicrogram to micrograms per liter. Also, commonly, the fungi will attenuate taxol production in culture, with some possibility for recovery, if certain activator compounds are added to the medium [66]. Efforts are being made to determine the feasibility of making microbial taxol a commercial possibility; the greatest prospect for this may be the discovery of endophytes that produce large quantities of one or more taxanes that could then be used as platforms for the organic synthesis of taxol or one of its anticancer relatives.

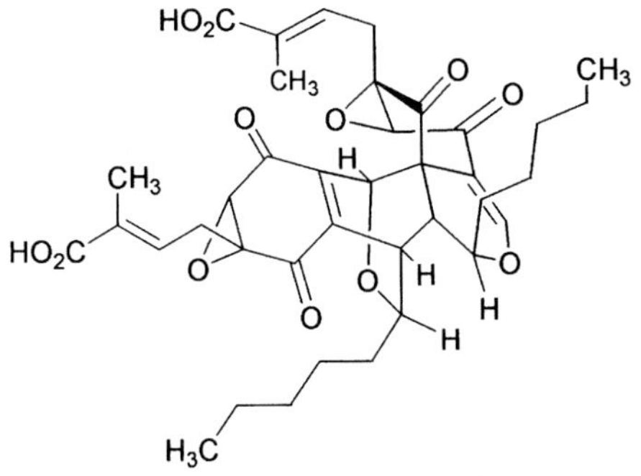

Fig. 4.8 Torreyanic acid, an anticancer compound, from Pestalotiopsis microspora

Torreyanic acid, a selectively cytotoxic quinone dimer and potential anticancer agent, was isolated from a *P. microspora* strain (Fig. 4.8). This strain was originally obtained as an endophyte associated with the endangered tree *Torreya taxifolia* (Florida torreya), as mentioned above [72]. Torreyanic acid was tested in several cancer cell lines, where it demonstrated five- to ten-fold more potent cytotoxicity in cell lines sensitive to protein kinase C agonists; cell death was caused by apoptosis. Recently, torreyanic acid was successfully synthesized by application of a biomimetic oxidation/dimerization cascade [73].

Alkaloids are also commonly found in endophytic fungi. Such fungal genera as xylaria, phoma, hypoxylon, and chalara are representative producers of a relatively large group of substances known as the cytochalasins, of which over 20 are now known. Many of these compounds possess antitumor and antibiotic activities, but because of their cellular toxicity they have not been developed into pharmaceuticals. Three novel cytochalasins have recently been reported from *Rhinocladiella* sp. as an endophyte on *Tripterygium wilfordii*. These compounds have antitumor activity and have been identified as 22-oxa-[12]-cytochalasins [74]. Thus, it is not uncommon to find one or more cytochalasins in endophytic fungi, and this provides an example of the fact that redundancy in discovery does occur, making dereplication an issue even for these under investigated sources.

4.7.2.7 Endophytic Fungal Products as Antioxidants

Two compounds, pestacin and isopestacin, have been obtained from culture fluids of *Pestalotiopsis microspora*, an endophyte isolated from a combretaceaous plant, *Terminalia morobensis*, growing in the Sepik River drainage system of Papua New Guinea [75,76]. Both pestacin and isopestacin display antimicrobial as well as antioxidant activity. Isopestacin was attributed with antioxidant activity based on its structural similarity to the flavonoids (Fig. 4.9). Electron spin resonance spectroscopy measurements confirmed this antioxidant activity; the compound is able to scavenge superoxide and hydroxyl free radicals in solution [75]. Pestacin was later described from the same culture fluid, occurring naturally as a racemic mixture and also possessing potent antioxidant activity [76]. The antioxidant activity of pestacin arises primarily via cleavage of an unusually reactive C–H bond and to a less-

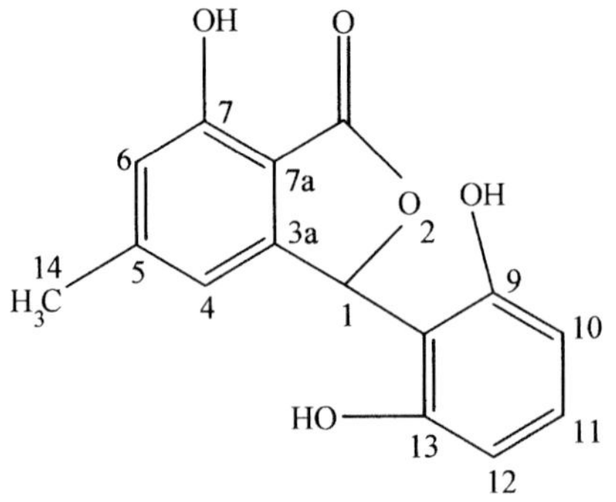

Fig. 4.9 Isopestacin, an antioxidant produced by an endophytic *Pestalotiopsis microspora* strain, isolated from *Terminalia morobensis* growing on the north coast of Papua New Guinea.

er extent, through O–H abstraction [76]. The antioxidant activity of pestacin is at least one order of magnitude more potent than that of trolox, a vitamin E derivative [76].

4.7.2.8 Endophytic Fungal Products as Immunosuppressive Compounds

Immunosuppressive drugs are used today to prevent allograft rejection in transplant patients, and in future could be used to treat autoimmune diseases such as rheumatoid arthritis and insulin-dependent diabetes. The endophytic fungus, *Fusarium subglutinans*, isolated from *T. wilfordii*, produces the immunosuppressive, but noncytotoxic diterpene pyrones subglutinols A and B [77]. Subglutinol A and B are equipotent in the mixed lymphocyte reaction (MLR) assay and thymocyte proliferation (TP) assay, with an IC_{50} of 0.1 µM. In the same assay systems, the famed immunosuppressant drug, cyclosporine A (also a fungal metabolite), was roughly as potent in the MLR assay, but 10^4-fold more potent in the TP assay. Nonetheless, the lack of toxicity associated with subglutinols A and B suggests that they should be explored in greater detail as potential immunosuppressants [77].

4.8
Surprising Results from Molecular Biology Studies on Pestalotiopsis microspora

Of some compelling interest is an explanation as to how the genes for taxol production may have been acquired by *P. microspora* [78]. Although the complete answer to this question is not at hand, some other relevant genetic studies have been performed on this organism. *Pestalotiopsis microspora* Ne 32 is one of the most easily genetically transformable fungi that has been studied to date. The *in-vivo* addition of telomeric repeats to foreign DNA generates extrachromosomal DNAs in this fungus [78]. Repeats of the telomeric sequence 5′-TTAGGG-3′ were appended to nontelomeric transforming DNA termini. The new DNAs, carrying foreign genes and the telomeric repeats, replicated independently of the chromosome and expressed the information carried by the foreign genes. The addition of telomeric repeats to foreign DNA is unusual among fungi. This finding may have important implications in the biology of *P. microspora* Ne 32 as it explains at least one mechanism as to how new DNA can be captured by this organism and eventually expressed and replicated. Such a mechanism may begin to explain how the enormous biochemical variation may have arisen in this fungus [28]. These initial studies also represent a framework to aid in the understanding of how this fungus might adapt itself to the environment of its plant hosts, and suggests that the uptake of plant DNA into its own genome may occur. In addition, the telomeric repeats have the same sequence as human telomeres, which points to the possibility that *P. microspora* might serve as a means to make artificial human chromosomes, a totally unexpected result.

4.9
Concluding Statements

Endophytes are a poorly investigated group of microorganisms that represent an abundant and dependable source of bioactive and chemically novel compounds with potential for exploitation in a wide variety of medical applications. The mechanisms through which endophytes exist and respond to their surroundings must be better understood in order to be more predictive about which higher plants to seek, study, and employ in isolating their microfloral components. This may facilitate the natural product discovery process.

Although investigations into the utilization of this vast resource of poorly understood microorganisms has only just begun, it is already clear that the enormous potential for organism, product, and utilitarian discovery in this field holds exciting promise. This is evidenced by the discovery of a wide range of products, and microorganisms that present potential, as mentioned above. It is important for all involved in these studies to realize the importance of acquiring the necessary permits from governmental, local, and other sources to pick, and transport plant materials (especially from abroad) from which endophytes are eventually to be isolated. In addition to this aspect of the work is the added activity of producing the necessary agreements and financial sharing arrangements with indigenous peoples or governments in case a product does develop an income stream.

Certainly, one of the major problems facing the future of endophyte biology and natural product discovery is the rapidly diminishing rainforests which hold the greatest possible resource for acquiring novel microorganisms and their products. The total land mass of the world that currently supports rainforests is about equal to the area of the United States [16]. Each year, an area the size of Vermont or greater is lost to clearing, harvesting, fire, agricultural development, mining, or other human oriented activities [16]. Presently, it is estimated that only a small fraction (10–20%) of what were the original rainforests existing 1000–2000 years ago, are currently present on the Earth [16]. The advent of major negative pressures on them from these human-related activities appears to be eliminating entire megalife forms at an alarming rate. Few have ever expressed information or opinions about what is happening to the potential loss of microbial diversity as entire plant species disappear. It can only be guessed that this loss is also happening, perhaps with the same frequency as the loss of mega life forms, especially since certain microorganisms may have developed unique specific symbiotic relationships with their plant hosts. Thus, when a plant species disappears, so too does its entire suite of associated endophytes and consequently all of the capabilities that they might possess to make natural products with medicinal potential. Multi-step processes are needed now to secure information and life forms before they continue to be lost. Areas of the planet that represent unique places housing biodiversity need immediate preservation. Countries must establish information bases of their biodiversity and, at the same time, begin to create national collections of microorganisms that live in these areas. Endophytes are just one example of a life form source that holds enormous promise to impact many aspects of human existence. The problem of the loss of biodiversity should be one of concern to the entire world.

Acknowledgments

The author thanks Dr. Gene Ford and Dr. David Ezra for helpful discussions. He also expresses his appreciation to the NSF, USDA, NIH, The BARD Foundation of Israel, The R&C Board of the State of Montana and the Montana Agricultural Experiment Station for providing financial support for some of the studies reviewed in this chapter.

References

1 Lane, G.A., Christensen, M.J., Miles, C.O. Coevolution of fungal endophytes with grasses: the significance of secondary metabolites. In: Bacon, C.W., White, J.F. (Eds.), *Microbial Endophytes*. Marcel Dekker, Inc., New York, **2000**.

2 Bacon, C., White, J.F. *Microbial Endophytes*. Marcel Dekker, Inc., New York, **2000**.

3 *NIAID Global Health Research Plan for HIV/AIDS, Malaria and Tuberculosis*. U.S. Department of Health and Human Services, Bethesda, MD, **2001**.

4 Newman, D.J., Cragg, G.M., Snader, K.M. *J. Nat. Prod.* **2003**, *66*, 1022–1037.

5 Grabley, S., Thiericke, R.. The impact of natural products on drug discovery. In: Grabley, S., Thiericke, R. (Eds.), *Drug Discovery from Nature*. Springer-Verlag, Berlin, **1999**, pp. 3–33.

6 Concepcion, G.P., Lazaro, J.E., Hyde, K.D. Screening for bioactive novel compounds. In: Pointing, S.B., Hyde,

K.D. (Eds.), *Bio-exploitation of Filamentous Fungi*. Fungal Diversity Press, Hong Kong, **2001**, pp. 93–130.

7 Suffness, M. (Ed.), *Taxol®: Science and Applications*. CRC Press, Boca Raton, **1995**.

8 Wani, M.C., Taylor, H.L., Wall, M.E., Goggon, P., McPhai, A.T. *J. Am. Chem. Soc.* **1971**, *93*, 2325–2327.

9 Rowinsky, E.K., Cazwnave, L.A., Donehower, F.C. *J. Natl. Cancer Inst.* **1990**, *82*, 1247–1259.

10 Demain, A.L.. *Science* **1981**, *214*, 987–994.

11 Schutz, B.. In: *Bioactive Fungal Metabolites – Impact and Exploitation*. British Mycological Society, International Symposium Proceedings, Swansea, University of Wales, UK, **2001**, p. 20.

12 Tan, R.X., Zou, W.X. *Nat. Prod. Rep.* **2000**, *18*, 448–459.

13 Redlin, S.C., Carris, L.M. (Eds.), *Endophytic Fungi in Grasses and Woody Plants*. APS Press, St. Paul, **1996**.

14 Hawksworth, D.C., Rossman, A.Y. *Phytopathology* **1987**, *87*, 888–891.

15 Dreyfuss, M.M., Chapela, I.H. Potential of fungi in the discovery of novel, low-molecular weight pharmaceuticals. In: Gullo, V.P. (Ed.), *The Discovery of Natural Products with Therapeutic Potential.* Butterworth-Heinemann, Boston, **1994**, pp. 49–80

16 Mittermeier, R.A., Myers, N., Gil, P.R., Mittermeier, C.G. *Hotspots: Earth's Biologically Richest and Most Endangered Ecoregions.* CEMEX Conservation International, Washington DC, **1999**.

17 Strobel, G.A., Li, J.Y., Sugawara, F., Koshino, H., Harper, J., Hess, W.M. *Microbiology* **1999**, *145*, 3557–3564.

18 Castillo, U.F., Giles, S., Browne, L., Strobel, G.A., Hess, W.M., Hanks, J., Reay, D. *Scanning Microscopy 27*, 305–311.

19 Castillo, U.F., Strobel, G.A., Ford, E.J., Hess, W.M., Porter, H., Jensen, J.B., Albert, H., Robison, R., Condron, M.A., Teplow, D.B., Stevens, D., Yaver, D. *Microbiolology* **2002**, *148*, 2675–2685.

20 Strobel, G.A., Stierle, A., Stierle, D., Hess, W.M. *Mycotaxon* **1993**, *47*, 71–78.

21 Redell, P., Gordon, V. Lessons from nature: can ecology provide new leads in the search for novel bioactive chemicals from rainforests? In: Wrigley, S.K., Hayes, M.A., Thomas, R., Chrystal, E.J.T., Nicholson, N. (Eds.), *Biodiversity: New Leads for Pharmaceutical and Agrochemical Industries.* The Royal Society of Chemistry, Cambridge, UK, **2000**, pp. 205–212.

22 Bills, G., Dombrowski, A., Pelaez, F., Polishook, J. Recent and future discoveries of pharmacologically active metabolites from tropical fungi. In: Watling, R., Frankland, J.C., Ainsworth, A.M., Issac, S., Robinson, C.H., Eda, Z. (Eds.), *Tropical Mycology: Micromycetes.* CABI Publishing, New York, **2002**, Vol. 2, pp. 165–194.

23 Stierle, A., Strobel, G.A., Stierle, D. *Science* **1993**, *260*, 214–216.

24 Strobel, G.A. *Can. J. Plant Pathol.* **2002**, *24*, 14–20.

25 Strobel, G.A. *Crit. Rev. Biotechnol.* **2002**, *22*, 315–333.

26 Strobel, G.A., Daisy, B. *Microbiol. Molec. Biol. Rev.* **2003**, *67*, 491–502.

27 Strobel, G.A., Daisy, B., Castillo, U.F., Harper, J. *J. Nat. Prod.* **2004**, *67*, 257–268.

28 Li, J.Y., Strobel, G.A., Sidhu, R., Hess, W.M., Ford, E. *Microbiology* **1996**, *142*, 2223–2226.

29 Strobel, G.A., Miller, R.V., Miller, C., Condron, M., Teplow, D.B., Hess, W.M.. *Microbiology* **1999**, *145*, 1919–1926.

30 Walsh, T.A. Inhibitors of β-glucan synthesis. In: Sutcliffe, J.A., Georgopapadakou, N.H. (Eds.), *Emerging Targets in Antibacterial and Antifungal Chemotherapy.* Chapman & Hall, London, 1992, pp. 349–373.

31 Li, J.Y., Strobel, G.A., Harper, J.K., Lobkovsky, E., Clardy, J. *Org. Lett.* **2000**, *2*, 767–770.

32 Li, J.Y., Harper, J.K., Grant, D.M., Tombe, B.O., Bashyal, B., Hess, W.M., Strobel, G.A. *Phytochemistry* **2001**, *56*, 463–468.

33 Harper, J.K., Mulgrew, A.E., Li, J.Y., Barich, D.H., Strobel, G.A., Grant, D.M. *J. Am. Chem. Soc.* **2001**, *123*, 9837–9842.

34 Harper, J.K., Arif, A.M., Li, J.Y., Strobel, G.A., Grant, D.M. *Acta Crystallogr.* **2000**, *C56*, e570.

35 Lee, J.C., Yang, X., Schwartz, M., Strobel, G.A., Clardy, J. *Chem. Biol.* **1995**, *2*, 721–727.

36 Pulici, M., Sugawara, F., Koshino, H., Uzawa, J., Yoshida, S., Lobkovsky, E., Clardy, J. *J. Org. Chem.* **1996**, *61*, 2122–2124.

37 Pulici, M., Sugawara, F., Koshino, H., Uzawa, J., Yoshida, S., Lobkovsky, E., Clardy, J. *J. Nat. Prod.* **1996**, *59*, 47–48.

38 Pulici, M., Sugawara, F., Koshino, H., Uzawa, J., Yoshida, S., Lobkovsky, E., Clardy, J. *J. Chem. Res.* **1996**, 378–379.

39 Li, J.Y., Strobel, G.A. Jesterone and hydroxy-jesterone antioomycete cyclohexenenone epoxides from the endophytic fungus, *Pestalotiopsis jesteri.* *Phytochemistry* **2001**, *57*, 261–265.

40 Hu, Y., Chaomin, L., Kulkarni, B., Strobel, G.A., Lobkovsky, E., Torczynski, R., Porco, J. *Org. Lett.* **2001**, *3*, 1649–1652.

41 Horn, W.S., Simmonds, M.S., Schwartz, R.E., Blaney, W.M. *Tetrahedron* **1995**, *14*, 3969–3978.

42 Brady, S.F., Clardy, J. *J. Nat. Prod.* **2000**, *63*, 1447–1448.

43 Zou, W.X., Meng, J.C., Lu, H., Chen, G.X., Shi, G.X., Zhang, T.Y., Tan, R.X. *J. Nat. Prod.* **2000**, *63*, 1529–1530.

44 Lu, H., Zou, W.X., Meng, J.C., Hu, J., Tan, R.X. *Plant Sci.* **2000**, *151*, 67–73.

45 Rodriguez-Heerklotz, K.F., Drandarov, K., Heerklotz, J., Hesse, M., Werner, C. *Helv. Chim. Acta.* **2001**, *84*, 3766–3772.

46 Miller, R.V., Miller, C.M., Garton-Kinney, D., Redgrave, G.B., Sears, J., Condron, M., Teplow, D., Strobel, G.A. *J. Appl. Microbiol.* **1998**, *84*, 937–944.

47 Harrison, L., Teplow, D., Rinaldi, M., Strobel, G.A. *J. Gen. Microbiol.* **1991**, *137*, 2857–2865.

48 Ballio, A., Bossa, F., DiGiogio, P., Ferranti, P.P, Paci, M., Pucci, P., Scaloni, A., Segre, A., Strobel, G.A. *FEBS Lett.* **1994**, *355*, 96–100.

49 Zhang, Y.Z., Sun, X., Zechner, D., Sachs, B., Current, W., Gidda, J., Rodriguez, M., Chen, S.H. *Bioorg. Med. Chem.* **2001**, *11*, 903–907.

50 Keiser, T., Bibb, M.J., Buttner, M.J., Charter, K.F., Hopwood, D.A. *Practical Streptomycetes Genetics.* The John Innes Foundation, Norwich, **2000**.

51 Guerny, K.A., Mantle, P.G. *J. Nat. Prod.* **1993**, *56*, 1194–1199.

52 Kunoh, H.J.. *Gen. Plant Pathol.* **2002**, *68*, 249–252.

53 Castillo, U., Harper, J.K., Strobel, G.A., Sears, J., Alesi, K., Ford, E., Lin, J., Hunter, M., Maranta, M., Ge, H., Yaver, D., Jensen, J.B., Porter, H., Robison, R., Millar, D., Hess, W.M., Condron, M., Teplow, D. *FEMS Lett.* **2003**, *224*, 183–190.

54 Ezra, D., Castillo, U., Strobel, G.A., Hess, W.M., Porter, H., Jensen, J., Condron, M., Teplow, D., Sears, J., Maranta, M., Hunter, M., Weber, B., Yaver, D. *Microbiology* **2004**, *150*, 785–793.

55 Worapong, J., Strobel, G.A., Ford, E.J., Li, J.Y., Baird, G., Hess, W.M. *Mycotaxon* **2001**, *79*, 67–79.

56 Strobel, G.A., Dirksie, E., Sears, J., Markworth, C. *Microbiology* **2001**, *147*, 2943– 2950.

57 Ezra, D., Strobel, G.A. *Plant Sci.* **2003**, *65*, 1229–1238.

58 Worapong, J., Strobel, G.A., Daisy, B., Castillo, U., Baird, G., Hess, W.M. *Muscodor roseus anna.* nov.: an endophyte from *Grevillea pteridifolia.* *Mycotaxon* **2002**, *81*, 463–475.

59 Ezra, D., Hess, W.M., Strobel, G.A. *Microbiology* **2004**, *150*, 4023–4031.

60 Atmosukarto, I., Castillo, U., Hess, W.M., Sears, J., Strobel, G.A. *Plant Sci.* *169*, 854–861.

61 Stinson, M., Ezra, D., Strobel, G.A. *Plant Sci.* **2003**, *165*, 913–922.

62 Guo, B., Dai, J., Ng, S., Huang, Y., Leong, C., Ong, W., Carte, B.K. *J. Nat. Prod.* **2000**, *63*, 602–604.

63 Strobel, G., Yang, X., Sears, J., Kramer, R., Sidhu, R.S., Hess, W.M. *Microbiology* **1996**, *142*, 435–440.

64 Strobel, G.A., Hess, W.M., Li, J.Y., Ford, E., Sears, J., Sidhu, R.S., Summerell, B. *Aust. J. Bot.* **1997**, *45*, 1073–1082.

65 Strobel, G.A., Ford, E., Li, J.Y., Sears, J., Sidhu, R., Hess, W.M. *System. Appl. Microbiol.* **1999**, *22*, 426–433.

66 Li, J.Y., Sidhu, R.S., Ford, E., Hess, W.M., Strobel, G.A. *J. Ind. Microbiol.* **1998**, *20*, 259–264.

67 Bashyal, B., Li, J.Y., Strobel, G.A., Hess, W.M. *Mycotaxon* **1999**, *72*, 33–42.

68 Young, D.H., Michelotti, E.J., Sivendell, C.S., Krauss, N.E. *Experientia* **1992**, *48*, 882–885.

69 Wang, J., Li, G., Lu, H., Zheng, Z., Huang, Y., Su, W. *FEMS Microbiol. Lett.* **2000**, *193*, 249–253.

70 Shrestha, K., Strobel, G.A., Prakash, S., Gewali, M. *Planta Med.* **2001**, *67*, 374 –376.

71 Hoffman, A., Khan, W., Worapong, J., Strobel, G., Griffin, D., Arbogast, B., Borofsky, D., Boone, R.B., Ning, L., Zheng, P., Daley, L. *Spectroscopy* **1998**, *13*, 22–32.

72 Lee, J.C., Strobel, G.A., Lobkovsky, E., Clardy, J.C. *J. Org. Chem.* **1996**, *61*, 3232–3233.

73 Li, C., Johnson, R.P., Porco, J.A. *J. Am. Chem. Soc.* **2003**, *125*, 5059–5106.

74 Wagenaar, M., Corwin, J., Strobel, G.A., Clardy, J. *J. Nat. Prod.* **2000**, *63*, 1692–1695.

75 Strobel, G.A., Ford, E., Worapong, J., Harper, J.K., Arif, A.M., Grant, D., Fung, P.C.W., Chan, K. *Phytochemistry* **2002**, *60*, 179–183.

76 Harper, J.K., Ford, E.J., Strobel, G.A., Arif, A., Grant, D.M., Porco, J., Tomer, D.P., O'Neill, K. *Tetrahedron* **2003**, *59*, 2471–2476.

77 Lee, J., Lobkovsky, E., Pliam, N.B., Strobel, G.A., Clardy, J. *J. Org. Chem.* **1995**, *60*, 7076–7077.

78 Long, D.E., Smidmansky, E.D., Archer, A.J., Strobel, G.A. *Fungal Gen. Biol.* **1998**, *24*, 335–344.

5
DNA Profiling of Plants

Hilde Nybom and Kurt Weising

5.1
Introduction

Mistaken or willful adulteration of medicinal plant material can be a major problem in a wide range of situations. These include, for example, the production of traditional Chinese medicine prescriptions that often consist of a complex mixture of ingredients, or the manufacture of certain herbal products, that are regarded as dietary supplements in Western countries and therefore not subjected to stringent safety controls. The problem is perhaps best exemplified by the major medicinal crop ginseng. Cultivated common ginseng *Panax ginseng* C. A. Mey. fetches only about 10–20% of the value accorded to American ginseng *Panax quinquefolius* L., and is therefore often marketed under the more "expensive" name [1]. Even worse, completely unrelated and/or even poisonous species from various genera with similar-looking roots (e.g., *Bletilla*, *Curcuma*, *Gynura*, *Mirabilis* and *Phytolacca*) are sometimes also marketed as ginseng.

In medicinal plant production, quality control is usually attempted at two different levels: (i) taxonomic authentication of the source material; and (ii) correct prediction and standardization of the concentration of active phytochemicals. These issues are closely interrelated, as many medicinally important species and species complexes are quite heterogeneous, with differences in both composition and concentration of phytochemical substances [2]. For example, the use of field-collected material from *Achillea millefolium* as a drug is problematic because of the heterogeneous distribution of medicinally important azulenes, flavonoids and essential oils among closely related species that are difficult to distinguish in the field [3]. In order to avoid undesirable variability in the efficacy of the commercial products, not only the correct species assignment but also the unambiguous discrimination among intraspecific breeding lines, strains, cultivars or accessions is of utmost importance.

Traditionally, the identification of medicinal plants has been based on the evaluation of phenotypic characteristics such as morphology, smell, taste, color, texture, and size. These characteristics however have certain limitations, including: (i) in-

Medicinal Plant Biotechnology. From Basic Research to Industrial Applications
Edited by Oliver Kayser and Wim J. Quax
Copyright © 2007 WILEY-VCH Verlag GmbH & Co. KGaA, Weinheim
ISBN 978-3-527-31443-0

sufficient variation among the samples; (ii) subjectivity in the analysis; (iii) plasticity of the character, which could be due to the influence of the environment and management practice; and (iv) ability to be scored only at certain stages of the plant development (e.g., during flowering and/or fruiting) which may not coincide with the commercially important stage or organ (e.g., roots in ginseng production). More recently, traditional means of plant identification have been complemented by advanced analytical methods of phytochemistry such as high-pressure liquid chromatography (HPLC) and mass spectrometry (MS), as well as by molecular biology. Among these novel techniques, molecular markers based on DNA sequence variation have become increasingly important for the identification and authentication of medicinal plants and for the estimation of genetic diversity in these crops (see also the review by Shaw et al. [4]).

In this chapter, we shall shortly describe some of the most commonly used DNA methods for identifying plant material and for analyzing plant genetic diversity and relatedness. We also list a number of recent studies that illustrate the potential application ranges of plant DNA profiling in the characterization of medicinally important plants. This survey is necessarily brief, but for more comprehensive descriptions of the theory and methodology underlying DNA markers, experimental protocols, strategies of data evaluation, and more applications, the reader is referred to the monographic treatments presented by Henry [5], Avise [6] and Weising et al. [7].

5.2
Methodology of Plant DNA Profiling

An obvious first step of most DNA profiling methods is the isolation of genomic DNA from the sample of interest. Different profiling techniques require different levels of purity of the DNA preparation. For example, high-quality DNA (and also quantity!) samples are needed for DNA fingerprinting based on restriction fragment length polymorphism (RFLP; see Section 5.2.2.1) and amplified fragment length polymorphism (AFLP; see Section 5.2.2.4), as well as for all other methods that involve the initial treatment of genomic DNA with restriction enzymes. These enzymes are sensitive to small amounts of impurities, and total or partial inhibition of their activity will result in irreproducible banding patterns. On the other hand, even highly degraded and/or contaminated preparations obtained from so-called "quick and dirty" methods of DNA isolation, may still give interpretable results when analyzed by direct DNA sequencing (see Section 5.2.1) or microsatellite markers (see Section 5.2.3).

Fresh leaves are the standard starting material for plant DNA isolation, but numerous other sources have been successfully used. These include flower petals [8], pollen grains [9], seeds [10], roots [11], rhizomes [12], tubers [13] and even dried wood [14, 15]. Samples that cannot be processed immediately should preferably be preserved by either freezing at –20 °C (or lower), freeze-drying, treatment with certain chemicals [16], or quick-drying in silica gel [17].

There are many factors that potentially interfere with the successful isolation of DNA from plant tissues. For example, endogenous nucleases may be activated during the isolation procedure, resulting in DNA degradation. Also, polysaccharides, polyphenols, latex, essential oils and other secondary compounds (which are common ingredients of medicinal plants; see [18]) may be co-isolated and cause damage to DNA and/or inhibit subsequent enzymatic reactions. Numerous strategies have been proposed to remove such impurities, including ultracentrifugation, preparative gel electrophoresis, affinity chromatography, and many more [7]. DNA preparations from forensic samples, herbarium specimens or processed plant products are particularly problematic, because they are often degraded, chemically modified, and/or contaminated by foreign DNA and by substances inhibitory to the polymerase chain reaction (PCR). Such specimens usually require special treatments [19, 20].

As an example, a DNA isolation protocol for market samples of dry tea leaves was presented by Singh and Ahuja [21]. Relatively intact DNA was obtained, but PCR was only successful when the leaves were preincubated in water and re-dried before DNA extraction. This procedure removed much of the brown color, and presumably also part of the secondary products that otherwise interfered with DNA amplification. Post-brewed tea samples, however, yielded heavily degraded DNA, which was unsuitable for PCR [21].

Because of the factors listed above, plant DNA isolation efficiencies vary from species to species, and from tissue to tissue. Accordingly, a huge number of plant DNA isolation protocols have been published. These methods differ in many respects, including the disruption of tissues and cells, the composition of extraction buffers, and in the way that DNA is purified from other cell ingredients. A number of common principles allow the distinction of certain "families" of methods. Perhaps most commonly used is the so-called cetyltrimethylammonium bromide (CTAB) strategy initially developed by Murray and Thompson [22], of which more than 100 variants have since been published. More recently, commercial plant DNA isolation kits and automated extraction have gained increasing attention and importance, but these are also more costly than the traditional techniques.

A comprehensive description and discussion of the wide variety of plant DNA extraction procedures was provided by Weising et al. [7]. Specific protocols have been designed for various medicinally important plant species, including *Achillea millefolium* (milfoil; [23]), *Allium sativum* (garlic; [18]), *Artemisia annua* [18, 24]; *Catharanthus roseus* [18], *Curcuma longa* (turmeric; [12]), *Eleutherococcus senticosus* (Siberian ginseng; [23]), *Mentha arvensis* [18], *Papaver somniferum* (opium poppy; [25]), *Taxus wallichiana* (yew; [18]), and *Zingiber officinale* (ginger; [12]).

5.2.1
DNA Sequencing

Determining the DNA sequence of one or more defined genomic regions is the most obvious strategy to obtain genotypic data from any organism. Having been somewhat tedious in the past, the collection of DNA sequence data for genotyping purposes has been facilitated considerably by methodological and computational improvements, as well as by the availability of commercial sequencing services. The invention of the PCR by Mullis and colleagues (reviewed in [26]) has been a key step in this development. The PCR is an enzymatic method that allows the exponential *in-vitro* multiplication (or "amplification") of any DNA sequence of interest, as defined by a pair of flanking primers that serve as starting points for a thermostable DNA polymerase. Typically, PCR products from the DNA region of interest are sequenced directly on an automated, fluorescence-based sequencing machine. Mutational differences are then detected by vertical alignment of the DNA sequences obtained from orthologous regions of all organisms to be compared.

Several issues should be considered when choosing a suitable DNA region for sequence analysis. First, the region should exhibit an adequate level of variation for solving the problem under investigation. For example, housekeeping genes tend to be highly conserved, whereas microsatellites (see Sections 5.2.2.3 and 5.2.3) are hypervariable. Second, the mode of inheritance of the targeted piece of DNA may have to be taken into account (inheritance is typically uniparental for organellar DNA and biparental for nuclear DNA). Finally, sequence data from flanking regions should be available for PCR primer design. Often, so-called universal primer pairs are used to generate the sequencing template. These are designed to bind to conserved DNA sequence motifs within genes, and are therefore transferable across species, genera and even families. Well-known examples are the so-called internal transcribed spacer (ITS) primers that bind to the conserved coding region of the nuclear 18S and 26S ribosomal RNA (rRNA) genes and amplify the often polymorphic internal transcribed spacer (ITS) region [27]. Sets of universal primers are also available for organellar DNAs (see Section 5.2.4). In medicinal plants, sequence data of the ITS or other ribosomal RNA gene regions have frequently been applied for species identification [28–31] (see Section 5.3.1.1).

5.2.2
Multilocus DNA Profiling

The techniques summarized in this section share a number of common characteristics. First, a more or less complex banding pattern of DNA fragments is generated that can serve as a type of bar code ("DNA fingerprint"). Depending on the technique used and the organisms investigated, this bar code can be specific to, for example, an individual, a clone, a population, or even a species. Second, it is generally possible to generate this bar code without previous DNA sequence information. Most multilocus methods are therefore universally applicable to all organisms, whether prokaryotic or eukaryotic. Third, the DNA fragments underlying the bar code presumably reflect a more or less random sample of the whole genome (unlike DNA sequencing, where only one or few genomic loci are analyzed). Fourth, individual bands within a pattern are mostly biallelic and can only be scored as present or absent. It is usually not possible to assign individual DNA fragments to a particular genomic locus, allelic states cannot be determined, and homo- and heterozygotes cannot be distinguished. Multilocus markers must therefore be treated as dominant, which reduces their potential for in-depth genetic analyses.

5.2.2.1 Hybridization-Based RFLP Fingerprinting

Based on Southern blot analysis with radioactive hybridization probes and the RFLP technique, the so-called "DNA fingerprinting" methodology was first introduced to plant genome analysis during the late 1980s (reviewed in [32]). With this method, it became possible for the first time ever to distinguish directly between closely related plant genotypes at the DNA level, using only a single analysis. Hybridization-based DNA fingerprinting has been used in many early studies on medicinal plants, for example for the identification of micropropagated *Achillea* clones [3]. The technique is highly reproducible and reliable, but involves many experimental steps and requires microgram quantities of DNA. This latter issue is of particular relevance for medicinal samples, where often only small amounts of material are available for examination. Today, hybridization-based fingerprinting is rarely used, mainly because virtually all subsequently developed methods benefit from the advantages of the PCR.

5.2.2.2 PCR with Arbitrary Primers

The introduction of PCR-based molecular marker methods during the early 1990s constituted a new milestone in the field of DNA fingerprinting; two methods using single primers with arbitrary sequence were published in 1990 [33, 34]), and a third method in 1991 [35]. Numerous variants followed (for a review, see [7]). These usually short, arbitrary primers were shown to generate anonymous PCR amplicons from genomic DNA, resulting in polymorphic banding patterns after gel electrophoresis and staining.

The RAPD (random amplified polymorphic DNA) approach developed by Williams et al. [34] has become the most popular variant of this "prototype" of PCR-based DNA profiling, mainly because of its technical simplicity. On the negative side, the reproducibility of RAPD banding patterns turned out to be relatively poor, especially when results were compared among laboratories [36]. The value of RAPD (and related) markers for genotype identification and genetic relatedness studies has therefore been controversial. Nevertheless, literally thousands of RAPD investigations on hundreds of species have been published during the past 15 years, also including medicinally important plant species from various genera (e.g., *Artemisia* [37], *Achillea* [3], *Echinacea* [38], and *Panax* [30, 39, 40]).

5.2.2.3 PCR with Microsatellite-Complementary Primers

Another variant of multilocus PCR markers was proposed a few years later, and was variously coined inter-simple sequence PCR (ISSR; [41]), single-primer amplification reactions (SPAR; [42]) or microsatellite-primed PCR (MP-PCR; [43, 44]). The primers employed in this strategy were designed to anneal at the 5′- or 3′-ends of so-called microsatellite DNA motifs, also known as simple sequence repeats (SSRs). Microsatellites consist of tracts of short (1–5 bp) direct DNA repeats, which are present at tens of thousands of sites in all eukaryotic and many prokaryotic genomes examined to date (for reviews, see [7, 45]; see also Section 5.2.3). The copy number of the tandemly repeated motifs within a microsatellite is often highly variable, resulting in frequent size polymorphisms.

If two inversely oriented microsatellites are located close enough to each other on opposite strands of the DNA, primer binding results in the amplification of the inter-repeat region. The use of so-called 5′-anchored primers ensures that part of the polymorphic repeat itself is included in the product, enhancing the chance of size variation [41, 46]. Like RAPDs, MP-PCR bands are separated by gel electrophoresis and visualized by staining, fluorescence, or radioactivity. The method appears to be under-utilized in medicinal plants, but has proven useful for studying genetic variation in the endangered tropical tree *Hagenia abyssinica* [47].

5.2.2.4 AFLP Analysis

A further new method, known as AFLP analysis, which incorporated elements of both RFLP and RAPD, was also developed during the mid-1990s [48, 49]. Although technically more demanding than RAPDs, AFLP analysis produced very high numbers of polymorphic bands in a single experiment. The procedure involves several steps. In the first step, genomic DNA is digested with restriction enzymes (usually a combination of two enzymes is used, one with a six-base and one with a four-base recognition sequence), and short adapters of a defined sequence are ligated to both ends of the resulting DNA fragments. A subset of these fragments is then amplified by PCR with a specially designed pair of primers. The 5′-portion of such an AFLP primer is complementary to the adapter sequence, whereas its 3′-end extends for one or a few, arbitrarily chosen, so-called "selective" nucleotides into the restriction fragment. Because exact matching of the 3′-end of a primer is required for successful PCR, only those fragments are amplified whose outmost nucleotides are able to base-pair with the selective nucleotide(s) of the primer(s). Statistically, this is the case for 1/16 of all restriction fragments if both primers carry a single base extension, but only for 1/4096 of all fragments if both primers carry a three-base extension. PCR products are eventually separated on polyacrylamide gels, and are visualized by radioisotopic labeling, fluorescence, or silver staining.

The complexity of an AFLP banding pattern can be adapted to organisms of different genome size by choosing appropriate numbers of selective nucleotides. For eukaryotic genomes, PCR is usually performed in two successive steps, with +1 selective nucleotide in the first PCR (so-called preamplification), and +3 in the second PCR. This procedure typically results in about 50 to 100 bands which are relatively easy to evaluate. Examples of the use of AFLPs for investigating medicinal plants include, among many others, the authentication of American ginseng, *Panax quinquefolius* [1], and the prediction of phytochemical contents in *Echinacea* [2].

5.2.3
Locus-Specific Microsatellite DNA Markers

Another popular DNA marker method is based on PCR with primers that are complementary to the DNA sequences *flanking* microsatellites, rather than binding to the repeats themselves (which is the case in MP-PCR; see Section 5.2.2.3). As a consequence of the variable number of tandem repeats within a microsatellite, the products amplified by such primer pairs often exhibit considerable length differences among individuals or populations of the same species. Microsatellite analysis, which was first introduced for plants in 1992 [50], produces locus-specific, multiallelic, easy-to-use, codominantly inherited bands with high levels of polymorphism and reproducibility. The resulting markers are excellent tools for genotype identification, population genetics, genetic mapping and gene tagging. Moreover, microsatellite data are easily managed and compared in databases, several marker loci can be analyzed simultaneously by multiplexing fluorescence-labeled primer panels [51, 52], reasonably well-interpretable results are obtained even with degraded DNA (when neither RAPDs nor AFLPs perform well), and the species-specificity of the markers reduces the sensitivity to contamination by foreign DNA.

Their species-specificity is also a major drawback of microsatellite markers. Contrary to the generic multilocus profiling approaches described in Section 5.2.2, *a priori* sequence information is needed for primer design, and the development of species-specific primer pairs can be costly and time-consuming (for reviews, see [53, 54]). Public databases have become an increasingly important source for microsatellite sequences, and database mining has become a valuable alternative to genomic cloning in many species. Furthermore, microsatellite-flanking primer sequences are often conserved in congeneric species, or occasionally, even in other genera (reviewed in [7]). Marker transferability is usually best for trinucleotide repeats that are often located in gene coding regions. Expressed sequence tag (EST) databases are increasingly used for "fishing" well-conserved microsatellite markers, as has been demonstrated, for example, in cereals [55, 56].

One potential problem for some applications is the fact that mutation rates can be very high, especially for dinucleotide repeat loci. Microsatellite markers are therefore not recommended at or above the species level. To date, only a few studies have appeared on the application of microsatellite DNA in medicinal plants [57, 58], but this is likely to increase in the near future.

5.2.4
PCR-Based RFLP Analysis of Organellar and Nuclear Genomes

Whereas the above-mentioned DNA marker methods focus on nuclear DNA, it may sometimes be more feasible to study the organellar genomes. A series of universal PCR primers have been developed, which allow the amplification of chloroplast and mitochondrial DNA introns and intergenic spacers in a wide array of plant species [59–61]. PCR products amplified with organelle-specific primer pairs are either sequenced directly (see Section 5.2.1), or digested with restriction enzymes in an approach known as PCR-RFLP. The lack of genetic recombination of the chloroplast genome allows the combination of polymorphisms observed at several loci to form a so-called haplotype. PCR-based RFLP analysis can also be applied to nuclear DNA, as exemplified in the CAPS (cleaved amplified polymorphic sequences) approach [62].

5.2.5

Other DNA Marker Methods

During the past decade, a plethora of additional, mostly PCR-based DNA marker methods have been developed, none of which has become as popular as RAPDs, MP-PCR, AFLPs and microsatellites. The PCR primers employed in these more exotic methods are targeted at a wide range of genomic DNA motifs, including pathogen resistance genes, intron-exon splice junctions, transposons, minisatellites, and many others. The diversity of methods has been further enhanced by various combinations of existing methods, as exemplified by SAMPL (selective amplification of polymorphic microsatellite loci [63]), which combines AFLP- and microsatellite-specific primers, and by S-SAP (sequence-specific amplification polymorphism [64]), which combines AFLP- and retrotransposon-specific primers (for a comprehensive survey on these techniques see [7]).

5.2.6

The Next Generation: SNPs and DNA Microarrays

The analysis of single nucleotide polymorphisms (SNPs) represents a relatively recent addition to the spectrum of plant DNA profiling methods (reviewed in [65–67]). The term "SNP" refers to a single base pair position in the DNA, at which different sequence alternatives exist at notable frequencies in a population or species. In principle, a SNP locus can have two to four alleles, but biallelic polymorphisms usually prevail. Because SNPs are caused by base substitutions, mutation rates at each individual marker locus are relatively low. Nevertheless, even closely related genotypes are discriminated by thousands of SNPs distributed throughout the genome. The average density of SNPs depends on the species and genomic region investigated. In the few plant species examined in detail to date, about one SNP was found per 200–500 bp [67].

Establishing SNP-based genomic profiling for a plant species of interest involves: (i) the discovery of a sufficient number of SNPs in the respective genomes; and (ii) the development of a suitable assay. For model organisms and major crops, database mining is a promising strategy for discovering SNPs *in silico* [68]. However, few database data are yet available for medicinal plant species, and SNPs will therefore have to be discovered experimentally. This can be done in a variety of ways, perhaps most economically by single-stranded conformation polymorphism (SSCP) analysis of PCR products (reviewed in [69]). A good option is also to convert other types of markers into SNPs, as has been demonstrated for AFLP products [70].

Many strategies have been developed for SNP detection, ranging from direct sequencing or SSCP analysis on the trivial side of the spectrum, to high-throughput allele-specific PCR, ligation or oligonucleotide hybridization methods on the more elaborate side (for reviews, see [71, 72]). "SNPology" is a rapidly expanding field, and new techniques and modifications of existing techniques are developed at high rates. SNP assays become most time-efficient in conjunction with DNA microarray technology, where multiple SNP-carrying DNA oligonucleotides are assembled and analyzed simultaneously on a small silicon chip (for recent reviews, see [73, 74]).

In plants, SNP markers are already well-established in the model organism *Arabidopsis thaliana* [e.g., 75] and in major crops such as the cereals [e.g., 68, 76]. However, SNPs are not yet in common practice in medicinal plants. The need to develop specific SNP-based marker systems for each species of interest is a major obstacle, and is certainly the main reason why generic markers such as RAPDs and AFLPs are still preferred. Nevertheless, chip-based DNA profiling of Chinese medicinal plant DNA has already been performed [77–79] (for details, see Section 5.3.1.1). Moreover, an electrochemistry-based approach was suggested by Lee and Hsing [80], who used a 16-base DNA oligonucleotide immobilized on an electrode surface for hybridization detection of PCR-amplified target DNA molecules. With this hand-held device, it was possible to discriminate between two Chinese herbal species of *Fritillaria* [80].

5.3
Applications

5.3.1
Genotype Identification

Most of us are aware of the importance of individual-specific human DNA fingerprints in, for example, forensics. For medicinal plant material, identification issues usually focus on species (or other taxonomic units) or accessions but, occasionally, also on individual genotypes when, for example, suspected of somaclonal variation after *in-vitro* propagation.

5.3.1.1 Plant Species
Unambiguous identification of a given plant species by DNA sequencing is relatively easy and straightforward when the target species is sufficiently distinct from all other relevant taxa. Proper identification is considerably more ambiguous when closely related species are compared. For example, Ngan et al. [81] attempted to discriminate the roots of six *Panax* species from each other and from two common adulterants, *Mirabilis jalapa* and *Phytolacca acinosa*, by DNA sequence analysis of the 5.8S rRNA gene. High levels of sequence homology (ca. 99%) were shared among the six *Panax* species, but also between *Panax* and the two adulterants (96–97%). Instead, better discrimination was achieved by sequencing the two internal transcribed spacers (ITS1 and ITS2) surrounding this gene. Whereas ITS sequence homology among the various *Panax* species was still high (93–99%), it fell to less than 62% when these were compared with the two putative adulterants from

other genera. The few ITS sequence differences between the closely related species *P. ginseng* and *P. quinquefolious* were utilized in developing an RFLP-based discrimination protocol; even as little as 10% contamination of *P. ginseng* in the *P. quinquefolius* sample could be detected in this way. High levels of sequence homology among *Panax* species were also reported for the nuclear 18S rRNA gene and the *mat*K gene, although minor intraspecific polymorphism was noted in *P. notoginseng* [28]. In contrast, the 5S rRNA genes were relatively divergent (76–97% sequence homology among *Panax* species [30]), though with such sequences there is an increased risk of confounding intraspecific variability.

Species identification and discrimination may be less reliable when DNA profiling methods are used instead of DNA sequencing. Because most of the commonly used DNA marker methods reveal considerable amounts of intraspecific polymorphism, so-called "species-specific" or even "species-diagnostic" marker bands may not necessarily hold up for the whole species. This is especially true when such claims are based on the analysis of a few samples only (as with studies on *Echinacea* [38] and *Panax* species [40]). Considerable intraspecific diversity was reported in a study of *P. ginseng* and *P. quinquefolius*, based on RAPD and allozymes [82]. The same two species were also analyzed using AFLP on a set of samples obtained from several different farms in China and Korea, and in Canada and the USA, respectively [1]. The latter species was shown to contain more variation between localities (similar index values for pairwise comparisons: 0.64–0.96) than the former (0.88–0.99). One AFLP band detected only in *P. ginseng* was excised from the polyacrylamide gel, and then cloned and sequenced. It proved to contain a minisatellite region (eight repeats of 22 nucleotides). When used as a single PCR primer, the minisatellite motif produced a profile discriminatory for the two *Panax* species [1].

Sequencing of the nuclear ITS1-5.8S-ITS2 rRNA gene region was successfully used to differentiate among 16 species in the orchid genus *Dendrobium*, and also between the orchids (some of which are used for the Chinese medicine "herba dendrobii" or "shihu") and some commonly used adulterants such as *Pholidota* [29]. In this study, species authentication was facilitated by low levels of intra-specific ITS variation. More recently, this ITS sequencing information was used to develop a glass slide-based oligonucleotide DNA microarray, where target sequences were labeled with a fluorescent dye, and the presence of hybridized target sequence was detected with a confocal laser scanner [79]. This microarray produced distinctive signals for the five *Dendrobium* species traditionally recognized as having medicinal value, and it was able specifically to detect the presence of *D. nobile* material in a medicinal formulation containing nine different herbal components.

A DNA microarray approach for species identification was also taken by Carles et al. [78]. These authors designed a silicon chip that could differentiate among a set of 20 common and toxic plant species currently used in traditional Chinese medicine. The probes immobilized on the chip were based on 5S rRNA gene sequences or, in two cases, on the sequence of a leucine-specific transfer RNA (tRNA) gene. Even the closely related species pair *Datura innoxia* and *D. metel* (Solanaceae) could be distinguished, in spite of differing by only two nucleotides.

5.3.1.2 Plant Cultivars and Accessions

Successful discrimination among cultivars or accessions depends, to a large extent, on the means of propagation, as well as the mating system of the crop species under study. All plants belonging to a particular cultivar of a vegetatively propagated crop are expected to share identical DNA fingerprints, except for rare mutations – so-called "sports". The identification of cultivar-specific band patterns is fairly easy when the cultivar itself is homogeneous, as demonstrated by a RAPD study in *Panax ginseng* [83]. One RAPD band was found that could identify a selected elite strain, Kaishusan, from other strains. By sequencing this band, primers for a more easily used locus-specific marker (see Section 5.4) were developed.

In most cases, DNA markers are unable to detect any difference between a sport and its original cultivar [e.g., 84]. This is because: (i) only a very small portion of the genome is sampled, even when a large number of markers are employed; and (ii) changes in phenotype are often a consequence of chimerism, caused by a mutation in only one of the three meristematic cell layers that differentiate into the various plant tissues [85]. However, there are also exceptions. Although garlic (*Allium sativum*) has been propagated vegetatively since time immemorial, a surprisingly high number of quite different genotypes were encountered by analysis with AFLP, RAPD and isozyme markers [86]. Possibly many of the clones have existed for a very long time and therefore acquired numerous mutations, although occasional influx from sexual forms may also have occurred.

In sexually propagated crops, considerable levels of genetic variability usually persist also *within* cultivars or accessions. Pooling (or bulking) of samples is sometimes undertaken for generating cultivar-specific DNA profiles, especially in outcrossing crops. For soybean *Glycine max*, Diwan and Cregan [87] suggested that analyses should be based on bulks of 30 to 50 plants. Alternatively, several plants from each cultivar can be analyzed individually, and intra- and intercultivar variability are then partitioned statistically, for example by an analysis of molecular variance (AMOVA) [88].

The choice of profiling method should take the type of plant into consideration as well as the total number of samples to be examined. For small tests with only a few simultaneously analyzed samples, reproducibility and documentation may be less important, and any method providing sufficiently variable markers would be acceptable. Numerous studies in this direction have been carried out using multilocus methods such as AFLP and RAPD profiling. For example, six commercial cultivars of thyme (*Thymus vulgaris* L.), all belonging to the thymol chemotype, were successfully differentiated using RAPD markers, and could be divided into two major groups [89]. The same two groups were retrieved in a multivariate analysis based on their oil content profiles, as determined by gas chromatography.

5.3.1.3 *In-Vitro*-Propagated Plant Material

Tissue culture techniques often produce regenerants that differ from their parental phenotype, a phenomenon termed "somaclonal variability". As with sports, such variation is rarely detectable with DNA markers. Thus, somatic embryogenesis induced in callus tissues of *Panax notoginseng* resulted in plantlets that could not be differentiated even with a large set of RAPD primers [39]. Similarly, no variation in RAPD banding patterns was encountered among first-regeneration cycle somaclones of *Hypericum perforatum*, whereas three out of 51 somaclones obtained from the second cycle differed by a single band [90].

5.3.2
Genetic Diversity

In many situations, it is necessary to estimate the amount of genetic diversity in the plant material from which samples are taken for authentication. Markers based on slowly evolving DNA sequences are adequate for the analysis of historical events on longer timescales, whereas markers derived from fast-evolving sequences are more suitable for analyzing recently diverged populations. For highly diverse entities – for example, populations or accessions in cross-pollinating and variable species – a larger number of samples may need to be taken, or a higher number of markers analyzed for each sample.

5.3.2.1 Variation and Relatedness Among Cultivars and Accessions
Inter-cultivar relationships can be studied at different scales. Sometimes, the only objective is to assign the most likely parents from a selection of candidates, whereas in other cases the intention is to reveal relationships across the whole set of cultivars or accessions available for the study. In the search for *Artemisia annua* plants with increased levels of artemisinin production, screening of genetic diversity was undertaken in a ten-year-old introduced population in India [37]. Since no less than 97 out of 101 RAPD markers were polymorphic, high levels of variation obviously occurs in this population, suggesting that genetic improvement for phytochemical traits should be possible.

As may be expected from its rarity in nature, very low levels of genetic diversity were encountered within and among accessions of the "living fossil" *Ginkgo biloba* introduced from China and now grown in eastern USA [91]. Thus, no variation at all was detected by a PCR-RFLP analysis of chloroplast DNA (see Section 5.2.4), and only five out of 99 RAPD fragments turned out to be polymorphic. One of the fragments was turned into a probe and used for hybridization with Southern-blotted RAPD gels to allow a more accurate analysis. Using the excised RAPD band directly as a probe yielded multibanded (but different) patterns in all the screened samples. When the probe was first purified, hybridization revealed a corresponding band only in those samples that appeared to have the band also with the initial analysis [91].

In a study on *Cannabis sativa*, Gilmore et al. [57] demonstrated the potential of microsatellite analysis for forensic investigations. According to an AMOVA, 25% of the total genetic variation existed between accessions, and 6% between the two major *Cannabis sativa* groups used for fiber and drug production, respectively.

These results showed that microsatellite DNA fingerprinting may aid in determining agronomic type, geographic origin and production locality of these clonally propagated drug crops. It was, however, not possible to find any consistency between the patterns of relatedness among the different accessions on the one hand, and their contents of tetrahydrocannabinol on the other hand.

5.3.2.2 Amount and Distribution of Variability in Wild-Growing Plants

Information concerning the amount and geographic distribution of intraspecific genetic variation can be important when harvesting plant tissue in the wild for medicinal drug production, but even more when setting up programs for genetic improvement of the crop itself. Both dominantly (e.g., AFLP and RAPD) and codominantly inherited markers (e.g., microsatellites) have frequently been used to study population structure. The overall patterns concerning the extent and partitioning of genetic variability appear to be quite similar regardless of marker type, provided that the numbers of analyzed markers are sufficiently high [92]. The general picture arising from large data compilations is that long-lived, outcrossing and late successional species retain most of their variation *within* populations, whereas annual, self-pollinating and early successional species allocate more variation *among* populations [92].

In order to evaluate genetic variability in the rapidly decreasing *Valeriana wallrothii* (a member of the medicinally important *V. officinalis* species complex) in Italy, eight populations were studied with AFLP and chloroplast microsatellite markers [93]. AFLP data were highly informative, and demonstrated pronounced population differentiation in spite of the outcrossing breeding behavior of the species, as well as strong effects of grazing and agriculture. By contrast, only one of six chloroplast loci was polymorphic, with two alleles. The amount and geographic partitioning of genetic variation have been determined in many other medicinal plants, such as foxglove (*Digitalis obscura*) studied with RAPD [94], drum-stick tree (*Moringa oleifera*) studied with AFLP [95]), endod (*Phytolacca dodecandra*) studied with AFLP and RAPD [96], neem tree (*Azadirachta indica*) studied with AFLP and SAMPL [97], and tea tree (*Melaleuca alternifolia*) studied with microsatellite markers [98]. Wide ranges of genetic variability and its spatial distribution were observed, in spite of all these species being outcrossing and rather widely distributed.

Some plant species are able to form large clones, due to vegetative propagation and/or apomixis – that is, seed set without prior fertilization. Because all members of a clone are genetically identical, the type of marker used for differentiation among clones is largely irrelevant provided that the discriminatory power is sufficiently high. Thus, apomictically and sexually derived offspring in the facultative apomict *Hypericum perforatum* could be discriminated by RAPD profiling [99] as well as by AFLPs and RFLP fingerprinting [100].

DNA fingerprinting is an important instrument for the characterization of germplasm; that is, the total genetic diversity present in the world for a certain crop, encompassing old and newly bred cultivars, landraces and related wild species. DNA markers have also provided valuable data for the identification of suitable material for *in-situ* preservation, the establishment of *ex-situ* gene banks and core collections with maximum diversity [101, 102], and for the detection of undesirable duplicates in germplasm collections [86, 103]. If drug production is based mainly on wild-collected material of rare plant species, conservation issues may become very important, as has been discussed in the frame of a RAPD study of variability and gene flow in the African tree *Prunus africana* [104], and for domesticated and wild species of Asian *Panax* studied with ITS sequencing and AFLP [105].

5.3.2.3 Plant Systematics

Knowledge concerning phylogenetic relationships (species origination) and phenetic patterns (present-day similarity among species) can help to develop methods for discrimination among species, as well as provide information about which species may constitute the most promising sources of certain biochemical compounds. Many phylogenetic studies have been based on DNA sequencing data, especially of cpDNA and nuclear ribosomal RNA genes and spacers [e.g., 27]. Sequencing the ITS region was, for example, undertaken in a worldwide collected set of 50 *Hypericum* taxa [31]. Three strongly supported monophyletic clades plus several secondary monophyletic groupings were revealed. The medicinally important *H. perforatum* could be distinguished from all other analyzed species. Another *Hypericum* study made use of PCR-RFLP with universal chloroplast DNA primers [106]. Polymorphism was found between three of the six taxa collected in Croatia, but not among three subspecies of *H. performatum*.

In another cpDNA-CAPS study, all 15 recognized taxa in the genus *Hippophae* were analyzed [107]. Two parsimony analyses were carried out; one based on PCR-RFLP of cpDNA, and one based on a combined data set (cpDNA plus morphological traits). The results of both analyses were congruent with each other, with those of a previous RAPD study, and in part also with some earlier taxonomic classifications. A PCR-RFLP study of five amplified cpDNA fragments was performed with 13 populations of the genus *Panax* in Nepal and China [108]. Distinct groupings were obtained, suggesting that most of the subspecies treated under *P. pseudoginseng* are quite different from each other and from subsp. *pseudoginseng* itself, and should be given species rank.

Multilocus DNA profiles are often exploited for systematic analyses of closely related species, where neither ITS nor cpDNA sequencing reveal sufficient polymorphism [109]. *Echinacea* is a genus with very restricted polymorphism in the ITS1, ITS2 and 5.8S regions [110], and may therefore be more suited to multilocus marker approaches. Kim et al. [111] initiated an AFLP study comprising 39 plants from 12 *Echinacea* taxa. Two species were represented by two populations each, the others by a single population. A neighbor-joining phenogram based on approximately 1000 polymorphic AFLP fragments demonstrated the existence of two major clades, with *E. pupurea* in the smaller clade together with *E. sanguinea* and *E. simulata*.

Potential problems with the use of dominant markers for phylogenetic reconstruction are mainly caused by the (usually unproven) assumptions that all DNA fragments within a pattern represent independent characters, and that fragments of equal length are homologous with each other (see also two recent reviews by Bussell et al. [112] and Koopman [113]). In general, homology is quite high when closely related taxa are compared, but drops sharply with increasing phylogenetic distance. Thus, Mechanda et al. [114] analyzed co-migrating AFLP fragments from several *Echinacea* species by DNA sequencing. Whereas pairwise sequence identities for an AFLP band that was monomorphic among all taxa ranged from ~76% to 99% *within* a species, it fell to an average of ~58% *among* species. Still lower sequence identities were observed for a polymorphic band that was present in 48 of 79 taxa. Nevertheless, the existence of a sizeable phylogenetic signal has been demonstrated in many AFLP-based data sets, and various methods to quantify the existence of phylogenetic information and to evaluate the support of internal branches in the resulting trees are available [113].

5.3.3
Gene Tagging

The breeding of medicinal plant crop plants has the potential to enhance both yield and quality of the ultimate product. In this context, one of the most important applications of molecular markers is the establishment of a genetic linkage between a trait of interest and an easy-to-detect marker. These markers can then be exploited for so-called marker-assisted selection, a strategy that allows the detection of and selection for specific traits that would otherwise require time-consuming and/or costly analyses. Such a character is sex in dioecious plants, which may not be determined until the plants are old enough to flower for the first time. For example, a sex-linked RAPD marker was developed and subsequently shown to occur in all the male plants of sea buckthorn *Hippophae rhamnoides* while lacking in all the female plants of the same seedling family [115]. Another well-known dioecious plant of medicinal interest is *Cannabis sativa*, for which RAPD analysis has also proved useful for developing male-specific markers [116].

The construction of genetic linkage maps is one of the most prominent applications of DNA markers [117]. Such maps are, in their turn, very useful for elucidating the genetic basis of complex traits, and for localization and cloning of genes. Genetic linkage maps based on various types of molecular markers have also been constructed for plants of medicinal relevance, such as garlic and Pacific yew (*Taxus brevifolia*). In the latter, 41 of a total of 102 RAPD bands were distributed into 17 linkage groups [118]. In garlic, SNPs, SSRs and RAPDs were used to define nine linkage groups, one of which also included a locus for male fertility [119]. In another garlic project, 216 and 143 AFLP markers, respectively, were mapped in two segregating populations obtained by selfing [120]. In addition, gene-specific markers for alliinase, chitinase, sucrose 1-fructosyltransferase and chalcone synthase could be placed on these maps.

In a different gene tagging approach, directed at the establishment of markers for multiple-gene inherited traits, Baum et al. [2] screened 52 *Echinacea purpurea* plants (both cultivated and wild) with a total of 232 AFLP markers. These same

plants were also subjected to HPLC analysis of two bioactive phytochemicals, ci-
choric acid and "tetraene". Regression analysis and canonical correlation analysis
were used to explore the relationships between DNA bands and chemical contents.
Thus, a set of 22 DNA bands were found, which can be used in the future for predicting the phytochemical profile of individual plants.

5.4
Conclusions

The rapidity with which large numbers of samples can be processed and the ubiquitous applicability to all organisms without previous sequence information has
made PCR-based multilocus profiling methods highly popular for many applications, including the authentication of medicinal plants. However, these methods
have their drawbacks. An important issue is the often limited reproducibility, especially in the case of RAPDs [36]. MP-PCR analyses can be performed at higher
stringency and were therefore initially claimed to be more reliable than RAPDs
[43], but this view has been challenged in later investigations [44]. Among multilocus markers, reproducibility appears to be best for AFLP [121] which is also regarded as the method of choice when high numbers of bands and maximum discriminatory potential are desired. Reproducibility can also be increased by converting individual DNA fragments of particular diagnostic importance into locus-specific
SCAR markers (sequence characterized amplified regions [83, 122, 123]). This is
done by sequencing the gel-purified fragment, followed by designing a locus-specific PCR primer pair.

Data sets produced by DNA sequencing are usually more reliable, as are locus-
specific methods such as microsatellites and SNPs. Because these methods are also less sensitive to contamination and DNA degradation, they should actually be
preferred over multilocus profiling, be it RAPD or AFLP. Unfortunately, DNA sequencing is not sufficiently sensitive for many applications, and the development
of microsatellites and SNPs is both expensive and time-consuming. However, the
actual assays are quite cheap for both types of markers, and the costs of microsatellite analysis can be further reduced by establishing multiplexed "genotype identification sets" [51, 52]. A few recent reports have already demonstrated that microarray-based DNA profiling of herbal plant material is feasible, also in conjunction
with portable microdevices [77–79]. It is safe to assume that SNP profiling will become more cost-effective by the availability of large, standardized DNA chips that
allow the simultaneous analysis of hundreds of genotypes. The importance of multiplexed microsatellites and SNP profiling for medicinal plant authentication is
therefore likely to increase in the near future.

References

1 Ha, W.Y., Shaw, P.C., Liu, J., Yau, F.C.F., Wang, J. Authentication of *Panax ginseng* and *Panax quinquefolius* using amplified fragment length polymorphism (AFLP) and directed amplification with microsatellite region DNA (DAMD). *J. Agric. Food Chem.* **2002**, *50*, 1871–1875. [Application of two types of DNA markers, AFLP and DAMD, in an important crop.]

2 Baum, B.R., Mechanda, S., Livesey, J.F., Binns, S.E., Arnason, J.T. Predicting quantitative phytochemical markers in single *Echinacea* plants or clones from their DNA fingerprints. *Phytochemistry* **2001**, *56*, 543–549. [Application of AFLP, and analysis of linkage between DNA markers and phytochemical contents.]

3 Wallner, E., Weising, K., Rompf, R., Kahl, G., Kopp, B. Oligonucleotide fingerprinting and RAPD analysis of *Achillea* species: characterization and long-term monitoring of micropropagated clones. *Plant Cell Rep.* **1996**, *15*, 647–652.

4 Shaw, P.C., Ngan, F.N., But, P.P.H., Wang, J. Authentication of Chinese medicinal materials by DNA technology. *J. Food Drug Anal.* **1997**, *5*, 273–283. [Informative review paper about DNA-based identification of traditional Chinese medicinal plants.]

5 Henry, R.J. (Ed.), *Plant Genotyping: The DNA Fingerprinting of Plants*. CAB International Publishing, Wallingford, **2001**. [Multi-authored book covering various DNA marker techniques (including SNPs) and their applications in plants.]

6 Avise, J.C. *Molecular Markers, Natural History, and Evolution*. Second edition. Sinauer Associates, Sunderland, **2004**. [Survey of theoretical and practical aspects of DNA marker techniques.]

7 Weising, K., Nybom, H., Wolff, K., Kahl, G. *DNA Fingerprinting in Plants: Principles, Methods and Applications*. CRC Press, **2005**. [In-depth descriptions of DNA-based marker methodology, data analyses and applications.]

8 Lin, J.-Z., Ritland, K. Flower petals allow simpler and better isolation of DNA for plant RAPD analyses. *Plant Mol. Biol. Rep.* **1995**, *13*, 210–213.

9 Simel, E.J., Saidak, L.R., Tuskan, G.A. Method for extracting genomic DNA from non-germinated gymnosperm and angiosperm pollen. *BioTechniques* **1997**, *22*, 390–394.

10 Kang, H.W., Cho, Y.G., Yoon, U.H., Eun, M.Y. A rapid DNA extraction method for RFLP and PCR analysis from a single dry seed. *Plant Mol. Biol. Rep.* **1998**, *16*, 90(1–9).

11 Kumar, A., Pushpangadan, P., Mehrotra, S. Extraction of high-molecular-weight DNA from dry root tissue of *Berberis lycium* suitable for RAPD. *Plant Mol. Biol. Rep.* **2003**, *21*, 309a–309d.

12 Syamkumar, S., Lowarence, B., Sasikumar, B. Isolation and amplification of DNA from rhizomes of turmeric and ginger. *Plant Mol. Biol. Rep.* **2003**, *21*, 171a–171e.

13 Wulff, E.G., Torres, S., Vigil, E.G. Protocol for DNA extraction from potato tubers. *Plant Mol. Biol. Rep.* **2002**, *20*, 187a–187e.

14 Deguilloux, M.F., Pemonge, M.H., Petit, R.J. Novel perspectives in wood certification and forensics: dry wood as a source of DNA. *Proc. R. Soc. Lond. B* **2002**, *269*, 1039–1046.

15 Reynolds, M.M., Williams, C.G. Extracting DNA from submerged pine wood. *Genome* **2004**, *47*, 994–997.

16 Rogstad, S.H. Saturated NaCl-CTAB solution as a means of field preservation of leaves for DNA analyses. *Taxon* **1992**, *41*, 701–708.

17 Chase, M.W., Hills, H.H. Silica gel: an ideal material for field preservation of leaf samples for DNA studies. *Taxon* **1991**, *40*, 215–220.

18 Khanuja, S.P.S., Shasany, A.K., Darokar, M.P., Kumar, S. Rapid isolation of DNA from dry and fresh samples of plants producing large amounts of secondary metabolites and essential oils. *Plant Mol. Biol. Rep.* **1999**, *17*, 74(1-7).

19 Savolainen, V., Cuénoud, P., Spichiger, R., Martinez, M.D., Crèvecoeur, M., Manen, J.-F. The use of herbarium specimens in DNA phylogenetics:

evaluation and improvement. *Plant Syst. Evol.* **1995**, *197*, 87–98.

20 Biss, P., Freeland, J., Silvertown, J., McConway, K., Lutman, P. Successful amplification of rice chloroplast microsatellites from century-old grass samples from the park grass experiment. *Plant Mol. Biol. Rep.* **2003**, *21*, 249–257.

21 Singh, B.M., Ahuja, P.S. Isolation and PCR amplification of genomic DNA from market samples of dry tea. *Plant Mol. Biol. Rep.* **1999**, *17*, 171–178.

22 Murray, M.G., Thompson, W.F. Rapid isolation of high molecular weight plant DNA. *Nucleic Acids Res.* **1980**, *8*, 4321–4325.

23 Pirttilä, A.M., Hirsikorpi, M., Kämäräinen, T., Jaakola, L., Hohtola, A. DNA isolation methods for medicinal and aromatic plants. *Plant Mol. Biol. Rep.* **2001**, *19*, 273a–273f.

24 Sangwan, N.S., Sangwan, R.S., Kumar, S. Isolation of genomic DNA from the antimalarial plant *Artemisia annua*. *Plant Mol. Biol. Rep.* **1998**, *16*, 365(18).

25 Sangwan, R.S., Yadav, U., Sangwan, N.S. Isolation of genomic DNA from defatted oil seed residue of opium poppy (*Papaver somniferum*). *Plant Mol. Biol. Rep.* **2000**, *18*, 265–270.

26 Mullis, K.B., Ferré, F., Gibbs, R.A. (Eds.), *The Polymerase Chain Reaction*. Birkhäuser, Basel, **1994**.

27 Baldwin, B.G., Sanderson, M.J., Porter, J.M., Wojciechowski, M.F., Campbell, C.S., Donoghue, M.J. The ITS region of nuclear ribosomal DNA: a valuable source of evidence on angiosperm phylogeny. *Ann. Missouri Bot. Gard.* **1995**, *82*, 247–277.

28 Fushimi, H., Komatsu, K., Namba, T., Isobe, M. Genetic heterogeneity of ribosomal RNA gene and *mat*K gene in *Panax notoginseng*. *Planta Med.* **2000**, *66*, 659–661.

29 Lau, D.T.W., Shaw, P.C., Wang, J., But, P.P.H. Authentication of medicinal *Dendrobium* species by the internal transcribed spacer of ribosomal DNA. *Planta Med.* **2001**, *67*, 456–460. [Good example of the application of DNA sequencing for species authentication.]

30 Cui, X.M., Lo, C.K., Yip, K.L., Dong, T.T.X., Tsim, K.W.K. Authentication of *Panax notoginseng* by 5S-rRNA spacer domain and random amplified polymorphic DNA (RAPD) analysis. *Planta Med.* **2003**, *69*, 584–586.

31 Crockett, S.L., Douglas, A.W., Scheffler, B.E., Khan, I.A. Genetic profiling of *Hypericum* (St. John's Wort) species by nuclear ribosomal ITS sequence analysis. *Planta Med.* **2004**, *70*, 929–935. [Interesting example of the application of DNA sequencing in a medicinal plant.]

32 Weising, K., Kahl, G. Hybridization-based microsatellite fingerprinting of plants and fungi. In: Caetano-Anollés. G., Gresshoff, P.M. (Eds.), *DNA Markers: Protocols, Applications, and Overviews*. Wiley-VCH, New York, **1998**, pp. 27–54.

33 Welsh, J., McClelland, M. Fingerprinting genomes using PCR with arbitrary primers. *Nucleic Acids Res.* **1990**, *18*, 7213–7218.

34 Williams, J.G.K., Kubelik, A.R., Livak, K.J., Rafalski, J.A., Tingey, S.V. DNA polymorphisms amplified by arbitrary primers are useful as genetic markers. *Nucleic Acids Res.* **1990**, *18*, 6231–6235.

35 Caetano-Anollés, G., Bassam, B.J., Gresshoff, P.M. High resolution DNA amplification fingerprinting using very short arbitrary oligonucleotide primers. *Bio/Technology* **1991**, *9*, 553–557.

36 Penner, G.A., Bush, A., Wise, R., Kim, W., Domier, L., Kasha, K., Laroche, A., Scoles, G., Molnar, S.J., Fedak, G. Reproducibility of random amplified polymorphic DNA (RAPD) analysis among laboratories. *PCR Methods and Applications* **1993**, *2*, 341–345.

37 Sangwan, R.S., Sangwan, N.S., Jain, D.C., Kumar, S., Ranade, S.A. RAPD profile based genetic characterization of chemotypic variants of *Artemisia annua* L. *Biochem. Molec. Biol. Int.* **1999**, *47*, 935–944. [Application of RAPD, and analysis of the relationship between markers and the chemical contents of the plant material.]

38 Nieri, P., Adinolfi, B., Morelli, I., Breschi, M.C., Simoni, G., Martinotti, E. Genetic characterization of the three medicinal *Echinacea* species using RAPD analysis. *Planta Med.* **2003**, *69*, 685–686.

39 Shoyama, Y., Zhu, X.X., Nakai, R., Shiraishi, S., Kohda, H. Micropropaga-

tion of *Panax notoginseng* by somatic embryogenesis and RAPD analysis of regenerated plantlets. *Plant Cell Rep.* **1997**, *16*, 450–453.

40 Shim, Y.-H., Choi, J.-H., Park, C.-D., Lim, C.-J., Cho, J.-H., Kim, H.-J. Molecular differentiation of *Panax* species by RAPD analysis. *Arch. Pharm. Res.* **2003**, *26*, 601–605.

41 Zietkiewicz, E., Rafalski, A., Labuda, D. Genome fingerprinting by simple sequence repeat (SSR)-anchored polymerase chain reaction amplification. *Genomics* **1994**, *20*, 176–183.

42 Gupta, M., Chyi, Y.-S., Romero-Severson, J., Owen, J.L. Amplification of DNA markers from evolutionary diverse genomes using single primers of simple-sequence repeats. *Theoret. Appl. Genet.* **1994**, *89*, 998–1006.

43 Meyer, W., Mitchell, T.G., Freedman, E.Z., Vilgalys, R. Hybridization probes for conventional DNA fingerprinting used as single primers in the polymerase chain reaction to distinguish strains of *Cryptococcus neoformans. J. Clin. Microbiol.* **1993**, *31*, 2274–2280.

44 Weising, K., Atkinson, R.G., Gardner, R.C. Genomic fingerprinting by microsatellite-primed PCR: a critical evaluation. *PCR Methods and Applications* **1995**, *4*, 249–255.

45 Goldstein, D.B., Schlötterer, C. (Eds.), *Microsatellites: Evolution and Applications.* Oxford University Press, Oxford, UK, **1999**.

46 Fisher, P., Gardner, R.C., Richardson, T.E. Single locus microsatellites isolated using 5′ anchored PCR. *Nucleic Acids Res.* **1996**, *24*, 4369–4371.

47 Feyissa, T., Nybom, H., Bartish, I., Welander, M. Analysis of genetic diversity in the endangered tropical tree species *Hagenia abyssinica* using ISSR markers. *Genet. Res. Crop Evol.* (in press).

48 Zabeau, M., Vos, P. Selective restriction fragment amplification: a general method for DNA fingerprinting. *European Patent Application EP 0534858,* **1993**.

49 Vos, P., Hogers, R., Bleeker, M., Reijans, M., Van de Lee, T., Hornes, M., Frijters, A., Pot, J., Peleman, J., Kuiper, M., Zabeau, M. AFLP: a new technique for DNA fingerprinting. *Nucleic Acids Res.* **1995**, *23*, 4407–4414.

50 Akkaya, M.S., Bhagwat, A.A., Cregan, P.B. Length polymorphisms of simple sequence repeat DNA in soybean. *Genetics* **1992**, *132*, 1131–1139.

51 Macaulay, M., Ramsay, L., Powell, W., Waugh, R. A representative, highly informative "genotyping set" of barley SSRs. *Theoret. Appl. Genet.* **2001**, *102*, 801–809.

52 Tang, S., Kishore, V.K., Knapp, S.J. PCR-multiplexes for a genome-wide frame-work of simple sequence repeat marker loci in cultivated sunflower. *Theoret. Appl. Genet.* **2003**, *107*, 6–19.

53 Zane, L., Bargelloni, L., Patarnello, T. Strategies for microsatellite isolation: a review. *Mol. Ecol.* **2002**, *11*, 1–16.

54 Squirrell, J., Hollingsworth, P.M., Woodhead, M., Russell, J., Lowe, A.J., Gibby, M., Powell, W. How much effort is required to isolate nuclear micro-satellites from plants? *Mol. Ecol.* **2003**, *12*, 1339–1348.

55 Holton, T.A., Christopher, J.T., McClure, L., Harker, N., Henry, R.J. Identification and mapping of polymorphic SSR markers from expressed gene sequences of barley and wheat. *Mol. Breed.* **2002**, *9*, 63–71.

56 Yu, J.-K., Dake, T.M., Singh, S., Benscher, D., Li, W., Gill, B., Sorrells, M.E. Development and mapping of EST-derived simple sequence repeat markers for hexaploid wheat. *Genome* **2004**, *47*, 805–818.

57 Gilmore, S., Peakall, R., Robertson, J. Short tandem repeat (STR) DNA markers are hypervariable and informative in *Cannabis sativa*: implications for forensic investigations. *Forens. Sci. Int.* **2003**, *131*, 65–74. [Application of microsatellite DNA markers, and an analysis of the relationship between these markers and chemical contents in the plant material.]

58 Shokeen, B., Sethy, N.K., Choudhary, S., Bhatia, S. Development of STMS markers from the medicinal plant Madagascar periwinkle [*Catharanthus roseus* (L.) G. Don]. *Mol. Ecol. Notes* **2005**, *5*, 818–820.

59 Demesure, B., Sozi, N., Petit, R.J. A set of universal primers for amplification of polymorphic non-coding regions of mitochondrial and chloroplast DNA in plants. *Mol. Ecol.* **1995**, *4*, 129–131.

60 Dumolin-Lapègue, S., Pemonge, M.-H., Petit, R.J. An enlarged set of consensus primers for the study of organelle DNA in plants. *Mol. Ecol.* **1997**, 6, 393–397.

61 Duminil, J., Pemonge, M.-H., Petit, R.J. A set of 35 consensus primer pairs amplifying genes and introns of plant mitochondrial DNA. *Mol. Ecol. Notes* **2002**, 2, 428–430.

62 Konieczny, A., Ausubel, F.M. A procedure for mapping *Arabidopsis* mutations using co-dominant ecotype-specific PCR-based markers. *Plant J.* **1993**, *4*, 403–410.

63 Witsenboer, H., Vogel, J., Michelmore, R.W. Identification, genetic localization, and allelic diversity of selectively amplified microsatellite polymorphic loci in lettuce and wild relatives (*Lactuca* spp.). *Genome* **1997**, *40*, 923–936.

64 Waugh, R., McLean, K., Flavell, A.J., Pearce, S.R., Kumar, A., Thomas, B.B.T., Powell, W. Genetic distribution of *Bare*-1-like retrotransposable elements in the barley genome revealed by sequence-specific amplification polymorphisms (S-SAP). *Mol. Gen. Genet.* **1997**, *253*, 687–694.

65 Gupta, P.K., Roy, J.K., Prasad, M. Single nucleotide polymorphisms: a new paradigm for molecular marker technology and DNA polymorphism detection with emphasis on their use in plants. *Curr. Sci.* **2001**, *80*, 524–535.

66 Brumfield, R.T., Beerli, P., Nickerson, D.A., Edwards, S.V. The utility of single nucleotide polymorphisms in inferences of population history. *Trends Ecol. Evol.* **2003**, *18*, 249–256.

67 Bhattramaki, D., Rafalski, A. Discovery and application of single nucleotide polymorphism markers in plants. In: Henry, R.J. (Ed.), *Plant Genotyping: The DNA Fingerprinting of Plants.* CABI Publishing, Wallingford, **2001**, pp. 179–192.

68 Somers, D.J., Kirkpatrick, R., Moniwa, M., Walsh, A. Mining single-nucleotide polymorphisms from hexaploid wheat ESTs. *Genome* **2003**, *49*, 431–437.

69 Sunnucks, P., Wilson, A.C.C., Beheregaray, L.B., Zenger, K., French, J., Taylor, A.C. SSCP is not so difficult: the application and utility of single-stranded conformation polymorphism in evolutionary biology and molecular ecology. *Mol. Ecol.* **2000**, *9*, 1699–1710.

70 Bensch, S., Akesson, S., Irwin, D.E. The use of AFLPs to find an informative SNP: genetic differences across a migratory divide in willow warblers. *Mol. Ecol.* **2002**, *11*, 2359–2366.

71 Kwok, P.-Y. Methods for genotyping single nucleotide polymorphisms. *Annu. Rev. Genomics Hum. Genet.* **2001**, *2*, 235–258.

72 Syvanen, A.C. Genotyping single nucleotide polymorphisms. *Nature Rev. Genet.* **2001**, *2*, 930–942.

73 Schena, M. *Microarray Analysis.* Wiley-Liss, Inc., USA, **2003**.

74 Mantripragada, K.K., Buckley, P.G., de Stahl, T.D., Dumanski, J.P. Genomic microarrays in the spotlight. *Trends Genet.* **2004**, *20*, 87–94.

75 Schmid, K.J., Sörensen, T.R., Stracke, R., Törjek, O., Altmann, T., Mitchell-Olds, T., Weisshaar, B. Large-scale identification and analysis of genome wide single-nucleotide polymorphisms for mapping in *Arabidopsis thaliana*. *Genome Res.* **2003**, *13*, 1250–1257.

76 Nasu, S., Suzuki, J., Ohta, R., Hasegawa, K., Yui, R., Kitazawa, N., Monna, L., Minobe, Y. Search for and analysis of single nucleotide polymorphisms (SNPs) in rice (*Oryza sativa, Oryza rufipogon*) and establishment of SNP markers. *DNA Res.* **2002**, *9*, 163–171.

77 Carles, M., Lee, T., Moganti, S., Lenigk, R., Tsim, K.W.K., Ip, N.Y., Hsing, I.-M., Sucher, N.J. Chips and Qi: microcomponent-based analysis in traditional Chinese medicine. *Fresenius J. Anal. Chem.* **2001**, *371*, 190–194. [One of the first studies demonstrating the applicability of DNA microarray technology for DNA sequence-based identification of medicinal plants.]

78 Carles, M., Cheung, M.K.L., Moganti, S., Dong, T.T.X., Tsim, K.W., Ip, N.Y.,

Sucher, N.J. A DNA microarray for the authentication of toxic traditional Chinese medicinal plants. *Planta Med.* **2005**, *71*, 580–584. [Application of silicon chip-based DNA microarray analysis.]

79 Zhang, Y.B., Wang, J., Wang, Z.T., But, P.P.H., Shaw, P.C. DNA microarray for identification of the herb of *Dendrobium* species from Chinese medicinal formulations. *Planta Med.* **2003**, *69*, 1172–1174.

80 Lee, T.M.H., Hsing, I.-M. Sequence-specific electrochemical detection of asymmetric PCR amplicons of traditional Chinese medicinal plant DNA. *Anal. Chem.* **2002**, *74*, 5057–5062. [Development of a hand-held device for SNP-based discrimination of plant material.]

81 Ngan, F., Shaw, P., But, P., Wang, J. Molecular authentication of *Panax* species. *Phytochemistry* **1999**, *50*, 787–791. [Application of 5.8 S rDNA sequencing for root discrimination.]

82 Artyukova, E.V., Kozyrenko, M.M., Koren, O.G., Muzarok, T.I., Reunova, G.D., Zhuravlev, Y.N. RAPD and allozyme analysis of genetic diversity in *Panax ginseng* C.A. Meyer and *P. quinquefolius* L. *Russian J. Genet.* **2004**, *40*, 178–185.

83 Tochika-Komatsu, Y., Asaka, I., Ii, I. A random amplified polymorphic DNA (RAPD) primer to assist the identification of a selected strain, Aizu K-111 of *Panax ginseng* and the sequence amplified. *Biol. Pharmaceut. Bull.* **2001**, *24*, 1210–1213.

84 Debener, T., Janakiram, T., Mattiesch, L. Sports and seedlings of rose varieties analysed with molecular markers. *Plant Breed.* **2000**, *119*, 71–74.

85 Franks, T., Botta, R., Thomas, M.R. Chimerism in grapevines: implications for cultivar identity, ancestry and genetic improvement. *Theoret. Appl. Genet.* **2002**, *104*, 192–199.

86 Ipek, M., Ipek, A., Simon, P.W. Comparison of AFLPs, RAPDs, and isozymes for diversity assessment of garlic and detection of putative duplicates in germplasm collections. *J. Am. Soc. Hort. Sci.* **2003**, *128*, 246–252. [Analysis of genetic identity determined with different kinds of molecular markers.]

87 Diwan, B., Cregan, P.B. Automated sizing of fluorescent-labeled simple sequence repeat (SSR) markers to assay genetic variation in soybean. *Theoret. Appl. Genet.* **1997**, *95*, 723–733.

88 Excoffier, L., Smouse, P.E., Quattro, J.M. Analysis of molecular variance inferred from metric distances among DNA haplotypes: applications to human mitochondrial DNA restriction data. *Genetics* **1992**, *131*, 479–491.

89 Echeverrigeray, S., Agostini, G., Atti-Sefini, L., Paroul, N., Pauletti, G.F., dos Santos, A.C.A. Correlation between the chemical and genetic relationships among commercial thyme cultivars. *J. Agric. Food Chem.* **2001**, *49*, 4220–4223. [Application of RAPD, and analysis of the relationship between DNA markers and oil content in the plant material.]

90 Haluskova, J., Kosuth, J. RAPD analysis of somaclonal and natural DNA variation in *Hypericum perforatum* L. *Acta Biol. Cracov Ser. Botanica.* **2003**, *45*, 101–104.

91 Kuddus, R.H., Kuddus, N.N., Dvorchik, I. DNA polymorphism in the living fossil *Ginkgo biloba* from the eastern United States. *Genome* **2002**, *45*, 8–12. [Application of RAPD, as well as the utilisation of a gel-excised DNA band for probing.]

92 Nybom, H. Comparison of different nuclear DNA markers for estimating intraspecific genetic diversity in plants. *Mol. Ecol.* **2004**, *13*, 1143–1155. [Compilation and analysis of population genetic studies carried out using various DNA marker methods, comparison of dominantly and codominantly inherited marker methods.]

93 Grassi, F., Imazio, S., Gomarasca, S., Citterio, S., Aina, R., Sgorbati, S., Sala, F., Patrignani, G., Labra, M. Population structure and genetic variation within *Valeriana wallrothii* Kreyer in relation to different ecological locations. *Plant Sci.* **2004**, *166*, 1437–1441. [Application of AFLP and chloroplast DNA microsatellite markers.]

94 Nebauer, S.G., del Castillo-Agudo, L., Segura, J. RAPD variation within and among natural populations of out-

crossing willow-leaved foxglove (*Digitalis obscura* L.). *Theoret. Appl. Genet.* **1999**, *98*, 985–994.

95 Muluvi, G.M., Sprent, J.I., Soranzo, N., Provan, J., Odee, D., Folkard, G., McNicol, J.W., Powell, W. Amplified fragment length polymorphism (AFLP) analysis of genetic variation in *Moringa oleifera* Lam. *Mol. Ecol.* **1999**, *8*, 463–470.

96 Semagn, K., Bjørnstad, A., Stedje, B. Genetic diversity and differentiation in Ethiopian populations of *Phytolacca dodecandra* as revealed by AFLP and RAPD analyses. *Genet. Crop Res. Evol.* **2003**, *50*, 649–661.

97 Singh, A., Chaudhary, A., Srivastava, P.S., Lakshmikumaran, M. Comparison of AFLP and SAMPL markers for assessment of intra-population genetic variation in *Azadirachta indica* A. Juss. *Plant Sci.* **2002**, *162*, 17–25.

98 Rossetto, M., Slade, R.W., Baverstock, P.R., Henry, R.J., Lee, L.S. Microsatellite variation and assessment of genetic structure in tea tree (*Melaleuca alternifolia* – Myrtaceae). *Mol. Ecol.* **1999**, *8*, 633–643.

99 Arnholdt-Schmitt, B. RAPD analysis: a method to investigate aspects of the reproductive biology of *Hypericum perforatum* L. *Theoret. Appl. Genet.* **2000**, *100*, 906–911.

100 Mayo, G.M., Langridge, P. Modes of reproduction in Australian populations of *Hypericum perforatum* L. (St. John's wort) revealed by DNA fingerprinting and cytological methods. *Genome* **2003**, *46*, 573–579.

101 Hu, J., Zhu, J., Xu, H.M. Methods of constructing core collections by stepwise clustering with three sampling strategies based on genotypic values of crops. *Theoret. Appl. Genet.* **2000**, *101*, 264–268.

102 Garkava-Gustavsson, L., Persson, H.A., Nybom, H., Rumpunen, K., Gustavsson, B.A., Bartish, I.V. RAPD-based analysis of genetic diversity and selection of lingonberry (*Vaccinium vitis-idaea* L.) material for *ex situ* conservation. *Genet. Res. Crop Evol.* **2005**, *52*, 723–735.

103 Zhang, D.P., Huaman, Z., Rodriguez, F., Rossel, G., Ghislain, M. Identifying duplicates in sweet potato (*Ipomoea*

batatas (L.) Lam) cultivars using RAPD. *Acta Hort.* **2001**, *546*, 535–541.

104 Dawson, I.K., Powell, W. Genetic variation in the Afromontane tree *Prunus africana*, an endangered medicinal species. *Mol. Ecol.* **1999**, *8*, 151–156.

105 Zhou, S.L., Xiong, G.M., Li, Z.Y., Wen, J. Loss of genetic diversity of domesticated *Panax notoginseng* F H Chen as evidenced by ITS sequence and AFLP polymorphism: a comparative study with *P. stipuleatus* H T Tsai et K M Feng. *J. Integrat. Plant Biol.* **2005**, *47*, 107–115.

106 Pilepic, K.H. Analysis of chloroplast DNA variability among some *Hypericum* species in Croatia. *Periodicum Biologorum* **2002**, *104*, 457–462.

107 Bartish, I.V., Jeppsson, N., Swenson, U., Nybom, H. Phylogeny of *Hippophae* (Elaeagnaceae) inferred from parsimony analysis of chloroplast DNA and morphology. *Syst. Bot.* **2002**, *27*, 41–54.

108 Yoo, K.O., Malla, K.J., Wen, J. Chloroplast DNA variation of *Panax* (Araliaceae) in Nepal and its taxonomic implications. *Brittonia* **2001**, *53*, 447–453.

109 Després, L., Gielly, L., Redoutet, B., Taberlet, P. Using AFLP to resolve phylogenetic relationships in a morphologically diversified plant species complex when nuclear and chloroplast sequences fail to reveal variability. *Mol. Phylogenet. Evol.* **2003**, *27*, 85–196.

110 Urbatsch, L.E., Baldwin, B.G., Donoghue, M.J. Phylogeny of the coneflowers and relatives (Heliantheae: Asteraceae) based on nuclear rDNA internal transcribed spacer (ITS) sequences and chloroplast DNA restriction site data. *Syst. Bot.* **2000**, *25*, 539–565.

111 Kim, D.-H., Heber, D., Still, D.W. Genetic diversity of *Echinacea* species based upon amplified fragment length polymorphism markers. *Genome* **2004**, *47*, 102–111.

112 Bussell, J.D., Waycott, M., Chappill, J.A. Arbitrarily amplified DNA markers as characters for phylogenetic inference. *Perspect. Plant Ecol. Evol. Syst.* **2005**, *7*, 3–26.

113 Koopman, W.J.M. Phylogenetic signal in AFLP data sets. *Syst. Biol.* **2005**, *54*, 197–217.

114 Mechanda, S.M., Baum, B.R., Johnson, D.A., Arnason, J.T. Sequence assessment of comigrating AFLP bands in *Echinacea* – implications for comparative biological studies. *Genome* **2004**, *47*, 15–25.

115 Persson, H.A., Nybom, H. Genetic sex determination and RAPD marker segregation in the dioecious species sea buckthorn (*Hippophae rhamnoides*). *Hereditas* **1998**, *129*, 45–51.

116 Mandolino, G., Carboni, A., Bagatta, M., Moliterni, V.M.C., Ranalli, P. Occurrence and frequency of putatively Y chromosome linked DNA markers of *Cannabis sativa* L. *Euphytica* **2002**, *126*, 211–218.

117 Meksem, K., Kahl, G. *The Handbook of Genome Mapping*. Wiley-VCH, USA, **2004**.

118 Göcmen, B., Jermstad, K.D., Neale, D.B., Kaya, Z. Development of random amplified polymorphic DNA markers for genetic mapping of Pacific yew (*Taxus brevifolia*). *Can. J. Forest Res.* **1996**, *26*, 497–503.

119 Zewdie, Y., Havey, M.J., Prince, J.P., Jenderek, M.M. The first genetic linkages among expressed regions in the garlic genome. *J. Am. Soc. Hort. Sci.* **2005**, *130*, 569–574. [Genetic mapping with different types of DNA markers and identification of a locus for male fertility.]

120 Ipek, M., Ipek, A., Almquist, S.G., Simon, P.W. Demonstration of linkage and development of the first low-density genetic map of garlic, based on AFLP markers. *Theoret. Appl. Genet.* **2005**, *110*, 228–236. [Genetic mapping with both DNA markers (AFLP) and genes for chemical contents.]

121 Hansen, M., Kraft, T., Christansson, M., Nilsson, N.O. Evaluation of AFLP in *Beta*. *Theoret. Appl. Genet.* **1999**, *98*, 845–852.

122 Paran, I., Michelmore, R.W. Development of reliable PCR-based markers linked to downy mildew resistance genes in lettuce. *Theoret. Appl. Genet.* **1993**, *85*, 985–993.

123 Xu, M., Huaracha, E., Korban, S.S. Development of sequence-characterized amplified regions (SCARs) from amplified fragment length polymorphism (AFLP) markers tightly linked to the *Vf* gene in apple. *Genome* **2001**, *44*, 63–70.

6

Bioprospecting: The Search for Bioactive Lead Structures from Nature

Michael Wink

Abstract

As a consequence of the Human Genome Project and further research in functional genomics and proteomics, many genes have been identified that are connected with health disorders and illnesses. Several of the products of these genes (i.e., proteins) have become new targets for the development of novel therapeutics. As a consequence, the demand for bioactive compounds from Nature or synthesis is increasing. Plants, bacteria, fungi and sessile marine organisms produce a wide variety of secondary metabolites that have been selected during evolution as defense substances against herbivores, predators or infective microbes and viruses, or as signal compounds. These secondary compounds have attracted great interest for drug development as they may represent lead structures for new or already existing drug targets. In this chapter, the biological and evolutionary background of secondary metabolites (mainly from plants) is discussed, and examples of known interactions of secondary metabolites with interesting molecular targets are described.

6.1
Introduction

Biomedicine and molecular biotechnology have made much progress during the past 10 years, with the sequencing of the human genome being one of the major achievements. Moreover, with the aid of functional genomics, proteomics and bioinformatics, the next step will be the identification of the genes, proteins, and their respective functions. Because of alternative splicing the number of proteins by far exceeds the number of genes (25 000) in humans; thus, the task to determine all protein functions and the potential cross-talk between individual proteins will be vast. Nonetheless, it is important that these data are understood in order to treat health disorders and diseases. As a consequence of these ongoing studies, a number of new targets have already been defined that are being used to develop new drugs, and more targets will undoubtedly emerge in the near future (Fig. 6.1). Two

Medicinal Plant Biotechnology. From Basic Research to Industrial Applications
Edited by Oliver Kayser and Wim J. Quax
Copyright © 2007 WILEY-VCH Verlag GmbH & Co. KGaA, Weinheim
ISBN 978-3-527-31443-0

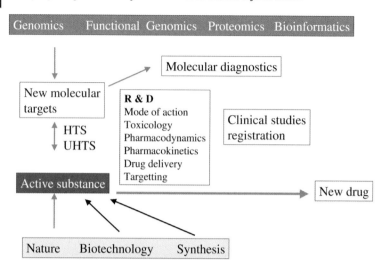

Fig. 6.1 The importance of bioactive compounds from Nature in drug development.

other technological advances that have changed the area of drug development during the past 20 years are those of automated high-throughput screening (HTS) and ultra-high-throughput screening (UHTS). By using HTS and UHTS, it is possible to screen up to one million compounds at one target, within one week.

This technology allows the screening of large substance libraries which consist of synthetic compounds, made by traditional or combinatorial synthesis, in addition to substance libraries from biotechnology and Nature. Nature produces an extreme diversity of bioactive compounds that, to date, have been exploited only to a minor degree. However, many of our medicinal drugs still derive from Nature either directly (e.g., many antibiotics, or the anticancer drugs paclitaxel, vinblastine, camptothecine or podophyllotoxin) or indirectly, in that they had served as lead compounds for the development of synthetic compounds with improved qualities. The search for new compounds from Nature – often referred to as "panning for biological gold" – is an active research field. Moreover, bioprospecting will become increasingly important in the future as large substance libraries with enriched drug candidates are required for screening at new molecular targets presently under development. Many natural products that have been shaped and selected during evolution provide a rich source of powerful drugs and lead structures, and these will be discussed in this chapter (Swain, 1977; Wink, 1988, 1993; Rosenthal and Berenbaum, 1991/1992; Harborne, 1993; Roberts and Wink, 1998; Seigler, 1998).

One typical feature of plants, fungi, sessile animals (especially from marine environments) and of bacteria is the production and accumulation of secondary metabolites that, in most cases, have a molecular weight of less than 1000 Dalton (Da). To date, more than 50 000 secondary metabolite structures have been determined using NMR, mass spectrometry (MS) and X-ray analysis, though only a small proportion of relevant organisms have been studied in some depth so far. Therefore, it is very likely that the actual number of secondary metabolites that exists in Nature is far in excess of 100 000. It is possible to distinguish between nitrogen (N)-

containing and N-free secondary metabolites; an overview of known classes and numbers of secondary metabolite structures is provided in Table 6.1, while the occurrences of secondary metabolites in the main groups of producers are detailed in Table 6.2. Although some groups of secondary metabolites are often defined by their bioactivities (e.g., antibiotics), they may also be grouped according to their

Table 6.1 Structural types of secondary metabolites and known structures.

Class	No. of structures
With nitrogen	
Alkaloids	21 000
Non-protein amino acids (NPAA)	700
Amines	100
Cyanogenic glycosides	60
Glucosinolates	100
Alkamides	150
Lectins, peptides	800
Without nitrogen	
Monoterpenes (incl. iridoids)	2500
Sesquiterpenes	5000
Diterpenes	2500
Triterpenes, steroids, saponins	5000
Tetraterpenes	500
Phenylpropanoids, coumarins, lignans	2000
Flavonoids, tannins	4000
Polyacetylenes, fatty acids, waxes	1500
Polyketides (anthraquinones)	750
Carbohydrates	200

Table 6.2 Occurrence of secondary metabolites in bacteria, plants, fungi, and animals.

Producer	Type of secondary metabolite
Blue-green algae (Cyanobacteria)	Polyketides (aplysiatoxin, brevetoxin B, dinophysistoxin-1), alkaloids (lyngbyatoxin, anatoxin-a)
Bacteria	"Antibiotics": polyketides, alkaloids, terpenoids, phenylpropanoids, peptides
Algae	Polyphenols, terpenoids, polysaccharides
Dinoflagellates (Dinophyceae)	Paralytic shellfish poisoning; alkaloids (saxitoxin, gonyautoxin); ciguatoxin
Lichen	Anthraquinones; polyphenols, phenylpropanoids
Plants (angiosperms, gymnosperms)	Alkaloids, amines, glucosinolates, cyanogenic glycosides, cyanolipids, glucosinolates, non-protein amino acids, terpenoids, saponins, phenylpropanoids, tannins, lignans, anthraquinones, fatty acids, polyines, phloroglucinols, alkylphenols, lectins

Table 6.2 Continued

Producer	Type of secondary metabolite
Fungi (Basidiomycetes)	Organic acids, alkaloids and peptides (hydrazine derivatives, amatoxines, bufotonin, ibotenic acid, muscarine, phallotoxins, psilocine, psilocybin, virotoxin), non-protein amino acids, cyanogenic glycosides, phenolics, sesquiterpenes, triterpenes
Molds (Deuteromycetes)	Mycotoxins (aflatoxins, phenolics, furanocoumarins, citrinin, citreoviridin, cytochalasin, *Penicillium* toxins, trichothecens, anthraquinones, ergot alkaloids, *Fusarium* toxins, *Aspergillus* toxins, ochratoxins, patulin, penitrem A, rugolosin, rubratoxins, zearalenone)
Animals	
Sponges (Porifera)	Terpenoids, sesquiterpenes, diterpenes, steroids, alkaloids, halogenated secondary metabolites, histamine derivatives
Corals, jelly fish (Cnidaria)	Toxic polypeptides in nematocysts (cytolysines; hemolysines; neurotoxins; physaliatoxin; esterase, hyaluronidase, proteases)
Sea anemones (Anthozoa)	Toxic polypeptides in nematocysts (neurotoxins)
Worms (Nermertini)	Alkaloids (anabaseine; nereistoxin)
Scorpions (Scorpiones)	Toxic polypeptides (neurotoxins: α-, β-, γ-toxin; phospholipase A, hyaluronidase)
Spiders (Araneae)	Polyamines (argiotoxin, argiopin, NSTX, ISTX), polypeptides (robustoxin, latrotoxin, sphingomyelinase D)
Scolopender (Chilopoda)	Toxic peptides, neurotransmitters (histamine, serotonin)
Insects (Insecta)	secondary metabolites acquired from host plants (alkaloids: pyrrolizidine alkaloids, quinolizidine alkaloids, aconitine, cardiac glycosides, cyanogenic glycosides, phorbol esters); quinones, terpenoids; toxic polypeptides (phospholipase A_2, hyaluronidase, hemolysins, melittin, apamin, wasp kinins, neurotransmitters (histamine), pyridine and piperidine alkaloids (solenopsine), iridoids (dolichodial), sesquiterpenes, diterpenes, cantharidin
Cone shells (Mollusca)	Toxic polypeptides, conotoxins; saxitoxin, gonyautoxin in *Turbo* and *Tectus* species
Octopus (Cephalopoda)	Toxic polypeptides (hyaluronidase; cephalotoxin, eledoisin); tetrodotoxin, neurotransmitters (serotonin, tyramine, octopamine, noradrenaline)
Starfish (Asteroidea)	Toxic polypeptides (phospholipase A_2), steroid saponins
Sea urchin (Echinoidea)	Toxic polypeptides, amines
Sea cucumber (Holothuroidea)	Steroidal saponins (holothurin A,B)
Stingray (Chondrichthyes)	Toxic polypeptides (phosphodiesterase)
Bony fish (Osteichthyes) (weever, scorpion, fire fish)	Toxic polypeptides, neurotransmitters (serotonin, acetylcholine), alkaloids (pahutoxin)

Table 6.2 Continued

Producer	Type of secondary metabolite
Puffer fish, mollusks, amphibia	Marine bacteria produce toxins, that accumulate in the food chain: tetrodotoxin, palytoxin
Toads, salamanders, frogs (Amphibia)	Alkaloids (bufotonin, samandarine, tetrodotoxin, batrachotoxin, pumiliotoxin, histrionicotoxin), bufadienolides (bufotoxin)
Gila monster (Reptilia)	Toxic polypeptides (hyaluronidase, kallikrein, gilatoxin)
Snakes (Reptilia)	Toxic polypeptides (neurotoxins: neurotoxic phospholipase A_2, α-neurotoxins, choline esterase-inhibiting proteins; enzymes: hyaluronidase, phosphatases, phospholipase A_2, proteases, oxidases)

structural types (as in this chapter). Whereas primary metabolites are present in all species, secondary metabolites occur in varying mixtures that differ between species and systematic units. Secondary metabolites are not essential for primary or energy metabolism but, as discussed below, they are important for the ecological fitness and survival of the organisms that produce them (Balandrin et al., 1985; Harborne, 1993; Wink, 1988, 1999a,b; Seigler, 1998).

Secondary metabolites are produced in specific pathways that involve substrate-specific biosynthetic enzymes (Luckner, 1990; Dewick, 2002). The sites of synthesis may differ between types of compounds and between species. Some compounds can be produced by all tissues, whereas most others are produced in a tissue- or even cell-specific fashion. It is likely that the corresponding genes are regulated by specific transcription factors similar to the situation of other genes that are differentially regulated. The site of synthesis is not necessarily the site of accumulation in plants. Indeed, several compounds have been shown to be transported within a plant either via the phloem or the xylem. Whereas hydrophilic compounds (e.g., alkaloids, amino acids, glucosinolates, cyanogenic glycosides, flavonoids, tannins and other polyphenols, carbohydrates and saponins) are stored in the vacuole, lipophilic secondary metabolites (e.g., many terpenoids) are sequestered in resin ducts, laticifers or in special (usually dead) cells, such as oil cells, trichomes, or in the cuticle. Often, epidermal cells, the initial role of which is to ward off enemies, are especially rich in secondary metabolites. In animals, secondary metabolites are often stored in special glandular structures commonly present in skin tissues.

It is a typical feature of secondary metabolites that many are stored at relatively high levels in sink tissues that are often important for the survival and reproduction of a plant, such as the flowers, seeds, seedlings, or the bark of perennial plants. Several secondary metabolites are not the end products of metabolism but can be recycled in plants. For example, nitrogen-containing secondary metabolites such as alkaloids, nonprotein amino acids (NPAA) or lectins are often accumulated as toxic nitrogen-storage compounds in the seeds of legumes in which nitrogen is remobilized during germination and seedling growth (Harborne, 1993; Wink, 1988, 1999a,b; Roberts and Wink, 1998).

6.2
The Function of Secondary Metabolites

Plants, fungi, sessile animals and bacteria cannot run away when attacked by herbivores or predators; neither do they have an immune system against invading bacteria, fungi, or viruses. Consequently, plants and other sessile organisms (e.g., marine animals) (see Table 6.2) have developed biologically active secondary metabolites during evolution that help them to defend themselves against predators (insects, mollusks, vertebrates), microbes, viruses, and other competing plants (Fig. 6.2). In order to be effective, secondary metabolites must be present at the correct site, time, and concentration. The biosynthesis of several secondary metabolites is constitutive, whereas in many plants it can be induced and enhanced by biological stress conditions, such as wounding or infection. This activation can be biochemical, for example through the hydrolysis of glycosides that are stored as "prodrugs" (Table 6.3) or via the activation of genes responsible for the synthesis, transport or storage of secondary metabolites. Signal transduction pathways that lead to gene activation in plants include those leading to jasmonic acid or salicylic acid that have been found to trigger defense reactions in plants (Harborne, 1993; Wink, 1988, 1999a,b).

Plants also use secondary metabolites (such as volatile essential oils and colored flavonoids or tetraterpenes) to attract insects for pollination or other animals for seed dispersion. In this case, secondary metabolites serve as *signal compounds*. Animals that store toxic secondary metabolites often advertise this property by warning colors (that are themselves secondary metabolites); this is termed *aposematism*.

In addition, some secondary metabolites concomitantly carry out physiological functions; for example alkaloids, NPAA, and peptides (lectins, protease inhibitors)

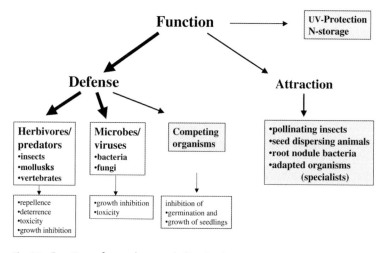

Fig. 6.2 Functions of secondary metabolites in plants.

Table 6.3 Typical "prodrugs" present in plants that are activated by wounding or infection.

Secondary metabolite of undamaged tissue	Active metabolite
Cyanogenic glycoside	HCN
Glucosinolate	isothiocyanate
Alliin	allicin
Coumaroylglycoside	coumarin
Arbutin	quinone
Salicin, methylsalicylate	saligenin, salicylic acid
Gein	eugenol
Bi-desmosidic saponins	mono-desmosidic saponins
Cycasin	methylazoxymethanol (MAM)
Ranunculin	protoanemonine

can serve as mobile and toxic nitrogen transport and storage compounds, while phenolics (e.g., flavonoids) can function as UV-protectants (Harborne, 1993; Wink, 1988, 1999a,b). In addition to chemical defense, a number of plants and marine animals use mechanical and morphological features for protection, including thorns, spikes, glandular and stinging hairs (often filled with noxious chemicals), or they may develop a barely penetrable bark (especially woody perennials) or armor (carapace).

The defense strategy usually works against generalist attackers, but often not against specialists that have adapted to their host plants or prey and their defense chemicals. A diverse collection of adaptations has already been detected and described (Harborne, 1993; Wink, 1988, 1999a,b). In general, it was observed that mono- and oligophagous insects can tolerate the defense chemicals of their particular host plants, but are susceptible to those of nonhost species.

The observed multiple functions of secondary metabolites are typical and do not contradict the main role of many secondary metabolites as *chemical defense* and *signal compounds*. If a costly trait can serve multiple functions (maintenance of the biochemical machinery to produce and store secondary metabolites is energetically costly; Wink, 1999a), it is more likely that it is maintained by natural selection as it provides a selective advantage for its carrier.

6.3
Modes of Action

In order to fulfill the role of defense substances against herbivores, predators and microbes, secondary metabolites must be able to interfere with molecular targets in the organs, tissues and cells of these organisms. The major types of molecular targets in prokaryotes and eukaryotes that are relevant in this context are listed in Table 6.4; these include biomembranes, proteins, and nucleic acids (DNA and RNA) (Wink 1993, 1999b, 2000, 2005; Teuscher and Lindequist, 1994; Roberts and Wink, 1998) (Fig. 6.3).

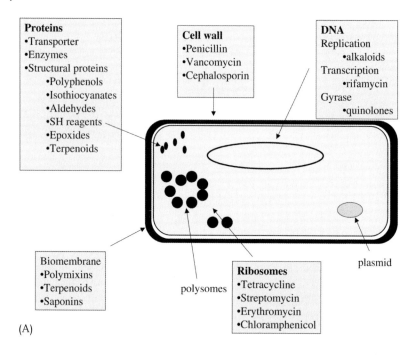

(A)

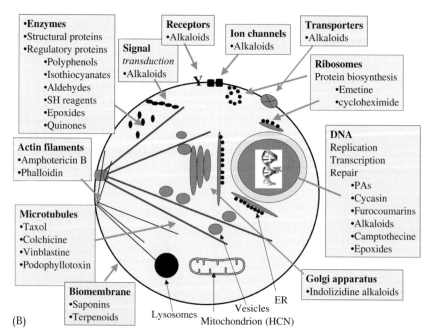

(B)

Fig. 6.3 Major molecular targets of cells. (A) Bacterial cells;
(B) animal cells. ER: endoplasmic reticulum.

Table 6.4 Interaction of secondary metabolites with molecular targets.

Target	Activity	Secondary metabolites (examples)
Unselective targets		
Biomembrane	Membrane disruption	Saponins
	Disturbance of membrane fluidity	Small lipophilic secondary metabolites
	Inhibition of membrane proteins (change of protein conformation)	Small lipophilic secondary metabolites
Proteins		
Nonselective interactions	Noncovalent bonding (change of conformation)	Polyphenols such as phenylpropanoids, flavonoids, catechins, tannins, lignans, quinones, anthraquinones, some isoquinoline alkaloids
	Covalent bonding (change of conformation)	Isothiocyanates sesquiterpene lactones, allicin, protoanemonine, furanocoumarins, iridoids (aldehydes), secondary metabolites with aldehydes, secondary metabolites with exocyclic CH_2 group, secondary metabolites with epoxide group
Specific interaction	Inhibition of enzymes	HCN from cyanogens, many structural mimics
	Modulation of regulatory proteins	Phorbol esters, caffeine
	Inhibition of ion pumps	Cardiac glycosides
	Inhibition of microtubule formation	Vinblastine, colchicines, podophyllotoxin, paclitaxel
	Inhibition of protein biosynthesis	Emetine, cycloheximide
	Inhibition of transporters	Non-protein amino acids
	Modulation of hormone receptors	Genistein, many other isoflavonoids
	Modulation of neuroreceptors	Nicotine, many alkaloids, conotoxins, nereistoxin, argiotoxin, argiopin
	Modulation of ion channels	Aconitine, many alkaloids; conotoxins, tetrodotoxin, saxitoxin, gonyautoxin, ciguatoxin, palytoxin
	Modulation of transcription factors	Cyclopamine, hormone mimics
DNA	Covalent modifications (point mutations)	Pyrrolizidine alkaloids, aristolochic acids, furanocoumarins, secondary metabolites with epoxy groups
	Intercalation (frameshift mutations)	Planar, aromatic and lipophilic secondary metabolites, sanguinarine, berberine, emetine, quinine, furanocoumarins, anthraquinones
	Inhibition of DNA topoisomerase I	Camptothecin
	Inhibition of transcription	Actinomycin D

6.3.1
Biomembranes

Biomembranes surround all living cells, and function as a permeation barrier. The barrier prevents polar molecules from leaking out of the cell, and unwanted molecules from entering the cell. Several secondary metabolites exist in Nature that interfere with membrane permeability (see Table 6.4). The most infamous of these are the saponins, which occur widely in the plant kingdom but less commonly in animals; mono-desmosidic saponins (see Table 6.3) are *amphiphilic* and function basically as detergents that can solubilize biomembranes. With their lipophilic moiety, they anchor in the lipophilic membrane bilayer, whereas the hydrophilic sugar moiety remains outside and interacts with other glycoproteins or glycolipids (Fig. 6.4). As a result, pores are generated in the membrane which causes it to leak. This effect can be easily demonstrated with red blood cells since, if saponins are present hemolysis occurs, and hemoglobin is able to flow out of the cell. Other lipophilic secondary metabolites, such as mono-, sesqui- and diterpenes, can also disturb membrane fluidity at higher concentrations. These compounds are able to interact with the lipophilic inner core of biomembranes represented by phospholipids and cholesterol (Fig. 6.4). This type of membrane disturbance is unselective, and therefore secondary metabolites with such properties are toxic to bacterial, fungal, and animal cells. Some may also affect viral membranes.

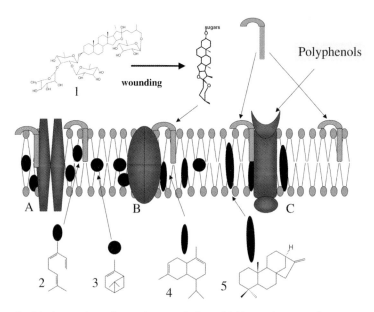

Fig. 6.4 Interactions of secondary metabolites with biomembranes and membrane proteins. 1, A steroidal saponin; sarsaparilloside. 2, Ocimene (linear monoterpene). 3, Alpha-pinene (bicyclic monoterpene). 4, Cadinene (sesquiterpene). 5, Phyllocladene (diterpene). A: ion channel; B: transporter; C: membrane receptor.

Biomembranes harbor a wide set of *membrane proteins,* including ion channels, transporters for nutrients and intermediates, receptors, and proteins of signal transduction and the cytoskeleton. These proteins can only function correctly if their structure is in the correct conformation. Membrane proteins with transmembrane domains are stabilized by the surrounding lipids. Hence, if lipophilic secondary metabolites are dissolved in the biomembrane they disturb the close interaction between the membrane lipids and proteins, changing the protein conformation (Table 6.4; Fig. 6.4), and a loss of function is the usual consequence. This mode of action is demonstrated by anesthetics, which are small, lipophilic compounds that inactivate ion channels and neuroreceptors, and thus block signal transduction. Several of the small terpenoids can react in a comparable manner; plants with essential oils are often used in medicine as carminative drugs (i.e., a drug that relieves intestinal spasms). In this respect it is possible that secondary metabolites inactivate ion-channels and receptors, leading to the relaxation of smooth muscles in the intestinal tissues.

6.3.2
Proteins and Protein Conformation

Proteins have multiple functions in a cell, ranging from catalytic enzymes, transporters, ion channels, receptors, microtubules, histones to regulatory proteins (e.g., signal molecules, transcription factors). Proteins can only function correctly if they have the correct shape and conformation. Conformational changes also alter the protein's properties and can prevent effective protein–protein cross-talk that is vital for intracellular communication. Protein activities are often regulated by phosphorylation or dephosphorylation, with the addition or subtraction of such a bulky group inducing a conformational change. It is likely that most secondary metabolites found in Nature interact with proteins in one way or another. Most secondary metabolites interfere with proteins in an *unselective* manner – that is, they affect any protein that they encounter (Fig. 6.5). Such unselective interactions can be subdivided into those that involve *noncovalent* bonding and those with *covalent* bond formation.

A major class of secondary metabolites, the *polyphenols,* includes structures such as phenylpropanoids, flavonoids, catechins, tannins, lignans, quinones, anthraquinones, and several alkaloids with one or several phenolic hydroxyl groups (Table 6.4; Fig. 6.5). The phenolic hydroxyl groups can partly dissociate under physiological conditions resulting in $-O^-$ ions (phenolates). The polyphenols have in common that they can interact with proteins by forming hydrogen bonds and ionic bonds with electronegative atoms of the peptide bond or the positively charged side chains of basic amino acids (lysine, histidine, arginine), respectively. A single noncovalent bond is quite weak, but because several such bonds are formed concomitantly when a polyphenols encounters a protein a change in protein conformation is likely to occur that commonly leads to protein inactivation.

However, the formation of *covalent bonds* also occurs (Fig. 6.6). Several types of secondary metabolites carry reactive functional groups that can bind to amino and

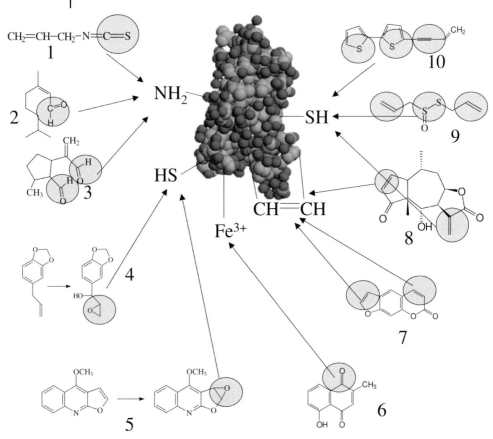

Fig. 6.5 Covalent modifications of proteins by secondary metabolites. 1. Allylisothiocyanate (mustard oil). 2. Citral (linear monoterpene). 3. Iridoid with opened lactol ring. 4. Safrole (phenylpropanoid). 5. Dictaminine (furoquinoline alkaloid). 6. Plumbagin (naphthoquinone). 7. Psoralen (furocoumarin). 8. Helenaline (sesquiterpene lactone). 9. Allicin (SH-reagent). 10. 5-(3-butenyl-1-ynyl)-2,2-bithienyl (BBT; polyine).

-SH groups or to double bonds in proteins. This covalent modification also leads to a conformational change and thus loss of activity. Secondary metabolites with reactive functional groups that are able to undergo electrophilic or nucleophilic substitutions are represented by isothiocyanates, allicin, protoanemonin, iridoid aldehydes, furanocoumarins, valepotriates, sesquiterpene lactones and secondary metabolites with active aldehydes, epoxide or terminal and/or exocyclic methylene groups (Table 6.4; Fig. 6.6). In several instances the reactive metabolites are not natively present in plants, but they can be converted to active metabolites either by the wounding process (releasing metabolizing enzymes) inside the producing organism or in the body of a herbivore/predator (after biotransformation in intestine or liver).

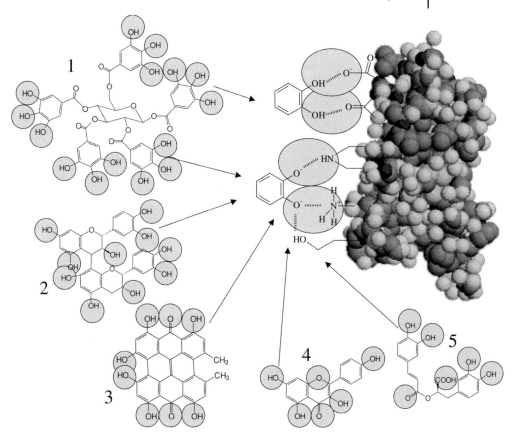

Fig. 6.6 Noncovalent modifications of proteins by secondary metabolites. 1, Pentagalloylglucose (gallotannin). 2, Dimeric procyanidin B4 (catechol tannin). 3, Hypericine (dimeric anthraquinone). 4, Kaempherol (flavonoid). 5, Rosmarinic acid (phenylpropanoid). Note: A color version of this figure is available in the color plate section.

In addition to the unselective interactions (broadband activities), several secondary metabolites are known to modulate protein activities in a specific manner in that they can bind as a ligand to the active site of a receptor or enzyme; this binding has been termed "induced fit". In this case, the structure of a given secondary metabolite is often a mimic of an endogenous ligand. Well-studied examples include several alkaloids that are structural analogues of neurotransmitters; for example, nicotine and hyoscyamine are mimics of acetylcholine, with nicotine binding to nicotinic acetylcholine receptors or hyoscyamine to muscarinic acetylcholine receptors. The various steps in neuronal signaling and signal transduction provide central targets that are affected by several amines and alkaloids (Table 6.4). The targets may be the neuroreceptor itself. Agonists mimic the function of a neurotransmitter (acetylcholine, dopamine, noradrenaline, adrenaline, serotonin, GABA, glutamate, glycine, endorphins, peptides) by binding to its receptor and causing the normal response. Antagonists (often called "blockers") also bind to the receptor but act as an inhibitor of the natural ligand by competing for binding sites on the receptor, thereby blocking the physiological response. Further targets are voltage-gated Na^+, K^+ and Ca^{2+} channels and the enzymes, which deactivate neurotransmitters after they have bound to a receptor, such as acetylcholine esterase, monoamine oxidase and catechol-*O*-methyltransferase. Also relevant are transport processes, which are important for the uptake and release of neurotransmitters in the presynapse or synaptic vesicles. Na^+, K^+, and Ca^{2+}-ATPases, which restore the ion gradients, must also be considered in this category. Furthermore, the modulation of key enzymes of signal pathways, including adenylyl cyclase (making cAMP), phosphodiesterase (inactivating cAMP or cGMP), phospholipase C (releasing inositol phosphates such as IP_3 and diacylglycerol (DAG) and several protein kinases, such as protein kinase C or tyrosine kinase (activating other regulatory proteins or ion channels) are important steps for which inhibitors from nature are known (see Table 6.3).

Another example of a more specific inhibitor is that of HCN released from cyanogenic glycosides common secondary metabolites in plants and some invertebrates. HCN is highly toxic for animals or microorganisms due to its inhibition of enzymes of the respiratory chain (i.e., cytochrome oxidase) and subsequent blockade of essential ATP production. HCN also binds to other enzymes containing heavy metal ions. In the case of emergency – that is, when plants are wounded by herbivores or other organisms – the cellular compartmentation breaks down and vacuolar cyanogenic glycosides come into contact with an active β-glucosidase of broad specificity, which hydrolyzes them to yield 2-hydroxynitrile (cyanohydrine). 2-Hydroxynitrile is further cleaved into the corresponding aldehyde or ketone and HCN by a hydroxynitrile lyase.

A number of diterpenes are infamous for their toxic properties (cytotoxicity, inflammation), such as phorbol esters of Euphorbiaceae and Thymelaeaceae (Fig. 6.7). These diterpenes specifically activate protein kinase C, which is an important key regulatory protein in animal cells. Another diterpene, forskolin, acts as a potent activator of adenynyl cyclase. Paclitaxel (Taxol®) is another diterpene that can be isolated from several yew species (including the north American *Taxus brevifolia* and the European *Taxus baccata*). Paclitaxel stabilizes microtubules and thus blocks cell division in the late G_2 phase; because of these properties, paclitaxel has been used for almost 10 years with great success in the chemotherapy of various tumors. Microtubule formation is a specific target for the alkaloids vinblastine (from *Catharanthus roseus*), colchicine (*Colchicum autumnale*) or the lignan podophyllotoxin (from *Podophyllum* and several *Linum* species) (Fig. 6.7).

One special case of steroidal saponins is that of cardiac glycosides that inhibit Na^+,K^+-ATPase, one of the most important targets in animal cells responsible for the maintenance of Na^+ and K^+ gradients. Cardiac glycosides can be divided into two classes:

- *Cardenolides* have been found in Scrophulariaceae (*Digitalis*), Apocynaceae (*Apocynum, Nerium, Strophanthus, Thevetia*), Asclepiadaceae (*Periploca, Xysmalobium*), Brassicaceae (*Erysimum, Cheiranthus*), Celastraceae (*Euonymus*), Convallariaceae (*Convallaria*) and Ranunculaceae (*Adonis*).

- *Bufadienolides* occur in Crassulaceae (*Kalanchoe*), Hyacinthaceae (*Urginea*) and Ranunculaceae (*Helleborus*).

Another example of specific protein inhibition can be found in the class of NPAA that often occur as anti-nutrients or anti-metabolites in many plants (e.g., in Fabaceae). Many NPAA mimic protein amino acids, and quite often can be considered to be their structural analogues that may interfere with the metabolism of a herbivore. For example, in ribosomal protein biosynthesis NPAA can be accepted in place of the normal amino acid, leading to defective proteins. NPAA may compet-

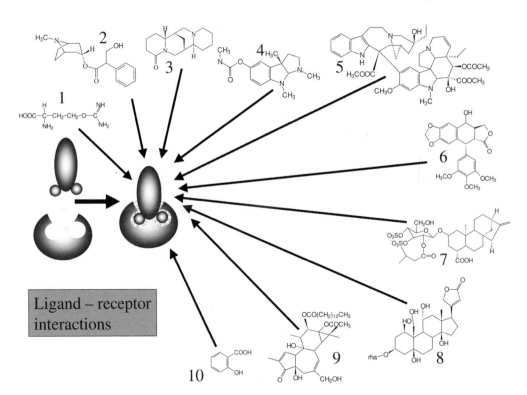

Fig. 6.7 Specific interactions (ligand–receptor relationships) of secondary metabolites with proteins. 1, Canavaline (non-protein amino acid). 2, Hyoscyamine (tropane alkaloid). 3, Lupanine (quinolizidine alkaloid). 4, Physostigmine (indole alkaloid). 5, Vinblastine (dimeric monoterpene indole alkaloid). 6, Podophyllotoxin (lignan). 7, Atractyloside (diterpenes). 8, Ouabain (cardiac glycoside). 9,12-Tetradecanoyl-phorbol-13-acetate (TPA; phorbolester). 10, Salicylic acid (phenolic acid). Note: A color version of this figure is available in the color plate section.

itively inhibit uptake systems (transporters) for amino acids; they may also inhibit amino acid biosynthesis by substrate competition or by mimicking end product-mediated feedback inhibition of earlier key enzymes in the pathway.

6.3.3
DNA and RNA

DNA is an important target in all organisms, as are the enzymes involved with DNA replication, DNA repair, DNA-topoisomerase and transcription. The translation of messenger RNA (mRNA) into protein in ribosomes is also a basic target that is present in all cells. Inhibitors of these systems are often active against a wide range of organisms, such as bacteria, fungi, and animal cells. The DNA itself can be modified by compounds with reactive groups, such as epoxides (Fig. 6.8). Infamous in this respect are the pyrrolizidine alkaloids, aristolochic acid, cycasin, fu-

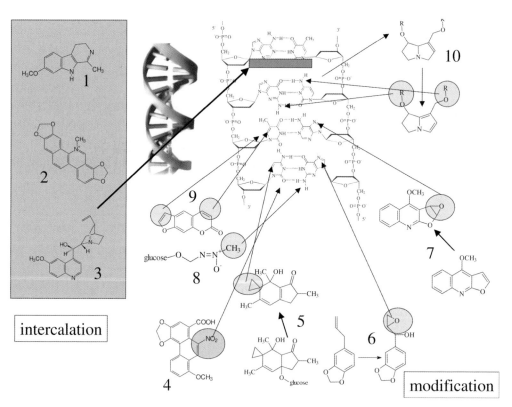

Fig. 6.8 Interactions of secondary metabolites with nucleic acids and corresponding proteins. 1, Harmaline (beta-carboline alkaloid). 2, Sanguinarine (protoberberine alkaloid). 3, Quinine (quinoline alkaloid). 4, Aristolochic acid.

5, Ptaquiloside (sesquiterpene).
6, Safrole (phenylpropanoid).
7, Dictamine (furoquinoline alkaloid).
8, Cycasine (azoxyglycoside).
9, Psoralen (furocoumarin).
10, Pyrrolizidine alkaloid.

ranocoumarins and secondary metabolites with epoxy groups (often produced in the liver). Covalent modifications can lead to point mutations and deletion of single bases or several bases if the converted bases are not exchanged by repair enzymes. Other secondary metabolites with aromatic rings and lipophilic properties intercalate DNA, which can lead to frameshift mutations (Fig. 6.8). Intercalating alkaloids include emetine, sanguinarine, berberine, quinine, β-carboline and furoquinoline alkaloids (Wink and Schimmer, 1999). Furocoumarins combine DNA alkylation with intercalation. Furocoumarins can intercalate DNA and, upon illumination with UV light, can form cross-links not only with DNA bases but also with proteins; they are therefore mutagenic and possibly carcinogenic. These compounds are abundant in Apiaceae (content up to 4%), but are also present in certain genera of the Fabaceae and Rutaceae.

Because frameshift mutations and nonsynonymous base exchanges in protein-coding genes alter the amino acid sequence in proteins, such mutations are usually deleterious for the corresponding cell. If they occur in germline cells, such as oocytes and sperm cells, even the next generation is negatively influenced either through malformations or protein malfunctions responsible for certain types of inheritable health disorders or illnesses.

Interference with DNA, protein biosynthesis and related enzymes can induce complex chain reactions in cells. Among these reactions is that of *apoptosis*, which leads to programmed cell death and is clearly an important process. Several alkaloids, flavonoids, allicin, saponins and cardiac glycosides have been shown to induce apoptosis in primary and tumor cell lines (Wink, 2006).

In summary, structures of allelochemicals appear to have been shaped during evolution in such a way that they can mimic the structures of endogenous substrates, hormones, neurotransmitters or other ligands; this process can be termed "evolutionary molecular modeling". Other metabolites intercalate or modify DNA, inhibit DNA and RNA related enzymes, protein biosynthesis, other metabolic enzymes and functional proteins, or they disturb membrane stability and membrane proteins in a less specific manner.

In general, we find a series of related compounds in a given organism; often, a few major metabolites and several minor components are present, which differ in the position of their chemical groups. The metabolic profile usually varies between plant organs, within developmental periods, and sometimes even diurnally. Marked differences can also usually be seen between individuals of a single population, and even more so between members of different populations. Even small changes in chemistry can serve as the base for a new pharmacological activity. It is evident that secondary metabolites are often multifunctional compounds, and most secondary metabolites carry more than one pharmacologically active chemical group (pharmacophor). In addition, secondary metabolites usually occur in complex mixtures of various types. In consequence, as the secondary metabolites present in a given plant or animal always affect several molecular targets, it is likely that several targets are modulated concomitantly in herbivores/predators or microbes and last – but not least – in patients treated with plant-based medicines.

6.4
The Utilization of Secondary Metabolites in Medicine

Since the early days of mankind, plants with secondary metabolites have been used to treat infections, health disorders, and illness (Mann, 1992; Wagner and Wiesenauer, 1995; Roberts and Wink, 1998; Wink, 1999b). Only during the past 100 years have natural medicines been replaced by synthetic drugs, for which plant structures were a lead in many instances (e.g., salicylic acid for aspirin). The use of plant drugs for medical treatment is possible since plants have evolved bioactive metabolites directed against microbes and herbivores (see above). The utilization of such compounds for medicinal purposes is, therefore, only the other side of the coin. Cardiac glycosides, for example, are very poisonous as they inhibit Na^+,K^+-ATPase, a central target in animals. However, if these compounds are used at lower concentrations the toxic effects are reduced and the medically beneficial activity is more pronounced. The Na^+,K^+-pumps are partly inactivated, leading to a higher Na^+ concentration in cardiac cells. This in turn stimulates a Na^+, Ca^{++} transporter, that increases the Ca^{++} concentration in the cell; the enhanced Ca^{++} concentration then stimulates muscular contraction.

Several secondary metabolites from plants are used medicinally as isolated compounds, including many alkaloids (Roberts and Wink, 1998; Wink, 1993, 2006) such as morphine (pain killer), codeine (antitussive), papaverine (phosphodiesterase inhibitor), ephedrine (stimulant), ajmaline (antiarrhythmic), quinidine (antiarrhythmic), quinine (antimalarial), paclitaxel (tumor therapy), vinblastine (tumor therapy), podophyllotoxin (tumor therapy), camptothecine (tumor therapy), reserpine (antihypertensive), galanthamine (acetylcholine esterase inhibitor; Alzheimer's disease), aconitine (pain killer), physostigmine (acetylcholine esterase inhibitor), atropine (spasmolytic; mydriatic), scopolamine (travel sickness), berberine (psoriasis), caffeine (stimulant), theophylline (antitussive), capsaicin (rheumatic pains), colchicines (gout), yohimbine (aphrodisiac), and pilocarpine (glaucoma). Other secondary metabolites include cardiac glycosides, genistein (tyrosine kinase inhibitor, phytoestrogen), khellin (angina pectoris), artemisinin (antimalarial), menthol (spasmolytic), and thymol (antiseptic).

In addition to individual compounds, plant extracts and even crude drugs are widely used in many parts of the world. In complex health disorders for which the targets are not known, the application of extracts with unselective ingredients that affect several proteins and biomembranes might even be the solution of choice. The likelihood is high that the relevant target will also be affected, even if it is not known.

One very large group of microbial secondary metabolites – the antibiotics – are most likely synthesized by microorganisms as defense compounds against other microbes, but have been used medicinally for the past 60 years. Today, because resistance against common antibiotics is increasing there is an urgent need for new antibiotics with new targets, and consequently bioprospecting for antibiotics and cytotoxic compounds from bacteria and fungi is an especially active field of research. Several bacteria cannot be cultivated and have therefore not been studied in the laboratory. However, by using genetic methods it might be possible to clone and express the corresponding genes that encode biosynthetic pathways for secondary metabolites, thereby overcoming the problem of cultivation.

Marine organisms produce secondary metabolites that have mostly been selected against microbes and lower animals. It is therefore not surprising that among the secondary metabolites isolated from these organisms, many have antibiotic and cytotoxic properties. Indeed, some of these have undergone extensive pharmacological trials and now provide interesting drug candidates.

6.5
Conclusions

Plants, sessile animals and bacteria produce a wide variety of bioactive metabolites (Table 6.2) that, to date, have been only partly studied. Understanding the physiology, biochemistry and ecology of secondary metabolism offers an opportunity to breed plants with a better protection against microbes and herbivores. Moreover, understanding their molecular pharmacology represents a key to exploit bioactive natural chemicals in a rational way in medicine and agriculture (biorational pesticides). The Human Genome Project will in time identify a large number of new molecular targets, and already today industry is employing high-throughput screening to seek interesting compounds that interact with such a target. Secondary metabolites from plants, animals and microbes which have been preselected and shaped during evolution offer a very interesting chance to obtain relevant "hits". In this respect, the search for new active compounds or leads has been aptly named as "bioprospecting" – the search for biological gold.

Key references

Balandrin, M.F., Klocke, J.A., Wurtele, E.S. and Bollinger, W.H. Natural plant chemicals: sources of industrial and medicinal materials. *Science* **1985**, *228*, 1154–1160.

Dewick, P.M. *Medicinal Natural Products. A biosynthetic approach*. Wiley, New York, **2002**.

Harborne, J.B. *Introduction to ecological biochemistry*. 4th edition. Academic Press, London, **1993**.

Luckner, M. *Secondary Metabolism in Microorganisms, Plants and Animals*. Springer, Heidelberg, **1990**.

Mann, J. *Murder, Magic and Medicine*. Oxford University Press, London, **1992**.

Roberts, M.F. and Wink, M. *Alkaloids-Biochemistry, ecological functions and medical applications*. Plenum, New York, **1998**.

Rosenthal, G.A. and Berenbaum, M.R. *Herbivores: Their interactions with secondary plant metabolites. Volume 1. The chemical participants; Volume 2. Ecological and evolutionary processes*. Academic Press, San Diego, **1991/1992**.

Seigler, D.S. *Plant Secondary Metabolism*. Kluwer, Boston, **1998**.

Swain, T. Secondary compounds as protective agents. *Annu. Rev. Plant Physiol.* **1977**, *28*, 479–501.

Teuscher, E. and Lindequist, U. *Biogene Gifte*. Fischer, Stuttgart, **1994**.

Wagner, H. and Wiesenauer M. *Phytotherapie*. Fischer, Stuttgart, **1995**.

Wink, M. Plant breeding: Importance of plant secondary metabolites for protection against pathogens and herbivores. *Theoret. Appl. Genet.* **1988**, *75*, 225–233.

Wink, M. Allelochemical properties and the raison d'être of alkaloids. In: Cordell, G. (Ed.), *The Alkaloids*. Academic Press, Orlando, Volume 43, **1993**, pp. 1–118.

Wink, M. *Biochemistry of Plant Secondary Metabolism*. Sheffield Academic Press and CRC Press, Annual Plant Reviews, Volume 2, **1999a**.

Wink, M. *Function of Plant Secondary Metabolites and their Exploitation in Biotechnology*. Sheffield Academic Press and CRC Press, Annual Plant Reviews, Volume 3, **1999b**.

Wink, M. Interference of alkaloids with neuroreceptors and ion channels. In: Atta-Ur-Rahman (Ed.), *Bioactive Natural Products*. Elsevier, Volume 11, **2000**, pp. 3–129.

Wink, M. and Schimmer, O. Modes of action of defensive secondary metabolites. In: Wink, M. (Ed.), *Function of Plant secondary metabolites and their exploitation in biotechnology*. Sheffield Academic Press and CRC Press, Annual Plant Reviews, Volume 3, **1999**, pp. 17–133.

Wink, M. Wie funktionieren Phytopharmaka? Wirkmechanismen der Vielstoffgemische. *Z. für Phytotherapie* **2005**, *26*, 271–274.

Wink, M. Molecular modes of action of cytotoxic alkaloids. From DNA intercalation, spindle poisoning, topoisomerase inhibition to apoptosis and multiple drug resistance. In: Cordell, G. (Ed.), *The Alkaloids*. Academic Press, **2006** (in press).

7
Biotechnological Approaches for the Production of some Promising Plant-Based Chemotherapeutics

Ashish Baldi, V.S. Bisaria, and A.K. Srivastava

Abstract

Today, natural products from plants provide better templates for the design of potential chemotherapeutic agents than synthetic drugs. Paclitaxel (Taxol), podophyllotoxin and camptothecin are some lead molecules which have proved to be Nature's boon in the treatment of cancer. To meet ever-increasing demands, biotechnological methods offer an excellent alternative, but the economy of such a production is the major hurdle to be overcome. The successful industrial application of plant cell cultivation for the production of these therapeutic compounds will trigger further research on other promising plant-based chemotherapeutics. As the initial hurdles for large-scale cultivation of plant cells have been overcome, the new areas of concern to produce desired products are synergistic product enhancement strategies, along with in-depth knowledge of the biosynthetic pathway. The complete pathway for the biosynthesis of podophyllotoxin has already been established, but alternative pathways of different metabolic fates of lignans, the precise sequence of the later steps in taxol biosynthesis, and the iridoid section of camptothecin biosynthesis have yet to be realized. Understanding of biochemistry, enzymology, physiology, bioreactor design and the application of proteomics and genomics are other areas on which to focus. Several points in a given metabolic pathway can be controlled simultaneously, either by overexpressing and/or suppressing several enzymes, or through the use of transcriptional regulators to control endogenous genes. Thus, multipoint metabolic engineering offers new perspectives for improvements in the production of plant-based chemotherapeutics. The aim of this chapter is to provide an overview of research on key anticancer drugs in order to elucidate the biotechnological approaches for their production in cell cultures. Special emphasis is placed on the biosynthetic pathway mapping and metabolic engineering.

Medicinal Plant Biotechnology. From Basic Research to Industrial Applications
Edited by Oliver Kayser and Wim J. Quax
Copyright © 2007 WILEY-VCH Verlag GmbH & Co. KGaA, Weinheim
ISBN 978-3-527-31443-0

7.1
Introduction

Cancer, an unrestrained proliferation and migration of cells, is a scourge that has afflicted mankind since time immemorial. In spite of the spectacular advances made by medical science during the past century, the treatment of cancer remains an enigma. Cancer represents the single largest cause of death in both men and women, and is a growing public health menace. Each year, about seven million new cases are diagnosed, and about five million people die as a result of cancer. Prevalence data indicate that, currently, about 14 million people are suffering from cancer. Bearing in mind the level of morbidity that is often affiliated with this disease, comprehension of such a high incidence is horrifying.

In recent years, efforts have been made to synthesize potential anticancer drugs; consequently, hundreds of chemical variants of known classes of anticancer therapeutic agents have been synthesized. It has been recognized that a successful anticancer drug should be one, which kills or incapacitates cancer cells without causing excessive damage to the normal cells. This criterion is difficult, or perhaps, impossible to attain, and that is why cancer patients suffer from unpleasant side effects while undergoing treatments. While vast amounts of synthetic chemistry have provided relatively small improvements over the prototype drugs, the synthesis of modified forms of known drugs continues as an important aspect of research. There exists a need for new prototypes and new templates for use in the design of potential chemotherapeutic agents. Natural products are capable of providing such templates.

In the United States, under the Cancer Chemotherapy National Service Centre (CCNSC), National Cancer Institute (NCI) programme, over 35 000 plants were screened for anticancer activity between 1960 and 1986, and over 2000 crystalline plant-derived compounds were isolated and tested for activity against P_{388} lymphocytic leukemia and KB carcinoma in cell culture. The screening program of the CCNSC brought to light hundreds of plant species, which were never used to treat cancer in any system of medicine. Some of the well-recognized anticancer plants are listed in Table 7.1. Chemical and biochemical investigations on some of these plants have yielded certain "lead" molecule as Nature's boon for cancer chemotherapeutic uses. With an understanding of the mechanism of action as well as the structure–activity relationship, several better analogues of these "lead" molecules have been prepared. Thus, it is now doubtless that plants are the most vital source of several compounds, which possess significant therapeutic values for combating cancer.

Plants are the most exclusive source of drugs for the majority of the world's population, and plant products constitute about 25% of prescribed medicines [1,2]. The impact of natural products upon anticancer drug discovery and design can be gauged by the fact that approximately 60% of all drugs, now in clinical trials for the treatment of cancer, are either natural products, compounds derived from natural products, or contain pharmacophores derived from natural products [3,4]. Some important antitumor compounds isolated from different parts of higher plants are listed in Table 7.2.

Table 7.1 The anticancer plants and their active constituents.

Plant	Family	Active constituent(s)
Acer negundo	Aceraceae	Acer saponin P (Saponin)
Acnistus arborescens	Solanaceae	Withalerin A (Withanolide)
Acronychia baueri	Rutaceae	Acronycine (Acrlaone alkaloid)
Allamanda cathartica	Apocynaceae	Allarnandin (Monoterpene)
Baccharis megapotamica	Asteraceae	Baccharin (Sesquiterpene)
Baileya multiradiata	Asteraceae	Pseudoguaianolide
Bersama abyssinica	Melianthaceae	Hellebrigenin acetate (Buladienullde)
Bouvardia temifolia	Rubiaceae	Bourvardin (Peptide)
Brucea antidysenterica	Simaroubaceae	Broceanlin (Simaroubolide)
Caesalpinia gilliesii	Fabaceae	Cesalin
Camptotheca acuminata	Nyssaceae	Camptothecin (Pyrraloquinolille alkaloid)
Catharanthus roseus	Apocynaceae	Vinblastine, Vincristine (Bis-indole alkaloid)
Cephalis acuminata	Rubiaceae	Emetine (Isoquinoline alkaloid)
C. ipeccacuanha	Rubiaceae	Emetine (Isoquinoline alkaloid)
Cephalotaxus harringtonia	Cephalotaxaceae	Harringtonine (Cephalotaxine alkaloid)
Cocculus sp.	Menispennaceae	Gocculine, Cocculidine (Bisclaurine alkaloid)
Colchicum autumnale	Liliaceae	Colchicine (alkaloid)
C. speciosum	Liliaceae	Colchicine (alkaloid)
Crocosmia crocosmiiflora	Iridaceae	Medicagenic acid (Saponin)
Crotalaria assamica	Leguminosae	Monocrotaline (Pyrrolizidine alkaloid)
C. spectabilis	Leguminosae	Monocrotaline (Pyrrolizidine alkaloid)
Croton macrostachys	Euphorbiaceae	Crotepoxide
C. tiglium	Euphorbiaceae	Phorbol derivatives (Terpenoid)
Cyclea peltata	Menisperrnaceae	Tetrandrine (Isoquinoline alkaloid)
Daphne mezereum	Thymelaeaccae	Mezerein (Diterpene)
Elephantopus elatus	Asteraceae	Elephantopin (Sesquiterpene)
E. mollis	Asteraceae	Molephantinin (Sesquiterpene)
Eupatorium hyssopifolium	Asteraceae	Eupahyssopin (Sesquiterpene)
Euphorbia escula	Euphorbiaceae	Ingenal dibenzoate
Fagara macrophylla	Rutaceae	Nilidine (Benzophenanthridine alkaloid)
F. zanthoxyloides	Rutaceae	Fagaronine (Benzophenanthridine alkaloid)
Gnidia lamprantha	Thymelaceaceae	Guidin (Diterpene)
Gossypium sp.	Malvaceae	Gossypol (Sesquiterpene dimer)
Helenium autumnale	Asteraceae	Helevalin (Sesquiterpene)
H. microcephaiam	Asteraceae	Microlenin (Sesquiterpene dimer)
Heliotropium indicum	Boraginaceae	Indicine-*N*-oxide (Pyrrolizidine alkaloid)
Holacantha emoryi	Simaroubaceae	Holacanthone (Sirnaroubalide)
Hymenoclea salsola	Asteraceae	Ambrosin (Sesquiterpene)
Ipomoea batatas	Convovulaceae	4-Ipomeanal (Monoterpene)
Jacaranda caucana	Bignoniaceae	Jacaranone (Quinone)
Jatropha gossypiifolia	Euphorbiaceae	Jatrophone (Diterpene)
Juniperus chinensis	Cupressaceae	Podophyllotoxin (Lignan)
Liatris chapmanii	Asteraceae	Liatrin (Sesquiterpene)
Linum album	Linaceae	Podophyllotoxin (Lignan)
Linum flavum	Linaceae	5-Methoxypodophyllotoxin (Lignan)
Mappia foetida	Olinaceae	Camptothecin (Pyrroloquinoline alkaloid)
Marah omganus	Cucurbitaceae	Cucurbitacin E
Maytenus buchananii	Celaslraceae	Maytansine (Ansa macrolide)

Table 7.1 Continued

Plant	Family	Active constituent(s)
M. ovatus	Celastraceae	Maytansine (Ansa macrolide)
M. serrata	Celastraceae	Maytansine (Ansa macrolide)
Montezuma speciasissima	Malvaceae	Gossypol (Sesquiterpene dimer)
Ochrosia elliptica	Apocynaceae	Ellipticine (Pyridocarbazole alkaloid)
O. maculata	Apocynaceae	9-Methoxyellipticine (Pyridocarbazole alkaloid)
O. moorei	Apocynaceae	Ellipticine (Pyridocarbazole alkaloid)
Parquetina nigrescens	Asclepiadaceae	Strophanthidin (Cardenolide)
Penstemon deutus	Scrophulariaceae	Penstimide (Monoterpene)
Phyllanthus acuminatus	Euphorbiaceae	Phyllathoside
Pierreodendron kerstingii	Sapotaceae	Glaucarbuinone (Simaroubolide)
Piper futokadzura	Piperaceae	Crotepoxide
Podtxatpus gracilori	Podocarpaceae	Podolide (Dilactone)
Podophyllum hexandrum	Podophyllaceae	Podophyllotoxin, Peltatin (Lignan)
P. peltatum	Podophyllaceae	Podophyllotoxin, (Lignan)
Putterlickia verrucosa	Celastraceae	Maytansine (Ansa macrolide)
Simarouba glauka	Simaroubaceae	Glaucarubinone (Simaroubolide)
Steganotaenia araliaceae	Umbelliferae	Liatrin (Sesquiterpene)
Stereospermum suaveolens	Bignoniaceae	Lapachol (Quinone)
Strophanthus sp.	Apocynaceae	Strophanthidin (Cardenolide)
Taxodium distichum	Taxodiaceae	Taxodione (Diterpene)
Taxus baccata	Taxaceae	Paclitaxel (Diterpene)
T. brevifolia	Taxaceae	Paclitaxel (Diterpene)
Thalictrum dasycatpum	Ranunculaceae	Thalicarpine (Isoquinoline alkaloid)
T. minus	Ranunculaceae	Thalicarpine (Isoquinoline alkaloid)
Tripterygium wilfordii	Celastraceae	Tripdiolide, Triptolide (Diterpene)
Tylophora asthmatica	Asclepiadaceae	Tylophorine
T. cerbiflora	Asclepiadaceae	Tylocrebine (Phenanthroidolizidine alkaloid)
Vernonia guineensis	Asteraceae	Vemolepin (Sequiterpene lactone)
V. hymenolepis	Asteraceae	Vemolepin (Sequiterpene lactone)
Withania somnifera	Solanaceae	Withaferin A, B (Withanolides)
Zaulzainia sp.	Rubiaceae	Zaluzonic C (Ridoid)
Zanthoxylum sp.	Rutaceae	Nitidine (Benzoohenathridine)

Table 7.2 Concentrations of antitumor compounds present in higher plants.

No.	Antitumor compound	Concn. [dry wt. %]
1	Baccharin	2.0×10^{-2}
2	Bruceantin	1.0×10^{-2}
3	Camptothecin	5.0×10^{-3}
4	Ellipticine	3.2×10^{-5}
5	Homoharringtonine	1.8×10^{-5}
6	Maytansine	2.0×10^{-5}
7	Podophyllotoxin	6.4×10^{-1}
8	Taxol	5.0×10^{-1}
9	Tripdiolide	1.0×10^{-3}
10	Vinblastine, Vincristine	5.0×10^{-3}

During the past decade, renewed interest in investigating plant-based products has led to the advent of several important anticancer substances. Most important are paclitaxel (taxol) from *Taxus brevifolia* L., podophyllotoxin from *Podophyllum peltatum* L., and camptothecin from *Camptotheca acuminata* Decne. Interestingly, these substances embrace some of the most exciting new chemotherapeutic agents currently available for use in clinical settings. The analogues of podophyllotoxin, taxol and camptothecin are illustrated in Figures 7.1, 7.2 and 7.3, respectively. The

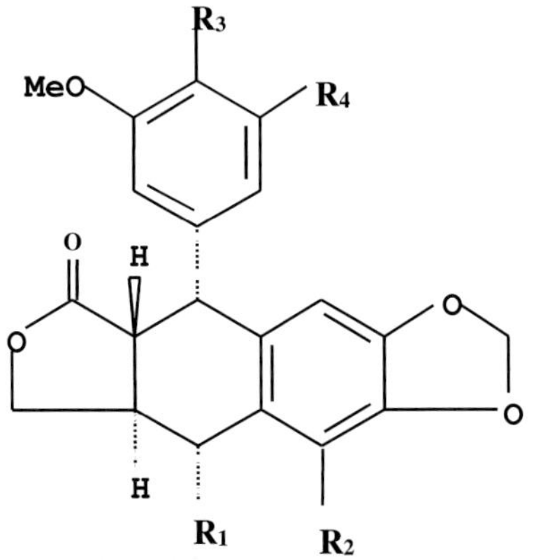

R_1	R_2	R_3	R_4	Other molecule	Name
- OH	H	- OCH$_3$	- OCH$_3$	--	Podophyllotoxin
- OH	- OCH$_3$	- OCH$_3$	- OCH$_3$	--	5-methoxypodophyllotoxin (5-MPT)
- OH	- OCH$_3$	- OCH$_3$	H	--	5'-demethoxy-5-MPT
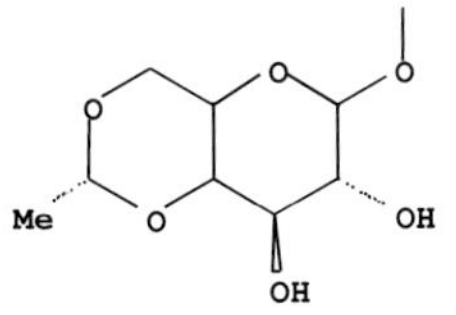	H	- OH	- OCH$_3$	--	Etoposide
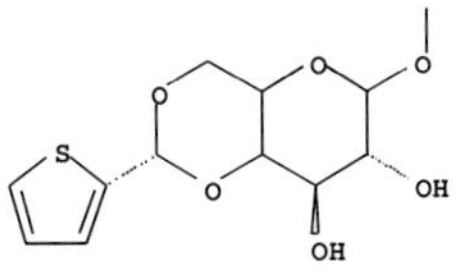	H	- OH	- OCH$_3$	--	Teniposide
H	- OH	- OH	- OCH$_3$	--	α-peltatin
H	- OH	- OCH$_3$	- OCH$_3$	--	β-peltatin
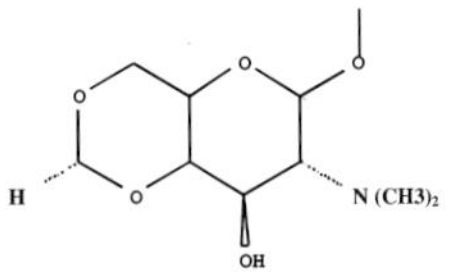	H	- OH	- OCH$_3$	.HCl. 2H$_2$O	NK611

Fig. 7.1 Podophyllotoxin and its analogues.

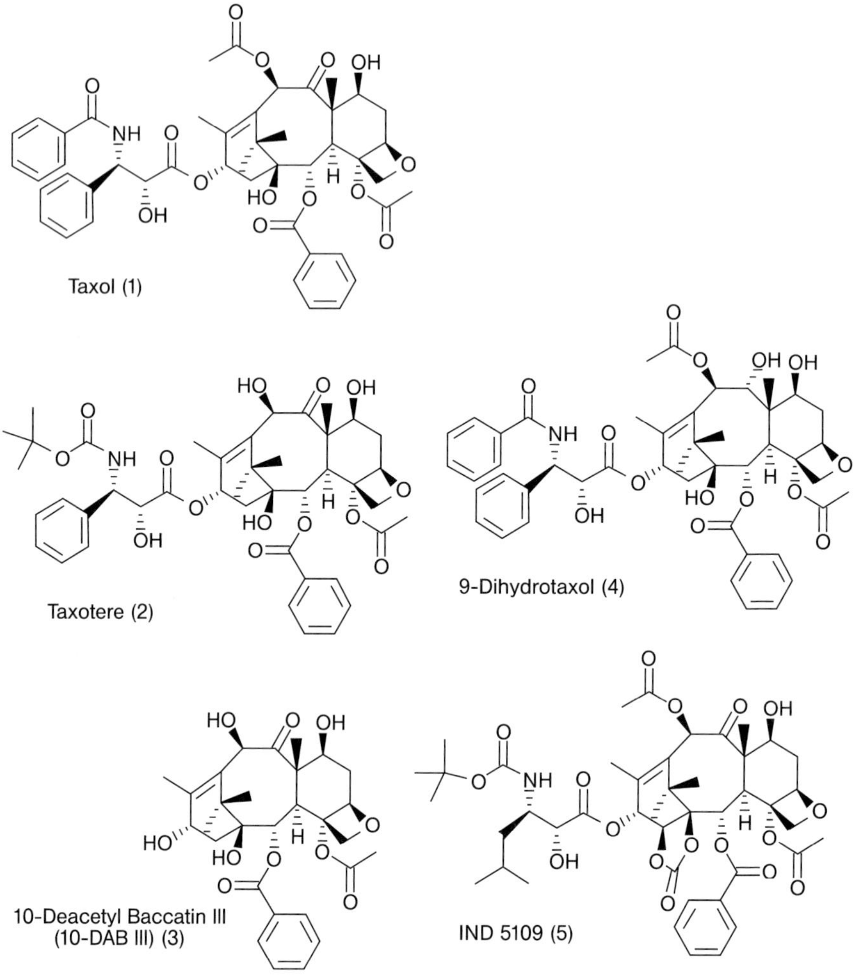

Fig. 7.2 Taxol and its analogues.

use of paclitaxel has been expanded to include a greater variety of cancers and more recently, Taxotere® has received approval by the FDA. Further, a podophyllotoxin analogue, teniposide (Vumon®), and a water-soluble camptothecin analogue, to-petocan hydrochloride (Hycamtin®), have been approved for human use during the past few years. Based on this impressive array of structures and activities, it is clear that plants and plant-derived drugs play a dominant role in contemporary cancer therapy [5]. The mode of action of some promising chemotherapeutics from natural origin and their derivatives is indicated in Table 7.3.

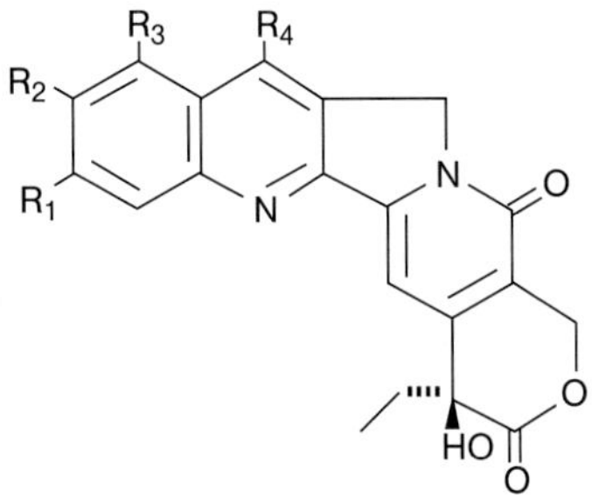

Fig. 7.3 Camptothecin and its analogues.

R$_1$	R$_2$	R$_3$	R$_4$	Analogue
H	H	H	H	Camptothecin
H	OH	CH$_2$N(CH$_3$)$_2$	H	Topetecan
H	H	NO$_2$	H	Rubitecan
H	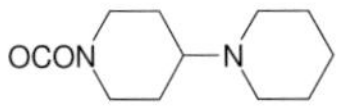 OCON⬡—N⬡	H	Ethyl	Irinotecan
F	Me	 [CH with NH$_2$ substituent]		Exatecan
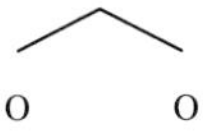 [O⌃O]		H	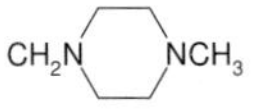 CH$_2$N⬡NCH$_3$	Lurtotecan

7.2
Production by Plant Cell Cultures

It is not an easy task to produce these compounds economically by extraction from intact plants and meet the ever-increasing demand. This may be due to very low concentrations of these active compounds in plants (Table 7.4), the slow growth rate of plants, complex accumulation patterns, and high susceptibility to geographical and environmental conditions. Other possible reasons are the non-availability of uniform and unadulterated quality plant material in quantities sufficient for in-

Table 7.3 Some chemotherapeutic products from natural sources, and their mode of action.

Source	Natural compounds or its derivatives	Mode of action	Cancer inhibited	Reference(s)
Podophyllum spp.	Podophyllotoxin (Natural)	As a mitotic spindle poison, binds the microtubule and causes mitotic arrest in metaphase	Lung	161
	Etoposide and teniposode (Semisynthetic derivatives)	Induce a premitotic blockage in the cell cycle, at two specific places, either in late S-phase or in early G-phase, by binding to and stabilizing the cleavable complex of DNA–topoisomerase II.	Lung, testicular, leukemias	161, 162
Taxus baccata	10-Deacetyl baccatin 111:Docetaxel (Semisynthetic derivatives)	Promotes tubulin assembly and inhibition of microtubule depolymerization; also acts as a mitotic spindle poison and induced mitotic block in proliferative cells	Breast, ovarian, nonsmall-cell lung, head and neck, colorectal melanoma	66, 163
Taxus brevifolia	Paclitaxel (Natural)	Promotes assembly of microtubules, stabilizes them against depolymerization, and inhibits cell replication; causes cell apoptosis.	Advanced breast, ovarian, adenocarcinoma, and other solid tumors	164–167
Camptotheca acuminata	10-hydroxy camptothecin, (Natural) Irinotecan (CPT-11), SN-38 (Semisynthetic derivatives)	Inhibits action of topoisomerase I, prevents religation of DNA strand, results in cell death	Liver, colorectal, head and neck cancer, leukemia	168–170

Table 7.4 Antitumor compounds from cell cultures versus intact plant.

Plant	Antitumor compound	Plant [% DW]	Cultured cells [% DW]	Reference(s)
Camptotheca acuminata	Camptothecin	5×10^{-3}	2.5×10^{-4}	160
Podophyllum spp.	Podophyllotoxin	6.4×10^{-1}	7.1×10^{-1}	8, 9
Taxus spp.	Taxol	5×10^{-1}	153 mg L^{-1}	25

DW, dry weight.

dustrial production and uneconomical chemical synthesis, particularly for large complex molecules. Therefore, biotechnological methods offer an excellent alternative for production of such compounds.

Some of the major limitations to meet the demand of podophyllotoxin are the inaccessible region from where the plant is obtained, its endangered status due to over-exploitation of natural resources, a long juvenile phase, poor fruit-setting ability, a long period of germination of the seeds, the continuing demand for the drug and very complicated and rather difficult chemical synthesis because of the presence of four chiral centers, a rigid trans lactone and an axially located 1-aryl substituent. The most difficult problems encountered in paclitaxel production are supply limitation arising from its very low concentration in the bark of the *T. brevifolia* (ca. 0.01% dry weight), and the very slow growth of the yew tree [6]. In order to isolate 1 kg of taxol, 10000 kg of dried bark from 3000 *T. brevifolia* trees must be extracted, whilst almost 2 g of taxol is needed to treat one patient. On the other hand, certain secondary metabolites such as camptothecin are accumulated only after a certain age or maturity of the plant. Hence, it is difficult to increase the area under plantation.

To overcome all these hurdles, the industry requires alternative methods of supply of uniform material throughout the year. Plant cell culture technology is undoubtedly one of the appropriate approaches to solve the above-described problems. However, before implementing this approach, the cost of the product and its demand should justify production by biotechnological means. For commercialization, one must consider the economic aspects and development of feasible bioprocess technologies and relevant reactor design and mode of operations for the production of these plant secondary metabolites. In this respect, batch, fed-batch and continuous cultivation of suspension culture are found to be the most economic methods [7]. The goal should therefore be to improve the productivity of cells in order to make the production of these compounds commercially viable. Commonly, no single growth-enhancing strategy can produce such a large increase, but in many cases the simultaneous application of multiple strategies results in synergistic interaction and thus an improvement of the yield. The biotechnological approaches used for production of anticancer drugs, podophyllotoxin, taxol, and camptothecin, will be addressed in the following sections.

7.3
Biotechnological Approaches for Production

7.3.1
Dedifferentiated Cultures

Today, the potential of plant cells to produce secondary metabolites in dedifferentiated cultures is used extensively to produce plant-derived drugs. Screening by selection of high-yielding cell lines is the most common method to enhance productivity.

Tissue cultures of *Podophyllum peltatum* for the production of podophyllotoxin were initiated from various explants such as rhizomes, roots and leaf segments of field-grown plants. The efficiency of various factors such as the source of explant, growth regulators and light conditions on callus growth and podophyllotoxin production were studied [8,9]. Callus was also induced from *P. hexandrum in-vitro*-grown seedlings [10]. When the callus derived from *P. hexandrum* was incubated in B5 medium containing 2,4-D, gibberellic acid and 6-benzylaminopurine, podophyllotoxin, 4'-demethyl-podophyllotoxin and podophyllotoxin-4-*O*-glucoside were produced, and the levels of podophyllotoxin and its derivatives were similar to those in the parent plant (0.3% dry weight) [11]. As the levels found in these cultures were quite low, the interest in podophyllotoxins was revived when root cultures of *Linum flavum* were found to contain high levels (1% dry weight) of 5-methylpodophyllotoxin and its glucosides [12]. Other podophyllotoxin-producing callus cultures are those from sterile leaves in *Juniperus chinensis*, needles of *Callitris drummondii*, and the stem and leaves of young seedlings of *Linum album* on different media [13–16]. Callus tissues and suspension culture cells of *Lilium album* also demonstrate podophyllotoxin production. One of the cell lines produced 0.3% podophyllotoxin of dried cells, together with small amounts of 5-methyl podophyllotoxin, lariciresinol and pinoresinol after three weeks of cultivation [14]. Cell suspension cultures of *Callitris drummondii* (conifer) also accumulated podophyllotoxin-β-D-glucoside. In the dark, the cells produced approximately 0.02% podophyllotoxin of the dry cell mass, and 85–90% of the lignans were of the β-D-glucoside form, while in light the yield of podophyllotoxin-β-D-glucoside increased to 0.11% [17].

Podophyllum hexandrum cell suspension was initiated under dark conditions on a rotary shaker at 26 °C using Gamborg's B5 salts supplemented with coconut milk, sucrose and naphthalene acetic acid (NAA) [10], on Gamborg's B5 salts supplemented with 2,4-D, gibberellin and benzyladenine [11], and on MS medium containing indole acetic acid (IAA) on a gyratory shaker at 20 °C in the dark in the authors' laboratory [18]. During culture growth, reduction in cell viability, biomass and product yield were found to be associated with browning of culture medium, clumping of cells, and a fall in medium pH. Supplementations of the medium with polyvinylpyrrollidone (PVP) stopped browning of the culture medium, as well as the fall in pH. Clumping of cells was reduced by the application of pectinase. *P. hexandrum* cells were found to be slow-growing and required 30 days to reach the maximum biomass of 8.3 g L^{-1}. Podophyllotoxin production in unoptimized medium was only 4.9 mg L^{-1} after 32 days [18]. The production of lignans by plant cell cultures at significant levels is listed in Table 7.5.

In the past, several investigators reported low paclitaxel levels in plant cell cultures [19–21], but today levels of 10–22 mg L^{-1} are common [22–24], significantly higher levels (i.e., 153 mg L^{-1} [25] and 110 mg L^{-1} [26]) have been reported by some research groups. Callus cultures and suspension cultures from *T. brevifolia* cv. Repandens, *T. cuspidata*, *T. media* cvs. Hicksii and Densiformis were established. Although some cell lines grew rapidly, with doubling times of 9–14 days, the levels of paclitaxel were too low for commercial production [20]. Twenty-seven different yew trees belonging to various genotypes and hybrids have been screened for their capacity to produce significantly high amounts of taxoids. *Taxus media* "Sargentii" proved capable of producing viable callus cultures from excised roots placed *in vitro*. In leaves and calluses, 0.069 and 0.032% paclitaxel contents were found, re-

Table 7.5 Production of lignans in various cell cultures.

(a) In callus culture

Species	Explant	Lignan content [% dry weight]	Culture strategy	Reference(s)
Callitris drummondii	Needle	0.02% (PT)	Callus	58
Juniperus chinensis	Leaf	0.005% (PT)	Callus	16
Linum album	Leaf and stem of young seedling	0.3% (PT)	Callus	14
Podophyllum peltatum	Rhizome	0.7% (PT)	Callus	9
Podophyllum hexandrum	Seedling	0.3% (PT)	Callus	11

(b) In shake flasks

Species	Culture time [days]	Lignan content [mg L^{-1}]	Culture Strategy	Reference(s)
Callitris drummondii	25	16.5 (PT)	Suspension	171
Linum album	11	18.0 (PT)	Suspension	172
L. flavum	14	5.4 (5-MPT)	Suspension	173
L. flavum	21	121.4 (5-MPT)	Root-like Suspension	37
L. nodiflorum	20	22.4 (PT)		
	12	76.6 (5-MPT)	Suspension	174
Podophyllum hexandrum	15	24.3 (PT)	Suspension	10
	21	48.8 (PT)	Suspension in optimized medium	47
P. peltatum	21	27.0 (PT)	Embryogenic Suspension	175

(c) In bioreactors

Species	Biomass [g L^{-1}]	Podophyllotoxin [mg L^{-1}]	Volumetric productivity [mg L$^{-1} \cdot$ d]	Culture strategy	Reference(s)
P. hexandrum	21.4	13.8	0.53	Batch	46
	48.0	43.2	0.72	Fed-batch	98
	53.0	48.8	0.81	Continuous	97

PT, podophyllotoxin; 5-MPT, 5-methoxy podophyllotoxin.

spectively; these were significantly higher than were previously reported for other genotypes [27]. The production of taxoids in various cell cultures of *Taxus* spp. on industrially interesting levels is provided in Table 7.6.

Table 7.6 Production of taxoids in various cell cultures.

(a) In shake flasks

Species	Culture time [days]	Taxoid content [mg L⁻¹]	Culture strategy	Reference
T. canadensis	12	117.0 (PcT)	Elicitation with methyl jasmonate	74
	14	506.0 (TT)		
T. media	14	110.3 (PcT)	Elicitation with methyl jasmonate	26
	14	145.3 (TT)		
T. chinensis	42	153.0 (PcT)	Medium optimization + elicitation + two-stage culture	25
	42	295.0 (TT)		
	35	67.0 (PcT)	Intermittent maltose feeding	49
	35	49.6 (PcT)	Silver nitrate elicitation	49
	35	885.9 (Tc)		
	35	82.4 (PcT)	Silver nitrate elicitation + temperature shift	49
	35	512.9 (Tc)		
	42	137.5 (PcT)	Silver nitrate elicitation + temperature shift	49
	23	274.4 (Tc)	Sucrose feeding	59
	15	527.0 (Tc)	Sucrose feeding + elicitation with methyl jasmonate	176

(b) In bioreactors

Species	Reactor	Taxoid content [mg L⁻¹]	Culture strategy	Reference
T. cuspidata	600-mL Wilson type	22.0 (PcT)	Batch study	22
	Balloon-type bubble reactor	74.0 (TT)	Media replacement	105
T. media	5-L stirred reactor	21.1 (PcT)	Two stage culture + Elicitation with methyl jasmonate + precursor	82
		56.0 (Bc)		
T. chinensis	1-L bubble column	229.0 (Tc)	Ethylene addition	107
	1-L airlift	336.0 (Tc)	Sucrose feeding + Elicitation with methyl jasmonate + ethylene addition	81
	1-L airlift	612.0 (Tc)	Sucrose feeding + Repeated elicitation with methyl jasmonate	48

PcT, paclitaxel; Tc, taxuyunnanine C; TT, total taxanes; Bc, Baccatin III.

Suspension cultures of *Nothapodytes foetida* were established, and the cell biomass was higher in the presence of NAA in comparison with 2,4-D. Culture medium supplemented with NAA and benzyl adenine (BA) attained 31.3 g L^{-1} dry weight during 20 days of cultivation in shake flasks. In the presence of NAA, maximum concentrations of camptothecin (0.035 mg mL^{-1}) and 9-methoxycamptothecin (0.026 mg mL^{-1}) were found in the medium. Alkaloid production was reduced in the presence of 2,4-D in the culture medium [28]. *Camptotheca acuminata* callus cultures were induced on MS medium containing 2,4-D, and kinetin and liquid cultures were developed in the presence of gibberellin, L-tryptophan, and conditioned medium, which yielded camptothecin at about 0.0025% on a dry weight basis [29]. When the cultures were grown on MS medium containing NAA, accumulation of camptothecin reached 0.998 mg L^{-1} [30]. Camptothecin and 10-hydroxy-camptothecin were also detected in the shoots and callus cultures of *C. acuminata*. Cultures maintained on MS medium supplemented with NAA contained 0.08% dry weight of 10-hydroxycamptothecin [31]. The culture strategies used to produce camptothecin and related compounds are listed in Table 7.7.

Table 7.7 Production of camptothecin in various cell cultures.

(a) In callus culture

Species	Alkaloid content [% dry weight]	Culture strategy	Reference
Camptotheca acuminata	0.0025% (CPT)	Gibberillin + Tryptophan + Conditioned medium	29
Camptotheca acuminata	0.08% (10 OH CPT)	MS + NAA	31

(b) In shake flasks

Species	Alkaloid content [mg L^{-1}]	Culture strategy	Reference
Nothapodytes foetida	0.035 (CPT) 0.026 (MCPT)	MS + NAA + BA	28

CPT, camptothecin; 10 OH CPT, 10-hydroxy camptothecin; MCPT, methoxy camptothecin.

7.3.2
Differentiated Cultures

Different plant parts such as roots, shoots and embryo have been cultured, and produce compounds similar to that from the whole plant. Embryogenic roots from a *Podophyllum peltatum* callus were induced in liquid MS medium supplemented with NAA, kinetin and casein hydrolyzate. The roots were then transferred to the medium without growth regulators, whereupon 1.6% of podophyllotoxin was detected in the dried tissues, which was sixfold higher than in the mother plant [32].

Strategies towards the *in-vitro* propagation and cultivation of *P. hexandrum* have been established by growing excised embryo on basal medium [33,34]. Rapidly producing and high-yielding *P. peltatum* plants were developed from *in-vitro* propagation protocol by growing rhizome tips on the basal MS medium containing sucrose, supplemented with benzyladenine and activated carbon. The podophyllotoxin contents of *in-vitro*-rooted bud and plantlet cultures were similar to the contents found in the wild [35].

Improvements in the productivity of podophyllotoxin required some morphological differentiation, for example root formation [37]. A threefold increase in podophyllotoxin content in comparison to controls was obtained in transformed calli of *P. hexandrum* developed by transformation of embryo using different strains of *Agrobacterium rhizogenes* viz. A4, 15834, and K599 [38]. Hairy root cultures of *Linum flavum* were also reported to produce up to 1% of 5-methoxypodophyllotoxin and its glucoside derivative on a dry cell weight basis. 5-Methoxypodophyllotoxin, isolated from the root cultures of *L. flavum* grown on vitamin-free MS basal medium supplemented with sucrose, appeared to have about the same cytotoxic potency as podophyllotoxins and its semisynthetic derivatives [36]. The levels of podophyllotoxins varied from 0.05% to 0.3% dry weight, depending on the culture conditions and the tendency to differentiate. Thus, hairy root cultures of *Linum flavum*, producing 1.5–3.5% podophyllotoxins in dry weight, seem to be most suitable for future research efforts [39].

Linum flavum hairy roots were also initiated from leaf discs with a success rate of approximately 50% using *A. rhizogenes* strains LBA9402 and TR105, while coniferin accumulation was significantly greater in LS medium containing 2,4-D and NAA than in MS medium without growth regulators. In LS medium, biomass and coniferin concentrations varied substantially by a factor of about 34 across eight different hairy root lines in shake-flask studies. The maximum coniferin concentration measured in the biomass was 58 mg g^{-1} dry weight, and little or no coniferin was released into the medium during culture [40].

The genetic transformation of *Taxus brevifolia* and *T. baccata* was also reported using two strains of *Agrobacterium tumefaciens* (Bo542 and C58) [41]. Very few reports exist of *Agrobacterium*-mediated genetic transformation of *Taxus*, as the botany hierarchy of the *Taxus* is classified within group Gymnosperm of the Taxaceae family and *Agrobacterium* strains are less or no virulence towards gymnosperms.

The embryos developed unorganized callus tissues in medium containing BA + 2,4-D. BA + NAA medium promoted multiple shoot development, while kinetin (Kn) + NAA and Kn + IAA in the medium favored plantlet formation. The differentiated plantlet cultures showed slightly higher amounts of 9-methoxycamptothecin (0.0007% dry weight) than the undifferentiated callus cultures (0.0001% dry weight) [43]. The hairy roots of *Ophiorrhiza pumila*, producing camptothecin and its related alkaloids (i.e., (3S)-pumiloside, (3S)- and (3R)- deoxypumilosides and strictosamide) were also reported [44].

7.3.3
Optimization of Culture Media Composition and Culture Conditions

Medium and culture condition optimization can increase the productivity in suspension culture by a factor of 20 to 30, and is very important in order to obtain an efficient system for the production of high levels of secondary metabolites from plant cells.

Podophyllotoxin accumulation was strongly affected by light, with red light stimulating its production in *P. pelatum* cell cultures [9]. Illumination stimulated the endogenous production of podophyllotoxin-β-D-glucoside, to a concentration of 0.11% (dry weight), in *L. flavum* cultures [44]. Dark-grown suspension cultures of *P. hexandrum* accumulated 0.1% podophyllotoxin, which was three- to fourfold that in light-grown cultures [44]. Similar results were also observed in the authors' laboratory [45]. *P. hexandrum* cells growing in shake flasks were found to be sensitive to hydrodynamic stresses generated by changing rotational speeds. A rotational speed of 150 rpm showed more than 80% viability of *P. hexandrum* cells. A medium pH of 6.0 was favorable for high biomass production and podophyllotoxin accumulation in *P. hexandrum* cell cultures [46]. MS medium with 60 mM nitrogen having an ammonium to nitrate salt ratio of 1:2, 60 g L^{-1} glucose and 1.25 mM phosphate were found to be optimum for podophyllotoxin production [47]. Statistical optimization methodology, such as Plackett–Burman design and Response Surface Methodology, have been employed to optimize the media and culture conditions for growth and podophyllotoxin production in *P. hexandrum* cultures. The optimized values of important nutritional parameters were found to be medium pH 6.0, 1.25 mg L^{-1} IAA concentration, 72 g L^{-1} glucose concentration, and an inoculum level of 8 g L^{-1}. When *P. hexandrum* was cultivated under statistically optimized culture conditions, a maximum of 20.2 g L^{-1} (dry weight) of biomass and 48.8 mg L^{-1} podophyllotoxin was obtained [46].

The addition of 50% (v/v) conditioned medium on day 0 resulted in a higher average growth rate compared to controls. In the case of bioreactor cultivation, taxuyunnanine (Tc) production and productivity of 43.0±17.6 mg L^{-1} and 1.0± 0.8 mg L^{-1} day^{-1} in the control culture increased to 78.9±20.6 mg L^{-1} and 4.5± 1.2 mg L^{-1} day^{-1}, respectively, when conditioned medium was added [48]. The effects of temperature shift on cell growth and paclitaxel production in suspension culture of *T. chinensis* was studied. Cell growth was optimum at 24 °C, while paclitaxel synthesis reached a maximum at 29 °C [49]. The most effective gas mixture composition in terms of paclitaxel production was 10% (v/v) oxygen, 0.5% (v/v) carbon dioxide, and 5 mg L^{-1} ethylene [50,51]. The manipulation of the amount and source of sugar in cell cultures was studied as a factor for enhanced growth and secondary metabolite production [52,53]. Elevated sucrose levels were favorable in some cultures [52,54], and the addition of fructose promoted taxol production in *Taxus* spp. cell cultures [48,55].

To date, no optimization studies have been reported on the production of camptothecin in plant cell cultures.

7.3.4
Immobilization

Although immobilization was originally proposed due to its bioprocessing advantages such as the re-use of expensive biomass and easier downstream processing, experimental evidence indicates that it may also have an impact on cell physiology and the production of secondary metabolites. The immobilization of plant cells often results in increased secondary metabolite production, most likely due to improved cell–cell adhesion and eventual aggregation.

Immobilization of *P. hexandrum* cells using calcium alginate in combination with precursor feeding of L-phenylalanine and L-tyrosine was attempted, but no improvement was obtained in the accumulation of podophyllotoxin, possibly because the stress environment [56].

Paclitaxel in the callus of *T. cuspidata* was detected at a level of $0.02 \pm 0.005\%$ (dry weight) after two months in culture. Suspension cultures of *T. cuspidata* were also established from the callus cultures and subsequently immobilized onto glass-fiber mats. The cells were maintained as immobilized cultures for six months, during which time the level of paclitaxel within the cells reached $0.012 \pm 0.007\%$ of dry weight [57].

7.3.5
Feeding of Precursors

The addition of precursors to the media in order to direct metabolic flux towards enhanced production of the desired products represents an interesting approach to exploit the biosynthetic potential of the enzymes present in plant cell cultures.

The addition of coniferyl alcohol complexed with β-cyclodextrin to *P. hexandrum* cell suspension cultures increased the concentration of podophyllotoxin fourfold, to 0.013% on a dry weight basis. Noncomplexed coniferyl alcohol, suspended in the medium, also enhanced podophyllotoxin production, albeit to a lower degree [17]. Coniferin, the β-D-glucoside of coniferyl alcohol, was found to be a more potent precursor in terms of yield of the anticancer compound (0.055%), but is not available commercially [17]. Various phenylpropanoid precursors (phenylalanine, tyrosine, cinnamic acid, caffeic acid, coumaric acid, ferrulic acid, coniferyl alcohol, coniferin, etc.) were utilized in cell cultivation for the improvement of podophyllotoxin levels in *P. hexandrum* cell cultures. Of these, only coniferin at a concentration of 2.1 mM was able significantly to increase podophyllotoxin accumulation on the tenth day of cultivation, by a factor of 12.8. However, most of the coniferin was transformed into unknown products [17,56]. Application of the amino acid, L-phenylalanine, to cell cultures of *Linum flavum* resulted in a three- to fivefold increase in levels of 6-methoxypodophyllotoxin [58]. Co-cultured hairy roots of *Linum flavum* and *Podophyllum hexandrum* suspensions resulted in increased podophyllotoxin concentrations by 240% in shake flasks, and by 72% in a dual bioreactor, as compared to *P. hexandrum* suspension cultured alone. The coniferin provided by

L. flavum hairy roots acted as a precursor for the production of podophyllotoxin by *P. hexandrum* suspension cultures [40].

Improved paclitaxel yields in both callus and cell suspension cultures of *T. cuspidata* were observed following the feeding of phenylalanine and other potential paclitaxel side-chain precursors such as benzoic acid and *N*-benzoylglycine. This might be related to the possible roles of phenylalanine and other compounds in the biosynthesis of the *N*-benzoylphenylisoserine paclitaxel side chain [19]. In suspension cultures of *T. chinensis*, a combination of an initial low sucrose concentration (20 g L^{-1}) and fed-batch mode improved both cell growth and taxane production [59]. A high level of paclitaxel production (26 mg L^{-1}) was achieved through intermittent feeding of 3%, 1%, and 2% sucrose on days 0, 7, and 21, respectively, while intermittent feeding of 1% and 2% maltose to the *T. chinesis* cultures on days 7 and 21 increased paclitaxel production to 67 mg L^{-1} [49]. A combination of sucrose feeding and dissolved O$_2$ tension control might represent a means of maximizing paclitaxel production [60].

Precursors such as phenylalanine, benzoic acid, serine and benzoylglycine were reported to increase the accumulation of paclitaxel in callus cultures of *T. cuspidata* [19,61,62]. In one study of *in-vitro* production of radiolabeled taxol by pacific yew, phenylalanine and leucine were shown to be the best precursors among the various compounds tested [63]. Feeding of acetate also enhanced the formation of taxol-like metabolites [23]. The addition of precursors such as sodium benzoate, hippuric acid, leucine and phenylalanine to cell cultures of *T. wallachiana* significantly improved the production of paclitaxel, baccatin, and 10-deacetyl baccatin [64].

Precursor studies using tryptamine and loganin (combined), secologanin and strictosidine were performed in *C. accuminata* cultures; feeding with strictosidine showed the precursor to be easily biotransformed by two enzymes (i.e., a hydroxylase and a dehydrogenase) to hydroxystrictosidine and didehydrostrictosidine, but camptothecin was never detected [65].

7.3.6
Elicitors

Elicitation is the induction of secondary metabolite production by molecules or treatments known as "elicitors". Elicitation is used to induce the expression of genes often associated with the enzymes responsible for synthesis of secondary metabolites by mimicking the pathogen defense or wound response in plants.

Attempts have been made to increase the accumulation of aryltetralin lignans in *Linum* spp. cultures by inducing a hypersensitive defense reaction with a variety of elicitors. However, nigeran, *Phytophthora megasperma* cell wall fractions, methyl jasmonate, hydrogen peroxide and salicylic acid were not found significantly to increase lignan accumulation in *Linum* spp. [66]. The addition of methyl jasmonate to cell cultures of *Forsythia intermedia* resulted in three- and sevenfold accumulations of the lignans, pinoresinol and matairesinol, predominantly as glucosides [67]. The effect of adding methyl jasmonate to various cell lines in suspension culture of *Linum album* has also been studied, with a twofold increase in podophyllotoxin (7.69 ± 1.45 mg g^{-1} dry weight) and 6-methoxypodophyllotoxin (1.11 ± 0.09 mg g^{-1} dry weight) being achieved as maximum in one of the cell lines [68]. The effect of yeast extract and abiotic elicitors (Ag^{2+}, Pb^{2+} and Cd^{2+}) was also studied on podophyllotoxin production in *L. album* suspension cultures, though only Ag^{2+} at 1 mM concentration was found to enhance production up to 0.24 mg g^{-1} cell dry weight [69].

It has been found that oxidative stress caused by treatment with a fungal elicitor leads to cell apoptosis and low paclitaxel production [70–72]. Methyl jasmonate treatment led to large improvements in both cost and efficiency over the use of undefined fungal extracts; 100 pM methyl jasmonate increased the paclitaxel level from 28 to 110 mg L^{-1} in a *T. media* cell line, and from 0.4 to 48 mg L^{-1} in *T. baccata* [26]. Other groups have reported similar effects of methyl jasmonate in cell cultures of *T. cuspidata* [73], *T. canadensis* and *T. cuspidata* [74] and *T. media* [75]. Significant enhancement of taxoid production by repeated elicitation using methyl jasmonate in bioreactor cell cultures of *T. chinensis* has also been reported [48]. Abiotic elicitors – especially the salts of some heavy metals such as lanthanum, cerium, and silver – were also effective in inducing paclitaxel production in plant cell cultures, generally resulting in several-fold increases in taxoid production [76,77].

The addition of cell extracts and culture filtrates of *Penicillium minioluteum*, *Botrytis cinerea*, *Verticillium dahliae*, and *Gilocladium deliqucescens* on the day 10 after transferring *Taxus* sp. cell suspensions into an induction medium improved taxoid production [78]. Enhanced paclitaxel production in a *T. chinensis* cell culture by a combination of fungal elicitation and medium renewal was also reported [79]. *T. chinensis* culture treated with 50 mg L^{-1} of a fungal elicitor with 50 mg L^{-1} salicylic acid not only gained more biomass than was the case with treatment by the elicitor alone, but also achieved higher paclitaxel production [80]. A higher level of paclitaxel (3.4 mg L^{-1}) was obtained when ethylene was added at a concentration of 5 ppm in the presence of 10 pM methyl jasmonate compared with treatment solely by 10 pM methyl jasmonate in suspension cultures of *T. cuspidata* [50]. In *T. chinensis* cells, conditioned medium addition combined with methyl jasmonate (100 pM) elicitation enhanced Taxuyunnanine C productivity in bioreactors by about 30% (from 9.2 ± 0.9 to 11.8 ± 0.9 mg L^{-1} day^{-1}) compared with methyl jasmonate elicitation alone [48]. Furthermore, it was shown that Taxuyunnanine C productivity could be increased by combining elicitation, sucrose feeding, and ethylene incorporation in bioreactors [81].

Taxus suspension cultures were elicited with cell extracts and culture filtrates of four different fungi and arachidonic acid [78]. Interestingly, a preferential increase in paclitaxel over other taxanes was achieved by arachidonic acid [78] and by the addition of methyl jasmonate [26]. Membrane-lipid peroxidation caused by fungal elicitor (F5), prepared from fungus isolated from the inner bark of *T. chinensis*, was decreased by the addition of salicylic acid, even if the latter also induced cell membrane-lipid peroxidation. F5 + salicylic acid also improved the activity of glucose-6-phosphate dehydrogenase compared to single F5 treatment, and achieved the greatest taxol production of 11.5 mg L^{-1}, this being 1.5-, 2.3- and 7.5-fold higher than that of F5, salicylic acid and the control [80]. The *Fusarium oxysporum* select-ed from seven species was found to be the most effective inducer of taxane accu-mulation. Suspension-cultured *Taxus chinesis* var. *mairei* Y901-L responded to crude elicitors from the fungus *Fusarium oxysporum* by influencing the general phenylpropanoid pathway and incorporating taxol synthesis. The maximum taxol concentration was eightfold that of controls, and increased amounts of phenolics were observed in the culture medium [71]. A two-stage culture for cell suspension was carried out with addition of methyl jasmonate (220 µg g^{-1} fresh weight) as elic-itor in combination with mevalonate (0.38 mM) and N-benzoylglycine (0.2 mM) as precursors. An approximately tenfold enhancement in taxol (21.12 mg L^{-1}) and 20-fold enhancement in baccatin III (56.03 mg L^{-1}) was found in suspension cultures of *Taxus media* [82].

Cobalt chloride and silver nitrate have also been reported as elicitors, either alone or in combination, for the production of taxol in suspension cultures of *Tax-us* spp. [71,79,83]. Synergistic enhancement of taxol content was also achieved by a mixture of ammonium citrate with salicylic acid [72]. Significant improvements in taxol content were also reported after the use of various elicitors such as methyl jas-monate, fungal elicitor and salicylic acid alone, or in a combination of two [26,71,84]. Statistical optimization of various precursors and elicitors for paclitaxel production using Central Composite Design was also reported. A maximum of 54 mg L^{-1} paclitaxel was produced in *T. chinesis* cell suspension cultures by adding 10 mg L^{-1} silver nitrate, 6 mg L^{-1} abscisic acid, 23 mg L^{-1} chitosan, 15 mg L^{-1} phen-ylalanine, 31 mg L^{-1} methyl jasmonate, 30 mg L^{-1} sodium benzoate, and 30 mg L^{-1} glycine on day 12, together with a feeding solution containing 20 g L^{-1} sucrose on day 16; this was twofold higher than paclitaxel production without optimization [85]. A combined strategy of adding various biotic and abiotic elicitors during dif-ferent stages of a two-stage cell suspension culture of *T. baccata* led to a 16-fold in-crease in taxol production (39.5 mg L^{-1}) than untreated cultures in B5 medium [86].

7.3.7
In-Situ Product Removal Strategies

Strategies such as permeabilization using solvents and the addition of adsorbants to cell cultures have been used to increase secondary metabolite production. The advantages of using these strategies include the stimulation of the secondary me-tabolite biosynthesis and the easy separation of products.

A combination of 0.5% isopropanol and coniferin led to an accumulation of to-tal extracellular podophyllotoxin of 12 mg L^{-1} due to partial release of intracellular product. A concentration of isopropanol exceeding 0.5% led to a reduction in dry cell weight of *P. hexandrum*, indicating a destructive effect of isopropanol on the cells [56].

The effects of adsorbent addition on the production of taxol and taxane com-pounds were investigated in *Taxus cuspidata* cell culture. As a suitable adsorbent, the non-ionic exchange resin XAD-4 (among several adsorbents tested) showed a maximum adsorptive capacity for taxol, due to hydrophobic interaction between them. The addition of XAD resin to the culture medium on day 16 after subculture enhanced taxol biosynthesis by 40–70% [87].

7.3.8
Biotransformation Studies

Plant enzymes are able to catalyze both regiospecific and stereospecific reactions. Thus, freely suspended and immobilized plant cells, as well as enzyme preparations, can be used for the production of pharmaceuticals by biotransformation, as such or in combination with chemical synthases.

The microbial transformation of deoxypodophyllotoxin to epipodophyllotoxin (a parent compound of etoposide) by *Penicillium* sp. has been reported whereas, in contrast, *Aspergillus niger* sp. transformed deoxypodophyllotoxin to podophyllotoxin [88]. The incubation of dibenzylbutanolides with cell-free extracts of *C. roseus* yielded an enzyme-catalyzed oxidative coupling of these compounds to picropodophyllotoxin analogues [89]. Four cyclodextrins – β-cyclodextrin, γ-cyclodextrin, dimethyl-β-cyclodextrin and hydroxypropyl-β-cyclodextrin – were investigated for the glucosylation of podophyllotoxin. A maximal bioconversion rate of 0.51 mmol L^{-1} day^{-1} was achieved with dimethyl-β-cyclodextrin complex at a final concentration of 1.35 mM in *L. flavum* cell suspension cultures [90]. A semi-continuous process for the biotransformation of butanolide to 4'-demethylepipodophyllotoxin at the multigram level by *P. peltatum* cell suspension cultures was also reported [91]. A synthetic dibenzylbutanolide was shown to be biotransformed into complex podophyllotoxin analogues by shoot cultures of *Hallophyllum patavinum* [92]. At low concentration (0.1 mM), deoxypodophyllotoxin was converted to 6-methyl podophyllotoxin-7-*O*-glycoside, 6-methyl podophyllotoxin, and traces of α-peltatin and podophyllotoxin in *L. album* cultures [93].

The bioconversion of taxol/cephalomannine by *Streptomyces* sp. MA 7065 resulted in hydroxylation on the 10-acetyl methyl group with 60% yield, and on the benzene ring at the *para* position of the phenylisoserine side chain with 10% yield [94]. The incubation of *Eucalyptus perriniana* cells in medium containing paclitaxel led to the isolation of three taxoid derivatives, which were subsequently identified by ^{1}H NMR and mass spectrometry as baccatin III, 10-deacetyl baccatin III, and 2-benzoyltaxol [95]. Similar results were also reported with *E. citridora* cell cultures [96].

7.3.9
Bioreactor and Scale-Up Studies

The environment in which plant cells grow usually changes when cultures are scaled-up from shake flasks to bioreactors, and this may result in reduced productivity. With the ultimate aim of implementing an industrial-scale process, the behavior of cell cultures in bioreactors has received much attention.

Submerged batch and fed-batch cultivation of *Podophyllum hexandrum* cells for podophyllotoxin production studied in a 3-L stirred-tank bioreactor. A 36% increase in volumetric productivity of podophyllotoxin was achieved in fed-batch cultivation of *P. hexandrum* cells over batch cultivation in a stirred-tank bioreactor [46, 97, 98]. Continuous cultivation of *P. hexandrum* with cell retention was also carried out in a 3-L bioreactor equipped with a spin filter mounted on the agitator shaft of the bioreactor. The result was an accumulation of 53 g L^{-1} biomass and 48.8 mg L^{-1} podophyllotoxin, with volumetric productivity of 0.8 mg L^{-1} day^{-1}) [97]. A maximal product yield of podophyllotoxin, up to 0.2% dry weight, was achieved when *Linum album* cell suspensions were cultured in a 20-L bioreactor [99].

Taxus has been cultivated successfully for paclitaxel production in a pneumatically mixed stirred tank [100], and in Wilson-type bioreactors [22]. A continuous production system for paclitaxel production with a mesh-net cell separator was developed that increased productivity by a factor of ten compared to batch operation [101]. Low-temperature treatment not only maintained a stable high taxol production but also further enhanced taxol production compared to elicited asynchronous cultures. This resulted in a highest taxol production of 27 mg L^{-1}, which was approximately two- and 11-fold higher than in asynchronous cultures and in elicited asynchronous cultures, respectively. Taxol production in a stirred bioreactor proved to be less than that in shake flasks [70].

Cell cultures of *T. chinensis*, when cultivated in a novel low-shear centrifugal impeller bioreactor (CIB), had a shorter lag phase and showed less cell adhesion to the reactor wall compared to a conventional cell-lift bioreactor [102, 103]. Pneumatically agitated bubble columns and airlift reactors have been more widely used for taxoid production than stirred-tank reactors [100]. Suspension cultures of *Taxus baccata* var. *fastigiata* and *Taxus wallichiana* were reported to grow in a 20-L airlift bioreactor operating in batch mode [104]. In cell cultures of *T. cuspidata*, a balloon-type bubble reactor was claimed to be more efficient in promoting cell growth than a bubble-column reactor (BCR), a BCR with a split-plate internal loop, a BCR with a concentric draught-tube internal loop, a BCR with a fluidized bed, and two different models of stirred-tank reactors. The cell growth pattern in pilot-scale balloon-type bubble reactors (100–500 L) was the same as that in smaller bioreactors (20 L) [105]. However, this study seemed not to have noted whether the inner structure of such reactors could affect reactor performance [106]. When cell cultures of *T. chinensis* were transferred from shake flasks to bioreactors, taxol production was greatly reduced [107]. However, the incorporation of 18 ppm ethylene into the inlet air of a BCR caused the taxuyunnanine C content and volumetric production in the reactor to recover to maximum values of 13.28 mg g^{-1} dry weight and 163.7 mg L^{-1}, respectively, which were very similar to values obtained in shake flasks. Using a low-shear CIB, the quantitative effects of mixing time on suspension cultures of *T. chinensis* were demonstrated, and the favorable effect of more rapid mixing on taxol production was confirmed [108].

To date, no studies on camptothecin production by plant cell cultures in bioreactors have been reported; hence, this represents an important area for future research.

7.3.10
Biosynthetic Pathway Mapping and Metabolic Engineering

In order to develop an improved biological process, it is desirable to understand the biosynthetic pathway to produce the required product, the enzymes catalyzing those sequences of reactions (especially the rate-limiting steps), and the genes encoding these enzymes. In this respect, a molecular understanding for complete knowledge of the biosynthesis of secondary metabolites is necessary. Multipoint metabolic engineering by overexpression of the enzymes required for synthesis of the desired product and/or down-regulation of the genes leading to undesired or side products would greatly help in improving the production of these chemotherapeutics.

In the very early biosynthetic studies, radiolabeled precursors were fed to whole plants, but more recently plant cell cultures fed with stable isotope-labeled precursors have been used as the preferred model system. This allows analyses to be performed of labeled metabolites using advanced methods such as nuclear magnetic resonance (NMR) and mass spectrometry. The final step in the analysis of biosynthetic pathways is the identification of enzymes catalyzing individual bioconversions in cell-free systems, and of the genes coding for these enzymes.

7.3.10.1   Biosynthesis of Podophyllotoxin
The biosynthesis of podophyllotoxin in *Linum album* is not yet known completely, but is thought to be similar to that in *Forsythia intermedia*, *Podophyllum peltatum*, and *Linum flavum* [109]. The initial step of this biosynthesis is the dirigent protein-mediated formation of (+)-pinoresinol from two molecules of coniferyl alcohol. (+)-Pinoresinol serves as a substrate for an NADPH-dependent pinoresinol-lariciresinol reductase, which sequentially converts (+)-pinoresinol into (+)-lariciresinol and (–)-secoisolariciresinol. The latter is dehydrogenated into (–)-matairesinol by NAD-dependent secoisolariciresinol dehydrogenase. The subsequent pathway from (–)-matairesinol to either 6-methoxypodophyllotoxin (*L. flavum*) or podophyllotoxin (*L. album*) is still under investigation. The detailed biosynthetic pathway for podophyllotoxin production is illustrated in Figure 7.4.

Extensive studies have been carried out on the isolation and overexpression of genes involved in the biosynthesis of podophyllotoxin, but the alternative pathways for different metabolic fates of lignans in complex biosynthesis are still unknown. The important enzymes involved in biosynthesis of podophyllotoxin and their functions are listed in Table 7.8.

A 1207-bp cDNA of cinnamoyl-CoA reductase (CCR) encoding a polypeptide of 344 amino acids with a predicted molecular mass of 37.4 kDa was isolated from a ryegrass (*Lolium perenne*) stem cDNA library [110]. A cDNA encoding CCR was isolated from a cDNA library of *Linum album* [111], while a PCR screening method was used to obtain laccase-gene-specific sequences from the white-rot fungus *Trametes sanguinea MU-2*. The *kc1* cDNA was inserted into yeast vectors for heterologous expression by *Saccharomyces cerevisiae* and *Pichia pastoris* [112]. The molecular cloning and characterization of a *Eucalyptus globulus* genomic fragment encoding cinnamyl alcohol dehydrogenase (CAD) and isolation of the corresponding full-length cDNA from young stem material has already been reported [113]. Two cDNA full-length clones probably encoding the phenylalanine ammonia lyase (PAL) were isolated from a cDNA library of *Linum album* [114]. The enzymes PAL,

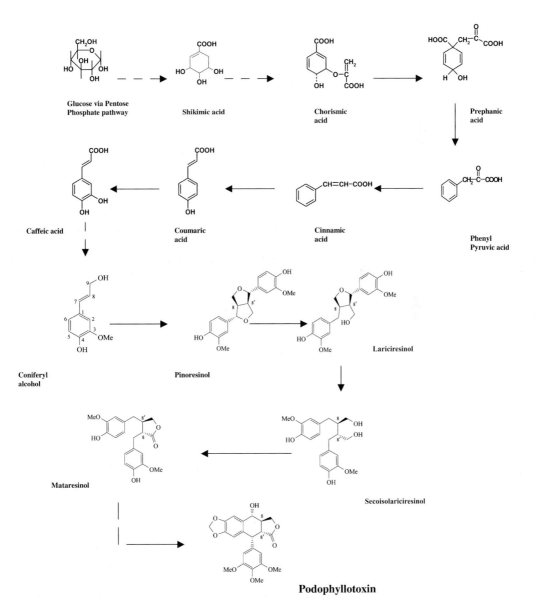

Fig. 7.4 Biosynthesis of podophyllotoxin (solid lines indicate single-step reactions; broken lines indicate multistep reactions).

Table 7.8 Enzymes involved (and function) in the biosynthesis of podophyllotoxin.

Serial no.	Enzyme involved	Function	Reference(s)
1	3-deoxy-D-arabino-heptulosonate 7-phosphate synthase (DAHPS)	Key point of regulation of the shikimate pathway, involved in DAHP synthesis	177
2	Chorismate synthase (CS)	Elimination of phosphate from EPSP to form chorismate, and introduces the second double bond towards aromatization	177
3	Phenylalanine ammonia lyase (PAL)	Connects the primary metabolism with the secondary metabolism by catalyzing the entry point reaction from phenylalanine to *trans*-cinnamic acid.	178
4	4-Hydroxycinnamate:CoA ligase	Attachment of CoA to hydroxycinnamic acid	179
5	Cinnamoyl-CoA reductase (CCR)	Conversion of cinnamoyl-CoAs into their corresponding cinnamaldehydes	180
6	*S*-Adenosyl-L-methionine: caffeic acid methyl transferase	Transfer of a methyl group from *S*-adenosyl-L-methionine to the 3-hydroxyl group of caffeic acid to form ferrulic acid; also catalyzes the methylation of 5-hydroxy ferrulic acid to sinnapic acid	117
7	Cinnamoyl alcohol dehydrogenase (CAD)	Reducing the cinnamyl aldehydes (*para*-coumaryl, coniferyl and sinapyl aldehydes) to the corresponding alcohols in the presence of NADPH	113
8	Caffeic acid 3-*O*-methyltransferase (COMT)	Methylation of caffeoylaldehyde to sinapyl alcohol formation	116
9	Caffeoyl CoA 3-*O*-methyltransferase (CCoAOMT)	3-*O*-methylation of caffeoyl CoA	116
10	Pinoresinol synthase	Stereo-specific phenoxy free-radical coupling reaction for pinoresinol synthesis	109
11	Pinoresinol-lariciresinol reductase	NADPH-dependent enzyme which sequentially converts (+)-pinoresinol into (+)-lariciresinol and (−)-secoisolariciresinol	116
12	Secoisolariciresinol dehydrogenase	Dehydrogenation of secoisolariciresinol into (−)-matairesinol via intermediate lactol	116
13	Cytochrome P450-dependent monooxygenase	Hydroxylation of deoxypodophyllotoxin (DOP) at the 5 position	179
14	Deoxypodophyllotoxin 7-hydroxylase	Conversion of deoxypodophyllotoxin to podophyllotoxin	181
15	Deoxypodophyllotoxin 6-hydroxylase	Conversion of deoxypodophyllotoxin to β-peltatin	82, 183
16	SAM: β-peltatin 6-*O* methyl transferase	Transfers a methyl group from *S*-adenosyl-L-methionine to the only free OH-group of β-peltatin in position 6 thus forming β-peltatin-A methylether	119
17	β-peltatin-A methylether 7-hydroxylase	Hydroxylation of β-peltatin-A methylether in position 7 to 6-methoxypodophyllotoxin	183
18	Laccases	Dehydrogenative polymerization of lignin	121

4-hydroxycinnamate: CoA ligase (4CL), cinnamoyl-CoA: NADP oxidoreductase and CAD were isolated from the cultures of *Linum album* [115]. Two different cDNAs with high homologies to known caffeic acid 3-*O*-methytransferases (COMTs) were isolated, although a cDNA with homology to caffeoyl CoA 3-*O*-methyltransferase (CCoAOMT) has not yet been found [116]. A number of c-DNA clones encoding *S*-adenosyl-L-methionine: caffeic acid methyl transferases were isolated by a heterologous probe screening of λZapII c-DNA library constructed from m-RNA isolated from kinetin-induced suspension cultures of *Vanila planifolia* [117]. Much of the overall biosynthetic pathway to podophyllotoxin in *Podophyllum peltatum* and *Linum flavum* has been established from the dirigent-mediated coupling of E-coniferyl alcohol and its subsequent conversions to 7-hydroxymatairesinol. Bifunctional pinoresinol/lariciresinol reductase genes were also isolated and cloned in the same study [109].

The ~32-kDa NAD-dependent secoisolariciresinol dehydrogenase gene was isolated from a *Forsythia intermedia* stem c-DNA library, and a homologous secoisolariciresinol dehydrogenase gene was also isolated from a *Podophyllum peltatum* rhizome cDNA library and expressed in *E. coli* [118]. *S*-Adenosyl-L-methionine: β-peltatin 6-*O*-methyltransferase was isolated and characterized from cell suspension cultures of *Linum nodiflorum* L. (Linaceae), a *Linum* species accumulating aryltetralin lignans such as 6-methoxypodophyllotoxin [119]. Characterization of deoxypodophyllotoxin 5-hydroxylase (dop5h), a cytochrome p450-dependent enzyme in lignan biosynthesis of *Linum* species cell cultures producing 5-methoxypodophyllotoxin (5-MPT), has been reported [120]. Five different laccase-encoding cDNA sequences were identified from ryegrass (*Lolium perenne*), four from the stem and one from meristematic tissue [121]. The down-regulation of cinnamoyl-CoA: NADP CCR in lignin biosynthesis leads to the reduction of lignin content [122].

7.3.10.2 Biosynthesis of Paclitaxel

The diterpenoid skeleton of paclitaxel was shown [123] to be derived via the mevalonate-independent (1-deoxy-D-xylulose-5-phosphate) pathway [124–126], The committed step in the biosynthesis of paclitaxel and other taxanes is represented by cyclization of the universal diterpenoid precursor geranylgeranyl diphosphate to taxa-4 (5), 11(12)-diene [127]. This slow (but apparently not rate-limiting) reaction is catalyzed by taxadiene synthase [128]. The second specific step in taxane biosynthesis is considered to be the cytochrome P450-dependent hydroxylation at the C-5 position of the taxane ring, which is accomplished by allylic rearrangement of the 4(5) double bond to the 4(20) position to yield taxa-4(20),11(12)-diene-5α-ol [129,130]. The taxadien-5α-ol-*O*-acetyltransferase likely represents the third specific step of the taxol biosynthetic pathway, and is responsible for generating the 4(20)-en-5α-acetoxy functional grouping that ultimately gives rise to the oxetane ring. Previous evaluations of the relative abundances of naturally occurring taxanes [131–133] have suggested that hydroxylations at positions C-5, C-10, C-9, and C-2 represent early steps of the paclitaxel pathway, which precede hydroxylation at C-13; and hydroxylation at the C-1 and C-7 positions of the taxane ring are considered to represent relatively late steps in paclitaxel biosynthesis. Such acylation re-actions appear to occur early, as well as late, in the paclitaxel pathway. Acetylation at the C-5 hydroxyl group of the taxane core represents an early step in the biosyn-thesis of paclitaxel, whereas acetylation at the C-10 hydroxyl and benzoylation at the C-2 hydroxyl represent later-stage transformations. As order of the oxygenation reactions of taxol biosynthesis, the precise sequence of the acylation reactions and timing of the epoxidation and ring expansion steps are not yet fully defined, and several routes to taxol synthesis may be possible. Hence, this remains an area for future exploration.

A detailed biogenesis of paclitaxel in plants is provided in Figure 7.5, and the en-zymes involved in various steps of pathway, and their functions, are listed in Table 7.9.

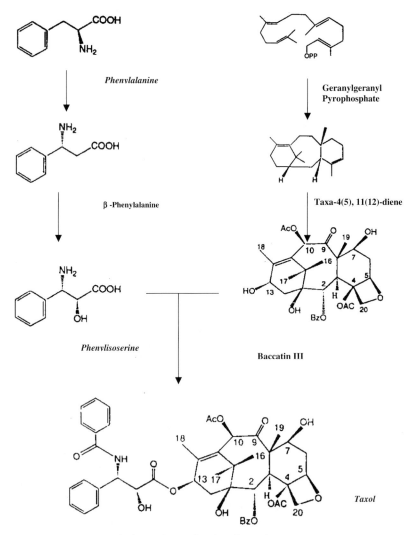

Fig. 7.5 Major steps in the biosynthesis of paclitaxel (Taxol).

Table 7.9 Enzymes involved (and their functions) in the biosynthesis of taxol.

Serial no.	Enzyme involved	Function	Reference(s)
1	Taxadiene synthase	Cyclization of the universal diterpenoid precursor geranylgeranyl diphosphate to taxa-4(5),11(12)-diene	127
2	Cytochrome P450 taxadiene 5α-hydroxylase	Hydroxylation at the C-5 position of the taxane ring, which is accomplished by allylic rearrangement of the 4(5) double bond to the 4(20) position to yield taxa-4(20),11(12)-diene-5α-ol	129, 130
3	Taxa-4(20),11(12)-dien-5α-o-acetyl transferase	Responsible for generating the 4(20)-en-5α-acetoxy functional grouping that ultimately gives rise to the oxetane ring	138
4	Cytochrome P450 taxane-13α-hydroxylase	Hydroxylation of taxane ring at C-13 position	184
5	Cytochrome P450 taxane-10β-hydroxylase	Hydroxylation at the C-10 position	184

A homology-based PCR cloning strategy led to the isolation of a cDNA encoding taxadiene synthase from *T. brevifolia* [134]. A cytochrome P450-specific differential display of the mRNA-PCR method and a PCR-based homology-cloning strategy led to the isolation of full-length cytochrome P450 cDNA from *T. cuspidata* [135–137]. More recent studies employing the baculovirus–insect cell expression system identified a second cytochrome P450 cDNA clone capable of hydroxylating taxadien-5α-ol at the C-13 position [137]. cDNA clones encoding all three of these taxane *O*-acyltransferases have been identified. The taxadien-5α-ol-*O*-acetyltransferase cDNA clone was obtained by a reverse genetic approach, based upon partial amino acid sequences derived from the corresponding purified protein [138–140]. The cDNAs encoding the 10-deacetylbaccatin-III-10β-*O*-acetyltransferase and the taxane-2α-*O*-benzoyltransferase were identified by a homology-based PCR cloning strategy [139–141]. A method for the heterologous overexpression of cDNA encoding taxadiene synthase in *E. coli* using thioredoxin fusion expression system, which increases the solubility of expressed protein, is described in Chapter 13.

7.3.10.3 Biosynthesis of Camptothecin

Camptothecin, which is structurally grouped in the quinoline alkaloids, is biogenetically a modified monoterpenoid indole alkaloid (TIA). The common intermediate, from which a variety of TIAs are formed, is strictosidine; this is formed by the condensation of tryptamine with the iridoid glucoside, secologanin [142]. This condensation is catalyzed by strictosidine synthase (STR) [143, 144]. The detailed pathway for the first part of the biosynthesis leading to strictosidine was discussed

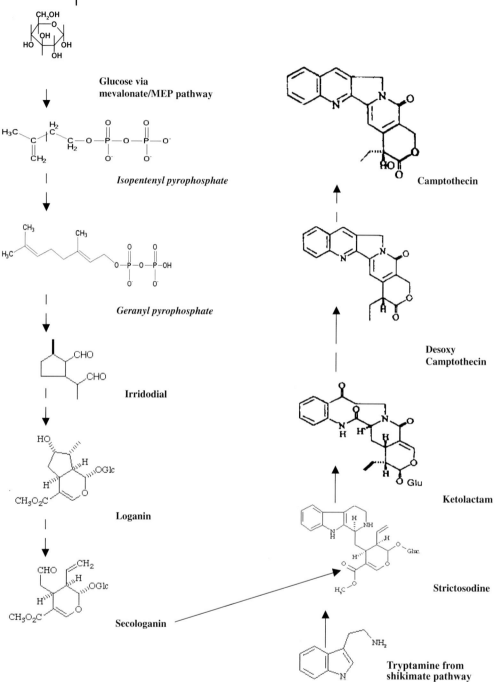

Glucose via mevalonate/MEP pathway

Isopentenyl pyrophosphate

Geranyl pyrophosphate

Irridodial

Loganin

Secologanin

Tryptamine from shikimate pathway

Strictosodine

Ketolactam

Desoxy Camptothecin

Camptothecin

Fig. 7.6 Biosynthesis of camptothecin (solid lines indicate single-step reactions; broken lines indicate multistep reactions).

earlier [145]. Subsequent intramolecular cyclization of strictosidine yields strictos-amide; a penultimate precursor of camptothecin formation in *C. acuminata* is also reported [146]. As an alternative pathway from strictosidine, a variety of TIAs are derived from deglucosylated strictosidine, and the resulting reactive dialdehyde produces different types of TIAs, such as cathenamine in *Catharanthus roseus* [147]. The iridoid section of the pathway has still not been completely elucidated, but details of the biogenetic pathway for camptothecin production are provided in Figure 7.6, with details of the enzymes involved listed in Table 7.10.

Tryptophan decarboxylase (TDC) [145,148] and strictosidine synthase (STR) [145,149] have been studied extensively. TDC channels tryptophan from primary metabolism into the terpenoid indole alkaloid pathway, while STR couples trypta-mine with secologanin, a product of the iridoid pathway. Both enzymes have been overexpressed in a number of plants and plant cells [150–154]. Expression of TDC and STR was coordinately regulated by elicitation, with jasmonate being the inter-mediate signal required, indicating that common regulators control these genes [155–158]. The cloning and characterization of cDNAs encoding STR (OpSTR) and TDC (OpTDC) – two key enzymes in the biosynthesis of monoterpenoid indole al-kaloids from hairy roots of *O. pumila* – has already been carried out, and the cDNA coding for NADPH: cytochrome P450 reductase (OpCPR) – which is presumed to be indirectly involved in camptothecin synthesis – has also been isolated [159].

Table 7.10 Enzymes involved (and their function) in the biosynthesis of camptothecin.

Serial no.	Enzyme involved	Function	Reference(s)
1	Tryptophan decarboxylase	Channels tryptophan from primary metabolism into terpenoid indole alkaloid pathway	145, 148
2	Geraniol 10 hydroxylase	Hydrolysis of geraniol at 10th carbon	185
3	Primary alcohol dehydrogenase	Conversion of 10-hydroxygeraniol to 10-oxogeranial	185
4	Cyclases	Cyclization of oxogeranial to irridodial	185
5	Loganic acid methyl transferase	Conversion of loganic acid to loganin	185
6	Secologanin synthase	Formation of secologanin from loganin	185
7	Stroctisidine synthase	Condensation of tryptamine and secologanin to yield stroctisidine	143, 144
8	Stroctosamide synthase	Intramolecular recyclization of stroctisidine to yield stroctosamide	146
9	NADPH: cytochrome P450 reductase	Indirect involvement in camptothecin biosynthesis	159

7.4
Future Prospects

Today, natural-product drugs play a dominant role in pharmaceutical care, especially in the case of antitumor agents. This increased demand for plant-based compounds for medicinal purposes, in association with low product yields and concerns about the destruction of plant-growing areas, has led to great strides being made in the large-scale production of pharmaceuticals using plant cell cultures. In this respect, the use of genetic tools has provided a clearer picture of biosynthetic pathway regulation, of secondary metabolism, and of signal transduction and elicitation. Hence, the optimization of production has moved from an empirical approach towards one of multiple productivity enhancements and synergistic effects. Ultimately, this approach should lead to significant reductions in the time taken to achieve optimal production of plant-based drugs at commercially acceptable and economic levels.

Metabolic engineering is important for the production of phytopharmaceuticals, and concise knowledge of the biosynthetic pathway and the responsible genes should allow cell cultures to be bioengineered to provide high and commercially sustainable production rates. This technology can be applied in two ways. First, the genes controlling the slow reaction steps can be overexpressed, thus raising production of the desired product(s). Second, by using antisense technology, the amounts of unwanted byproducts can be suppressed. This would result not only in an improved flux to the desired end product, but also a simplification of downstream purification processes. Moreover, pathways could be redirected to produce novel derivatives with a greater range and potency, and fewer adverse side effects.

Clearly, the successful industrial production of paclitaxel by plant cell cultures will trigger research into other plant-based chemotherapeutics such as podophyllotoxin and camptothecin. Today, the metabolic fate of complexed lignans and details of multi-branched biosynthesis form the major area of concern for the production of chemotherapeutics on a commercial scale. Notably, the production of camptothecin by plant cell cultures will require much attention.

In conclusion, improvements in bioreactor design, the use of synergism, the development of *in-situ* product removal strategies, and an insight into the mechanisms that regulate the flux of precursors to desired end-products are the focal points of future research. Thus, the up-regulation of biosynthetic pathways using regulatory genes, and the development of short bioconversion pathways in microbes are the two main areas likely to be exploited for the production of plant-based chemotherapeutics.

References

1 Farnsworth, N.R., Bingel, A.S. Problems and prospects of discovering new drugs from higher plants by pharmacological screening, Wagner, H., Wolff, P. (Eds.), *New natural products with pharmacological, biological or therapeutic activity.* Springer-Verlag, New York, **1977**, pp. 1–22.

2 Principe, P.P. Moneterizing the pharmacological benefits of plants. In: Balick, M.J., Elisabetsky, E., Laird, S.A. (Eds.), *Medicinal Plant Resources of the Tropical Forest.* Columbia University Press, New York, **1996**, pp. 191–218.

3 Cragg, G.M., Newman, D.J., Snader, K.M. Natural products in drug discovery and development. *J. Nat. Prod.* **1997**, *60*, 52–60.

4 Cragg, G.M., Newman, D.J. Antineoplastic agents from natural sources: Achievements and future directions. *Expert Opin. Investig. Drugs* **2000**, *9*, 2783–2797.

5 Pezzuto, J.M. Plant-derived anticancer agents. *Biochem. Pharmacol.* **1997**, *53*, 121–133.

6 Cragg, G.M., Scherpartz, S.A., Suffness, M., Grever, M.R. The Taxol supply crisis. New NCI policies for handling the large-scale production of novel natural product anticancer and anti-HIV agents. *J. Nat. Prod.* **1993**, *56*, 1657–1668.

7 Verpoorte, R., Contin, A., Memelink, J. Biotechnology for the production of plant secondary metabolites. *Phytochem. Rev.* **2002**, *1*, 13–25.

8 Kadkade, P.G. Formation of podophyllotoxin by *Podophyllum peltatum* tissue cultures. *Naturwissenschaften* **1981**, *68*, 481–482.

9 Kadkade, P.G. Growth and podophyllotoxin production in callus tissues of *Podophyllum peltatum. Plant Sci. Lett.* **1982**, *25*, 107–115.

10 Van Uden, W., Pras, N., Visser, J.F., Malingre, T.M. Detection and identification of podophyllotoxin produced by cell cultures derived from *Podophyllum hexandrum* Royle. *Plant Cell Rep.* **1989**, *8*, 165–168.

11 Heyenga, A.G., Lucas, J.A., Dewick, P.M. Production of tumor-inhibitory lignans in callus cultures of *Podophyllum hexandrum. Plant Cell Rep.* **1990**, *9*, 382–385.

12 Berlin, J., Bedorf, N., Molleschott, C., Wray, V., Sasse, F., Hofle, G. On the podophyllotoxins of root cultures of *Linum flavum. Planta Med.* **1988**, *54*, 204–206.

13 Van Uden, W., Pras, N., Batterman, S., Viser, J.F., Malingre, T.M. The accumulation and isolation of coniferin from a high-producing cell suspension of *Linum flavum* L. *Planta* **1990**, *183*, 25–30.

14 Smollny, T., Wichers, H., De-Rijk, T., Van-Zwam, A., Shahsavari, A., Alfermann, A.W. Formation of lignans in suspension cultures of *Linum album. Planta Med.* **1992**, *58*, A622–A624.

15 Wichers, H.J., Versluis-de-Haan, G.G., Marsman, J.W., Harkes, M.P. Podophyllotoxin related lignans in plants and cell cultures of *Linum flavum. Phytochemistry* **1991**, *30*, 3601–3604.

16 Muranaka, T., Miyata, M., Kazutaka, I., Tachibana, S. Production of podophyllotoxin in *Juniperus chinensis* callus cultures treated with oligosaccharides and a biogenetic precursor. *Phytochemistry* **1998**, *49*, 491–496.

17 Woerdenbag, H.J., van Uden, W., Frijlink, H.W., Lerk, C.F., Pras, N., Malingre, T.M. Increased podophyllotoxin production in *Podophyllum hexandrum* cell suspension cultures after feeding coniferyl alcohol as a beta-cyclodextrin complex. *Plant Cell Rep.* **1990**, *9*, 97–100.

18 Chattopadhyay, S., Srivastava, A.K., Bhojwani, S.S., Bisaria, V.S. Development of suspension culture of *Podophyllum hexandrum* for the production of podophyllotoxin. *Biotechnol Lett.* **2001**, *23*, 2063–2066.

19 Fett-Neto, A.G., Melanson, S.J., Nicholson, S.A., Pennington, J.J., Dicosmo, F. Improved taxol yield by aromatic carboxylic acid and amino acid feeding to cell cultures of *Taxus cuspidata. Biotechnol Bioeng.* **1994**, *44*, 967–971.

20 Wickremesinhe, E.R.M., Arteca, R.N. *Taxus* cell suspension cultures:

optimizing growth and production of Taxol. *J. Plant Physiol.* **1994**, *144*, 183–188.

21 Kim, J.H., Yun, J.J., Hwang, Y.S., Byun, S.Y., Kim, D.I. Production of Taxol and related taxanes in *Taxus brevifolia* cell cultures: effect of sugar. *Biotechnol Lett.* **1995**, *17*, 101–106.

22 Pestchanker, L., Roberts, S.C., Shuler, M.L. Kinetics of taxol production and nutrient use in suspension cultures of *Taxus cuspidata* in shake flasks and a Wilson-type bioreactor. *Enzyme Microb. Tech.* **1996**, *19*, 256–260.

23 Hirasuna, T.J., Pestchanker, L.J., Srinivasan, V., Shuler, M.L. Taxol production in suspension cultures of *Taxus baccata*. *Plant Cell Tissue Organ Cult.* **1996**, *44*, 95–102.

24 Ketchum, R.E.B., Gibson, D.M. Paclitaxel production in suspension cell cultures of *Taxus*. *Plant Cell Tissue Organ Cult.* **1996**, *46*, 9–16.

25 Bringi, V., Kadkade, P.G., Prince, C.L., Schubmehl, B.F., Kane, E.J., Roach, B. Enhanced production of Taxol and taxanes by cell cultures of *Taxus* species (**1995**) US Patent 5407816.

26 Yakimune, Y., Tabata, H., Higashi, Y., Hara, Y. Methyl jasmonate-induced overproduction of paclitaxel and baccatin III in *Taxus* cell suspension cultures. *Nat. Biotechnol.* **1996**, *14*, 1129–1132.

27 Parc, G., Canaguier, A., Landre, P., Hocquemiller, R., Chriqui, D., Meyer, M. Production of taxoids with biological activity by plants and callus culture from selected *Taxus* genotypes. *Phytochemistry* **2002**, *59*, 725–730.

28 Fulzele, D.P., Satdive, R.K., Pol, B.B. Growth and production of camptothecin by cell suspension cultures of *Nothapodytes foetida*. *Planta Med.* **2001**, *67*, 150–152.

29 Sakato, K., Misawa, M. Effects of chemical and physical conditions on growth of *Camptotheca acuminata* cell cultures. *Agric. Biol. Chem.* **1974**, *38*, 491–497.

30 Van Hengal, A.J., Harkes, M.P., Witchers, H.J., Hesselinic, P.G.M., Buitglaar, R.M. Characterization of callus formation and camptothecin production by cell lines of *Camptotheca acuminata*. *Plant Cell Tissue Organ Cult.* **1992**, *28*, 11–18.

31 Wiedenfeld, H., Furmanowa, M., Roeder, E., Guzewska, J., Gustowski, W. Camptothecin and 10-hydroxycamptothecin in callus and plantlets of *Camptotheca acuminata*. *Plant Cell Tissue Organ Cult.* **1997**, *49*, 213–218.

32 Sakata, K., Morita, E., Takezono, T. Annual Meeting of the Society of Agricultural Chemistry. Fukuoka, Japan, March 30, **1990**.

33 Arumugam, N., Bhojawani, S.S. Somatic embryogenesis in tissue cultures of *Podophyllum hexandrum*. *Can. J. Bot.* **1990**, *68*, 487–491.

34 Nadeem, M., Palni, L.M.S., Purohit, A.N., Pandey, H., Nandi, S.K. Propagation and conservation of *Podophyllum hexandrum* Royle: an important medicinal herb. *Biol. Conserv.* **2000**, *92*, 121–129

35 Moraes-Cerdeira, R.M., Burandt, C.L., Jr, Bastos, J.K., Nanayakkara, N.P.D., McChesney, J.D. *In vitro* propagation of *Podophyllum peltatum*. *Planta Med.* **1998**, *64*, 42–46.

36 Van Uden, W., Homan, B., Woerdenbag, H.J., Pras, N., Malingre, T.M., Wichers, H.J. Isolation, purification and cytotoxicity of 5-methoxypodophyllotoxin, a lignan from a root culture of *Linum flavum*. *J. Nat. Prod.* **1992**, *55*, 102–110.

37 Van Uden, W. , Pras, N., Homan, B., Malingre, T.M. Improvement of the production of 5-methoxypodophyllotoxin using a new selected root culture of *Linum flavum* L. *Plant Cell Tissue Organ Cult.* **1991**, *27*, 115–121.

38 Giri, A., Giri, C.C., Dhingra, V., Narasu, M.L. Enhanced podophyllotoxin production from *Agrobacterium rhizogenes* transformed cultures of *Podophyllum hexandrum*. *Nat. Prod. Lett.* **2001**, *15*, 229–235.

39 Oostdam, A., Mol, J.N.M., van der Plas, L.H.W. Establishment of hairy root cultures of *Linum flavum* producing the lignan 5-methoxypodophyllotoxin. *Plant Cell Rep.* **1993**, *12*, 474–477.

40 Lin, H., Kwok, K.H., Doran, P.M. Production of podophyllotoxin using cross-species coculture of *Linum flavum* hairy roots and *Podophyllum hexandrum* cell suspensions. *Biotechnol. Prog.* **2003**, *19*, 1417–1426.

41 Han, K.H., Fleming, P., Walker, K., Loper, M., Chilton, W.S., Mocek, U., Gordon, M.P., Floss, H.G. Genetic transformation of mature *Taxus*: an approach to genetically control the *in vitro* production of the anticancer drug, taxol. *Plant Sci.* **1994**, *95*, 187–196.

42 Roja, G., Heble, M.R. The quinoline alkaloids camptothecin and 9-methoxy-camptothecin from tissue cultures and mature trees of *Nothapodytes foetida*. *Phytochemistry* **1994**, *36*, 65–66.

43 Kitajima, M., Fischer, U., Nakamura, M., Ohsawa, M., Ueno, M., Takayama, H., Unger, M., Stöckigt, J., Aimi, N. Anthraquinones from *Ophiorriza pumila* tissue and cell cultures. *Phytochemistry* **1998**, *48*, 107–111.

44 Van Uden, W., The biotechnological production of podophyllotoxin and related cytotoxic lignans by plant cell cultures. *Pharm. World Sci.* **1993**, *15*, 41–43.

45 Chattopadhyay, S., Srivastava, A.K., Bisaria, V.S. Optimization of culture parameters for production of podophyllotoxin in suspension culture of *Podophyllum hexandrum*. *Appl. Biochem. Biotechnol.* **2002**, *102/103*, 381–393.

46 Chattopadhyay, S., Srivastava, A.K., Bhojwani, S.S., Bisaria, V.S. Production of podophyllotoxin by plant cell cultures of *Podophyllum hexandrum* in bioreactor. *J. Ferment. Bioeng.* **2002**, *93*, 215–220.

47 Chattopadhyay, S., Mehra, R.S., Srivastava, A.K., Bhojwani, S.S., Bisaria, V.S. Effect of major nutrients on podophyllotoxin production in *Podophyllum hexandrum* suspension cultures. *Appl. Microbiol. Biotechnol.* **2003**, *60*, 541–546.

48 Wang, Z.Y., Zhong, J.J. Combination of conditioned medium and elicitation enhances taxoid production in bioreactor cultures of *Taxus chinensis* cells. *Biochem. Eng. J.* **2002**, *12*, 93–97.

49 Choi, H.K., Kim, S.I., Son, J.S., Hong, S.S., Lee, H.S., Chung, I.S., Lee, H.J. Intermittent maltose feeding enhances paclitaxel production in suspension culture of *Taxus chinensis* cells. *Biotechnol. Lett.* **2000**, *22*, 1793–1796.

50 Mirjalili, N., Linden, J.C. Gas phase composition effects on suspension

cultures of *Taxus cuspidata*. *Biotechnol. Bioeng.* **1995**, *48*, 123–132.

51 Linden, J.C., Haigh, J.R., Mirjalili, N., Phisaphalong, M. Gas concentration effects on secondary metabolite production by plant cell cultures. *Adv. Biochem. Eng. Biotechnol.* **2001**, *72*, 28–62.

52 Smith, M.A.L. Large-scale production of secondary metabolites. In: Terzi, M., Cella, R., Falavigna, A. (Eds.), *Current Issues in Plant Molecular and Cellular Biology*. Kluwer Academic Publishers, Dordrecht, The Netherlands, **1995**, pp. 669–674.

53 Stierle, A., Strobel, G., Stierle, D. Taxol and taxane production by *Taxomyces andreanae*, an endophytic fungus of pacific yew. *Science* **1993**, *260*, 214–216.

54 Altstadt, T.J., Fairchild, C.R., Golik, J., Johnston, K.A., Kadow, J.F., Lee, F.Y., Long, B.H., Rose, W.C., Vyas, D.M., Wong, H., Wu, M.J., Wittman, M.D. Synthesis and antitumor activity of novel C-7 paclitaxel ethers: discovery of BMS-184476. *J. Med. Chem.* **2001**, *44*, 4577–4583.

55 Pezzutto, J.M. Taxol production in plant cell culture comes of age. *Nat. Biotechnol.* **1996**, *14*, 1083.

56 Van Uden, W., Pras, N., Malingre, T.M. On the improvement of the podophyllotoxin production by phenylpropanoid precursor feeding to cell cultures of *Podophyllum hexandrum* Royle. *Plant Cell Tissue Organ Cult.* **1990**, *23*, 217–224.

57 Fett-Neto, A.G., Dicosmo, F., Reynolds, W.F., Sakata, K. Cell culture of *Taxus* as a source of the antineoplastic drug taxol and related taxanes. *Biotechnology* **1992**, *10*, 1572–1575.

58 Van Uden, W., Pras, N., Malingre, T.M. The accumulation of podophyllotoxin-β-D-glucoside by cell suspension cultures derived from the conifer *Callitris drummondii*. *Plant Cell Rep.* **1990**, *9*, 257–260.

59 Wang, H.Q., Yu, J.T., Zhong, J.J. Significant improvement of taxane production in suspension cultures of *Taxus chinensis* by sucrose feeding strategy. *Process Biochem.* **2000**, *35*, 479–483.

60 Luo, J., Mei, X., Liu, L., Hu, D.W. Improved paclitaxel production by fed-

batch suspension cultures of *Taxus chinensis* in bioreactors. *Biotechnol. Lett.* **2002**, *24*, 561–565.

61 Fleming, P.E., Knaggs, A.R., He, X.G., Mocek, U., Floss, H.G. Biosynthesis of taxoids. Mode of attachment of the taxol side chain. *J. Am. Chem. Soc.* **1994**, *116*, 4137.

62 Srinivasan, V., Ciddi, V., Bringi, V., Shuler, M.L., Metabolic inhibitors, elicitors, and precursors as tools for probing yield limitation in taxane production by *Taxus chinensis* cell cultures. *Biotechnol. Prog.* **1996**, *12*, 457–465.

63 Strobel, G.A., Stierle, A., Vankujik, F.J.G.M. Factors influencing the *in vitro* production of radiolabelled taxol by pacific yew, *Taxus brevifolia*. *Plant Sci.* **1992**, *84*, 65–74.

64 Mamatha, R. Studies in tissue cultures of taxus species and other higher plants. PhD thesis, Kakatiya University, Warangal, A.P., India, **2000**.

65 Silvestrine, A., Pasqua, G., Botta, B., Monacelli, B., Van der Heijden, R., Verpoorte, R. Effects of alkaloid precursor feeding on a *Camptotheca acuminata* cell line. *Plant Physiol. Biochem.* **2002**, *40*, 749–753.

66 Aapro, M. Docetaxel versus doxorubicin in patients with metastatic breast cancer who have failed alkylating chemotherapy: a preliminary report of the randomized phase III trial. *Semin. Oncol.* **1998**, *25*, 7–11.

67 Schmitt, J., Petersen, M. Influence of methyl jasmonate and coniferyl alcohol on pinoresinol and matairesinol accumulation in a *Forsythia intermedia* suspension culture. *Plant Cell Rep.* **2002**, *20*, 885–889.

68 Van Furden, B., Humburg, A., Fuss, E. Influence of methyl jasmonate on podophyllotoxin and 6-methoxypodophyllotoxin accumulation in *Linum album* cell suspension cultures. *Plant Cell Rep.* **2005**, *24*, 312–317.

69 Shams-Ardakani, M., Hemmati, S., Mohagheghzadeh, A. Effect of elicitors on the enhancement of podophyllotoxin biosynthesis in suspension cultures of *Linum album*. *DARU* **2005**, *13*, 56–60.

70 Yu, L.J., Lan, W.Z., Qin, W.M., Jin, W.W., Xu, H.B. Oxidative stress and taxol production induced by fungal elicitor in cell suspension cultures of *Taxus chinensis*. *Biol. Plant. (Prague)* **2002**, *45*, 459–461.

71 Yuan, Y.J., Li, C., Hu, Z.D., Wu, J.C. A double oxidative burst for taxol production in suspension cultures of *Taxus chinensis* var. *mairei* induced by oligosaccharide from *Fusarium oxysprum*. *Enzyme Microb. Technol.* **2002**, *30*, 774–778.

72 Yuan, Y.J., Li, C., Hu, Z.D., Wu, J.C., Zeng, A.P. Fungal elicitor-induced cell apoptosis in suspension cultures of *Taxus chinensis* var. *mairei* for taxol production. *Process Biochem.* **2002**, *38*, 193–198.

73 Mirajilili, N., Linden, J.C. Methyl jasmonate induced production of taxol in suspension cultures of *Taxus cuspidate*: ethylene interaction and induction models. *Biotechnol. Prog.* **1996**, *12*, 110–118.

74 Ketchum, R.E.B., Gibson, D.M., Croteau, R.B., Shuler, M.L. The kinetics of taxoid accumulation in cell suspension cultures of *Taxus* following elicitation with methyl jasmonate. *Biotechnol. Bioeng.* **1999**, *62*, 97–105.

75 Baebler, S., Camloh, M., Kovac, M., Ravnikar, M., Zel, J. Jasmonic acid stimulates taxane production in cell suspension culture of yew (*Taxus media*). *Planta Med.* **2002**, *68*, 475–476.

76 Yuan, Y.J., Hu, G.W., Wang, C.G., Jing, Y., Zhou, Y.Q., Shen, P.W. Effect of rare earth compounds on the growth, taxol biosynthesis and release in *Taxus cuspidata* cell culture. *J. Chin. Rare Earth Soc.* **1998**, *16*, 56–60.

77 Wu, J., Wang, C., Mei, X. Stimulation of taxol production and excretion in *Taxus* spp. cell cultures by rare earth chemical lanthanum. *J. Biotechnol.* **2001**, *85*, 67–73.

78 Ciddi, V., Srinivasan, V., Shuler, M.L. Elicitation of *Taxus* sp. cell cultures for production of taxol. *Biotechnol. Lett.* **1995**, *17*, 1343–1346.

79 Wang, C., Wu, J., Mei, X. Enhanced taxol production and release in *Taxus chinensis* cell suspension cultures with selected organic solvents and sucrose feeding. *Biotechnol. Prog.* **2001**, *17*, 89–94.

80 Yu, L.J., Lan, W.Z., Qin, W.M., Xu, B.H. Effects of salicylic acid on fungal elicitor induced membrane-lipid peroxidation and taxol production in cell suspension cultures of *Taxus chinensis*. *Process Biochem.* **2001**, *37*, 477–482.

81 Dong, H.D., Zhong, J.J. Enhanced taxane production in bioreactor cultivation of *Taxus chinensis* cells by combining elicitation, sucrose feeding and ethylene incorporation. *Enzyme Microb. Technol.* **2002**, *31*, 116–121.

82 Cusido, R.M., Palazo, M., Bonfill, M., Navia-Osorio, A., Morales, M., Pinol, M.T. Improved paclitaxel and baccatin III production in suspension cultures of *Taxus media*. *Biotechnol. Prog.* **2002**, *18*, 418–423.

83 Zhang, C.H., Wu, J.Y. Ethylene inhibitors enhance elicitor-induced paclitaxel production in suspension cultures of *Taxus* spp. cells. *Enzyme Microb. Technol.* **2003**, *32*, 71–77.

84 Laskaris, G., Bounkhay, M., Theodoridis, G., Van der Heijden, R., Verpoorte, R., Jaziri, M. Induction of geranylgeranyl diphosphate synthase activity and taxane accumulation in *Taxus baccata* cell cultures after elicitation by methyl jasmonate. *Plant Sci.* **1999**, *147*, 1–8.

85 Luo, J., He, G.Y. Optimization of elicitors and precursors for paclitaxel production in cell suspension culture of *Taxus chinensis* in the presence of nutrient feeding. *Process Biochem.* **2004**, *39*, 1073–1079.

86 Khosroushahi, A.Y., Valizadeh, M., Ghasempour, A., Khosrowshahli, M., Naghdibadi, M., Dadpour, M.R., Omidi, Y. Improved Taxol production by combination of inducing factors in suspension cell culture of *Taxus baccata*. *Cell Biol. Inter.* **2005**, *30*, 262–269.

87 Kwon, I.C., Yoo, Y.J., Lee, J.H., Hyun, J.O. Enhancement of taxol production by *in situ* recovery of product. *Process Biochem.* **1998**, *33*, 701–707.

88 Kondo, O., Honda, H., Taya, M., Kobayashi, T. Comparison of growth properties of carrot hairy root in various bioreactors. *Appl. Microbiol. Biotechnol.* **1989**, *32*, 291–294.

89 Kutney, J.P., Hewitt, G., Jarvis, T.C., Palaty, J., Retting, S.J. Studies with plant cell cultures of *Catharanthus roseus*. Oxidative coupling of dibenzyl-butanolides catalyzed by plant cell culture extracts. *Can. J. Chem.* **1992**, *70*, 2115–2133.

90 Van Uden, W., Holidi Oeij, K., Woerdenbag, H.J., Pras, N. Glucosylation of cyclodextrin-complexed podo-phyllotoxin by cell cultures of *Linum flavum*. *Plant Cell Tissue Organ Cult.* **1993**, *34*, 169–175.

91 Kutney, J.P. Biotechnology and synthetic chemistry routes to clinically important compounds. *Pure Appl. Chem.* **1999**, *71*, 1025–1032.

92 Puricelli, L., Caniato, R., Delle Monache, G. Biotransformation of a dibenzyl-butanolide to podophyllotoxin analogues by shoot cultures of *Haplophyllum patavinum*. *Chem. Pharm. Bull.* **2003**, *51*, 848–850.

93 Koulman, A., Beekman, A.C., Pras, N., Quax, J.W. The bioconversion process of deoxypodophyllotoxin with *Linum flavum* cell cultures. *Planta Med.* **2003**, *69*, 739–744.

94 Chen, T.S., Li, X., Bollag, D., Liu, Y.C., Chang, C.J. Biotransformation of taxol. *Tetrahedron Lett.* **2001**, *42*, 3787–3789.

95 Hamada, H., Furaya, T. Recent advances in plant biotransformation. *Plant Tiss. Cult. Biotech.* **1996**, *2*, 52–60.

96 Veeresham, C., Rao, M.A., Babu, P.C., Mamatha, R., Kokate, C.K. Biotrans-formation of paclitaxel (Taxol) by plant cell cultures. *Indian Drugs* **2000**, *37*, 86–89.

97 Chattopadhyay, S., Bisaria, V.S., Srivastava, A.K. Enhanced production of podophyllotoxin by *Podophyllum hexandrum* using in situ cell retention bioreactor. *Biotechnol. Prog.* **2003**, *19*, 1026–1028.

98 Chattopadhyay, S., Bisaria, V.S., Bhojwani, S.S., Srivastava, A.K. Enhanced production of podophyllotoxin by fed-batch cultivation of *Podophyllum hexandrum*. *Can. J. Chem. Eng.* **2003**, *81*, 1–8.

99 Arroo, R.R.J., Alfermann, A.W., Medarde, M., Petersen, M., Pras, N., Woolley, J.G. Plant cell factories as a source for anti-cancer lignans. *Phytochem. Rev.* **2002**, *1*, 27–35.

100 Srinivasan, V.L., Pestchanker, L., Moser, S., Hirasuna, T., Tatlcek, R.A., Schular, M.L. Taxol production in bioreactors: Kinetics of biomass accumulation, nutrient uptake, and taxol production by cell suspensions of *Taxus baccata*. *Biotech. Bioeng.* **1995**, *47*, 666–676.

101 Seki, M., Ohzora, C., Takede, M., Furasaki, S. Taxol (paclitaxel) production using free and immobilized cells of *Taxus cuspidata*. *Biotech. Bioeng.* **1997**, *53*, 214–219.

102 Wang, S.J., Zhong, J.J. A novel centri-fugal impeller bioreactor. I. Fluid circula-tion, mixing, and liquid velocity profiles. *Biotechnol. Bioeng.* **1996**, *51*, 511–519.

103 Zhong, J.J., Wang, S.J., Wang, H.Q. A new centrifugal impeller reactor and its application to plant cell cultures. In: *Horizon of biochemical engineering.* The Society of Chemical Engineers, Japan, Tokyo, **1997**, pp. 175–179.

104 Navia-Osorio, A., Garden, H., Cusido, R.M., Palazon, J., Alfermann, A.W., Pinol, M.T. Taxol and baccatin III production in suspension cultures of *Taxus baccata* and *Taxus wallichiana* in an airlift bioreactor. *J. Plant Physiol.* **2002**, *159*, 97–102.

105 Son, S.H., Choi, S.M., Lee, Y.H., Choi, K.B., Yun, S.R., Kim, J.K., Park, H.J., Kwon, O.W., Noh, E.W., Seon, J.H., Park, Y.G. Large-scale growth and taxane production in cell cultures of *Taxus cuspidata* (Japanese yew) using a novel bioreactor. *Plant Cell Rep.* **2000**, *19*, 628–633.

106 Hu, W.W., Zhong, J.J. Effect of bottom clearance on performance of airlift bioreactor in high-density culture of *Panax notoginseng* cells. *J. Biosci. Bioeng.* **2001**, *92*, 389–392.

107 Pan, Z.W., Wang, H.Q., Zhong, J.J. Scale-up study on suspension cultures of *Taxus chinensis* cells for production of taxane diterpene. *Enzyme Microb. Technol.* **2000**, *27*, 714–723.

108 Zhong, J.J., Pan, Z.W., Wang, Z.Y., Wu, J.Y., Chen, F., Takagi, M., Yoshida, A.T.

Effect of mixing time on taxoid produc-tion in suspension cultures of *Taxus chinensis* in a centrifugal impeller bioreactor. *J. Biosci. Bioeng.* **2002**, *94*, 244–250.

109 Xia, Z.Q., Costa, M.A., Proctor, J., Davin, L.B., Lewis, N.G. Dirigent-mediated podophyllotoxin biosynthesis in *Linum flavum* and *Podophyllum peltatum*. *Phytochemistry* **2000**, *55*, 537–549.

110 Larsen, K. Cloning and characterization of a ryegrass (*Lolium perenne*) gene encoding cinnamoyl-CoA reductase (CCR). *Plant Sci.* **2004**, *166*, 569–581.

111 Windhövel, J., Fuss, E., Alfermann, A.W. The biosynthesis of lignans in *Linum album*: Cinnamoyl-CoA: NADP oxido-reductase: Abstract (Biochemistry) (**2001**) www.biologie.uni-erlangen.de/ pharmabiol/Abstract/Biochemistry.

112 Hoshida, H., Nakao, M., Kazava, N., Kubo, K., Hakukawa, T., Morimasa, K., Akada, R., Nishizawa, Y. Isolation of five lactase gene sequences from the white-rot fungus *Trametes sanguinea* by PCR, and cloning, characterization and expression of the laccase cDNA in yeasts. *J. Biosci. Bioeng.* **2001**, *92*, 372–380.

113 De Melis, L.E., Whiteman, P.H., Stevenson, T.W. Isolation and character-isation of a cDNA clone encoding cinnamyl alcohol dehydrogenase in *Eucalyptus globulus* Labill. *Plant Sci.* **1999**, *143*, 173–182.

114 Schwelm, A., Fuss, E., Schönell, B., Windhövel, J., Alfermann, A.W. The biosynthesis of cytotoxic lignans in *Linum album*: phenylalanine ammonia lyase and cinnamoyl alcohol dehydro-genase: Abstract (Biochemistry) (**2001**) www.biologie.uni-erlangen.de/ pharmabiol/Abstract/Biochemistry.

115 Seidel, V., Windhövel, J., Eaton, G., Alfermann, A.W., Arroo, R.R.J., Woolley, J.G. Podophyllotoxin biosynthesis in *Linum album* cell cultures – the role of phenylpropanoid ring substitution patterns: Abstract (Biochemistry) (**2001**) www.biologie.uni-erlangen.de/ pharmabiol/Abstract/Biochemistry.

116 Hegener, U., Fuss, E., Alfermann, A.W. Biosynthesis of cytotoxic lignans in *Linum album*: caffeic acid 3-*O*-methyl-transferase: Abstract (Biochemistry)

(2001) www.biologie.uni-erlangen.de/pharmabiol/Abstract/Biochemistry.

117 Xue, Z., Brodellius, P. Kinetin induced caffeic acid O-methyltransferases in cell suspension cultures of *Vanila planifolia* Andr. and isolation of caffeic acid O-methyltransferases cDNAs. *Plant Physiol. Biochem.* **1998**, *36*, 779–788.

118 Xia, Z.Q., Costa, M.A., Pellisier, H., Davin, L.B., Lewis, N.G. Secoisolariciresinol dehydrogenase purification, cloning and functional expression. *J. Biol. Chem.* **2001**, *276*, 12614–12623.

119 Kranz, K., Petersen, M. β-Peltatin 6-O-methyltransferase from suspension cultures of *Linum nodiflorum*. *Phytochemistry* **2003**, *64*, 453–458.

120 Kuhlmann, S., Molog, G., Empt, U., Van Uden, W., Pras, N., Alfermann, A.W., Petersen, M. characterisation of deoxypodophyllotoxin 5-hydroxylase (dop5h), a cytochrome p450-dependent enzyme in lignan biosynthesis of *Linum* species: Abstract (Biochemistry) (**2001**) www.biologie.uni-erlangen.de/pharmabiol/Abstract/Biochemistry.

121 Gavnholt, B., Larsen, K., Rasmussen, S.K. Isolation and characterisation of laccase cDNAs from meristematic and stem tissues of ryegrass (*Lolium perenne*). *Plant Sci.* **2002**, *162*, 873–885.

122 Piquemal, J., Lapierre, C., Myton, K., O'Connell, A., Schuch, W., Grima-Pettenati, J., Boudet, A.M. Down-regulation of cinnamoyl-CoA reductase induces significant changes of lignin profiles in transgenic tobacco plants. *Plant J.* **1998**, *13*, 71–83.

123 Eisenreich, W., Menhard, B., Hylands, P.J., Zenk, M.H., Bacher, A. Studies on the biosynthesis of Taxol: the taxane carbon skeleton is not of mevalonoid origin of special interest. *Proc. Natl. Acad. Sci. USA* **1996**, *93*, 6431–6436.

124 Rohmer, M., Knani, M., Simonin, P., Sutter, B., Sahm, H. Isoprenoid biosynthesis in bacteria: a novel pathway for the early steps leading to isopentenyl diphosphate. *Biochem. J.* **1993**, *295*, 517–524.

125 Lichtenthaler, H.K., Rohmer, M., Schwender, J. Two independent biochemical pathways for isopentenyl diphosphate and isoprenoid biosynthesis

in higher plants. *Physiol. Plant.* **1997**, *101*, 643–652.

126 Lichtenthaler, H.K. The 1-deoxy-D-xylulose-5-phosphate pathway of isoprenoid biosynthesis in plants. *Annu. Rev. Plant Physiol. Plant Mol. Biol.* **1999**, *50*, 47–65.

127 Koepp, A.E., Hezari, M., Zajicek, J., Vogel, B.S., Lafever, R.E., Lewis, N.G., Croteau, R. Cyclization of geranylgeranyl diphosphate to taxa-4(5),11(12)-diene is the first committed step of Taxol biosynthesis in Pacific yew. *J. Biol. Chem.* **1995**, *270*, 8686–8690.

128 Hezari, M., Lewis, N.G., Croteau, R. Purification and characterization of tax-4(5),11(12)-diene synthase from Pacific yew (*Taxus brevifolia*) that catalyzes the first committed step of Taxol biosynthesis. *Arch. Biochem. Biophys.* **1995**, *322*, 437–444.

129 Hefner, J., Rubenstein, S.M., Ketchum, R.E.B., Gibson, D.M., Williams, R.M., Croteau, R. Cytochrome P450-catalyzed hydroxylation of taxa-4(5),11(12)-diene to taxa-4(20),11(12)-dien-5a-ol: the first oxygenation step in Taxol biosynthesis. *Chem. Biol.* **1996**, *3*, 479–489.

130 Vazquez, A., Williams, R.M. Studies on the biosynthesis of taxol. Synthesis of taxa-4(20),11(12)-diene-2,5-diol. *J. Org. Chem.* **2000**, *65*, 7865–7869.

131 Floss, H.G., Mocek, U. Biosynthesis of taxol. In: Suffness, M. (Ed.), *Taxol science and applications*. CRC Press, Boca Raton, **1995**, pp. 191–298.

132 Kingston, D.G.I., Molinero, A.A., Rimoldi, J.M. The taxane diterpenoids. *Prog. Chem. Org. Nat. Prod.* **1993**, *61*, 1–206.

133 Baloglu, E., Kingston, D.G.I. The taxane diterpenoids. *J. Nat. Prod.* **1999**, *62*, 1448–1472.

134 Wildung, M.R., Croteau, R. A cDNA clone for taxadiene synthase, the diterpene cyclase that catalyzes the committed step of Taxol biosynthesis. *J. Biol. Chem.* **1996**, *271*, 9201–9204.

135 Schoendorf, A., Rithner, C.D., Williams, R.M., Croteau, R.B. Molecular cloning of a cytochrome P450 taxane 10-hydroxylase cDNA from *Taxus* and functional expression in yeast. *Proc. Natl. Acad. Sci. USA* **2001**, *98*, 1501–1506.

136 Jennewein, S., Croteau, R. Taxol: biosynthesis, molecular genetics, and biotechnological applications. *Appl. Microbiol. Biotechnol.* **2001**, *57*, 13–19.

137 Jennewein, S., Rithner, C.D., Williams, R.M., Croteau, R.B. Taxol biosynthesis: taxane 13 alpha-hydroxylase is a cytochrome P450-dependent monooxygenase. *Proc. Natl. Acad. Sci. USA* **2001**, *98*, 13595–13600.

138 Walker, K., Ketchum, R.E., Hezari, M., Gatfield, D., Goleniowski, M., Barthol, A., Croteau, R. Partial purification and characterization of acetyl coenzyme A: taxa-4(20),11(12)-dien-5-alpha-ol O-acetyl transferase that catalyzes the first acylation step of taxol biosynthesis. *Arch. Biochem. Biophys.* **1999**, *364*, 273–279.

139 Walker, K., Croteau, R. Molecular cloning of a 10-deacetylbaccatin III-10-O-acetyl transferase cDNA from Taxus and functional expression in *Escherichia coli. Proc. Natl. Acad. Sci. USA* **2000**, *97*, 583–587.

140 Walker, K., Scboendorf, A., Croteau, R. Molecular cloning of a taxa-4(20),11(12)-dien-5-ol-O-acetyl transferase cDNA from *Taxus* and functional expression in *Escherichia coli. Arch. Biochem. Biophys.* **2000**, *374*, 371–380.

141 Huang, K.X., Huang, Q.L., Wildung, M.R., Croteau, R., Scott, A.I. Overproduction, in *Escherichia coli*, of soluble taxadiene synthase, a key enzyme in the Taxol biosynthetic pathway. *Protein Expr. Purif.* **1998**, *13*, 90–96.

142 Stockigt, J., Zenk, M.H. Strictosidine (isovincoside): the key intermediate in the biosynthesis of monoterpenoid indole alkaloids. *J. Chem. Soc. Chem. Commun.* 1977, 646–648.

143 Scott, A.I., Lee, S.L. Biosynthesis of the indole alkaloids – A cell-free system from *Catharanthus roseus. J. Am. Chem. Soc.* **1975**, *97*, 6906–6908.

144 Stockigt, J., Ruppert, M. Strictosidine – the biosynthetic key to monoterpenoid indole alkaloids. In: Barton, S.D., Nakanishi, K., Meth-Cohn, O. (Eds.), *Comprehensive Natural Products Chemistry*, Volume 4. Pergamon, Oxford, **1999**, pp. 109–138.

145 Verpoorte, R., Van der Heijden, R., Moreno, P.R.H. Biosynthesis of terpenoid indole alkaloids in *Catharanthus roseus* cells. In: Cordell, G.A. (Ed.), *The Alkaloids.* Academic Press, San Diego, **1997**, pp. 221–299.

146 Hutchinson, C.R., Heckendorf, A.H., Straughn, J.L., Daddona, P.E., Cane, D.E. Biosynthesis of camptothecin: Definition of strictosamide as the penultimate biosynthetic precursor assisted by 13C and 1H NMR spectroscopy. *J. Am. Chem. Soc.* **1979**, *101*, 3358–3369.

147 Hemscheidt, T., Zenk, M.H. Glucosidases involved in indole alkaloid biosynthesis of *Catharanthus* cell cultures. *FEBS Lett.* **1980**, *110*, 187–191.

148 De Luca, V. Metabolic engineering of crops with the tryptophan decarboxylase of *Catharanthus roseus.* In: Verpoorte, R., Alfermann, A.W. (Eds.), *Metabolic engineering of plant secondary metabolism.* Kluwer Academic Press, Dordrecht, **2000**, pp. 179–194.

149 Kutchan, T.M. Alkaloid biosynthesis – The basis for metabolic engineering of medicinal plants. *Plant Cell* **1995**, *7*, 1059–1070.

150 Canel, C., Lopez-Cardoso, M.I., Whitmer, S., Van der Fits, L., Pasquali, G., Van der Heijden, R., Hoge, J.H.C., Verpoorte, R. Effects of over-expression of strictosidine synthase and tryptophan decarboxylase on alkaloid production by cell cultures of *Catharanthus roseus. Planta* **1998**, *205*, 414–419.

151 Whitmer, S., Canel, C., Hallard, D., Goncalves, C., Verpoorte, R. Influence of precursor availability on alkaloid accumulation by transgenic cell lines of *Catharanthus roseus. Plant Physiol.* **1998**, *116*, 853–857.

152 Geerlings, A., Hallard, D., Martinez, C.A., Lopez, C.I., Van der Heijden, R., Verpoorte, R. Alkaloid production by a *Cinchona officinalis* 'Ledgeriana' hairy root culture containing constitutive expression constructs of tryptophan decarboxylase and strictosidine synthase cDNAs from *Catharanthus roseus. Plant Cell Rep.* **1999**, *19*, 191–196.

153 Whitmer, S. *Aspects of terpenoid indole alkaloid formation by transgenic cell lines of* Catharanthus roseus *over-expressing tryptophan decarboxylase and strictosidine synthase.* PhD Thesis, University of Leiden, The Netherlands, **1999**.

154 Hallard, D. *Transgenic plant cells for the production of indole alkaloids*. PhD Thesis, University of Leiden, The Netherlands, **2000**.

155 Menke, F.L.H., Champion, A., Kijne, J.W., Memelink, J. A novel jasmonate- and elicitor-responsive element in the periwinkle secondary metabolite biosynthetic gene *Str* interacts with a jasmonate- and elicitor-inducible *AP2*-domain transcription factor, *ORCA2*. *EMBO J.* **1999**, *18*, 4455–4463.

156 Menke, F.L.H., Parchmann, S., Mueller, M.J., Kijne, J.W., Memelink, J. Involvement of the octadecanoid pathway and protein phosphorylation in fungal elicitor-induced expression of terpenoid indole alkaloid biosynthetic genes in *Catharanthus roseus*. *Plant Physiol.* **1999**, *119*, 1289–1296.

157 Menke, F.L.H. *Elicitor signal transduction in* Catharanthus roseus *leading to terpenoid indole alkaloid biosynthetic gene expression*. PhD Thesis, University of Leiden, The Netherlands, **1999**.

158 Van der Fits, L. *Transcriptional regulation of stress-induced plant secondary metabolism*. PhD Thesis, University of Leiden, The Netherlands, **2000**.

159 Yamazaki, Y., Urano, A., Sudo, H., Kitajima, M., Takayama, H., Yamazaki, M., Aimi, N., Saito, K. Metabolite profiling of alkaloids and strictosidine synthase activity in camptothecin-producing plants. *Phytochemistry* **2003**, *62*, 461–470.

160 Misawa, M., Sakato, K., Tanaka, H., Hayashi, M., Samejima, M., Production of physiologically active substances by plant cell suspension cultures. In: Street, H.E. (Ed.), *Tissue culture and Plant Science*. Academic Press, London, New York, **1974**, pp. 235–246.

161 Stahelin, H.F., von Wartburg, A.V. The chemical and biological route from podophyllotoxin glucoside to etoposide. *Cancer Res.* **1991**, *51*, 5–15.

162 Imbert, T.F. Discovery of podophyllotoxins. *Biochimie* **1998**, *80*, 207–222.

163 Sjostrom, J., Blomqvist, C., Mouridsen, H., Pluzanska, A., Ottosson-Lonn, S., Bengtsson, N.O., Ostenstad, B., Mjaaland, I., Palm-Sjovall, M., Wist, E., Valvere, V., Anderson, H., Bergh, J. Docetaxel compared with sequential methotrexate and 5-fluorouracil in patients with advanced breast cancer after anthracycline failure: a randomized phase III study with crossover on progression by the Scandinavian Breast Group. *Eur. J. Cancer* **1999**, *35*, 1194–1201.

164 Fan, W. Possible mechanisms of paclitaxel-induced apoptosis. *Biochem. Pharmacol.* **1999**, *57*, 1215–1221.

165 Huang, Y., Fan, W. IkB kinase activation is involved in regulation of paclitaxel-induced apoptosis in human tumour cell lines. *Mol. Pharmacol.* **2002**, *61*, 105–113.

166 Johnson, K.R., Fan, W. Reduced expression of p^{53} and $p^{21WAF1/CIP1}$ sensitizes human breast cancer cells to paclitaxel and its combination with 5-fluorouracil. *Anticancer Res.* **2002**, *22*, 1–8.

167 Johnson, K.R., Wang, L., Miller, M.C., Willingham, M.C., Fan, W. 5-Fluoro-uracil interferes with paclitaxel cyto-toxicity against human solid tumour cells. *Clin. Cancer Res.* **1997**, *3*, 1739–1745.

168 Chabot, G.G. Clinical pharmacokinetics of irinotecan. *Clin. Pharmacokinet.* **1997**, *33*, 245–259.

169 Friedman, H.S., Petros, W.P., Friedman, A.H., Schaaf, L.J., Kerby, T., Lawyer, J., Parry, M., Houghton, P.J., Lovell, S., Rasheed, K., Cloughsey, T., Stewart, E.S., Colvin, O.M., Provenzale, J.M., Mclendon, R.E., Bigner, D.D., Cokgor, I., Haglund, M., Rich, J., Ashley, D., Malczyn, J., Elfring, G.L., Miller, L.L. Irinotecan therapy in adults with recurrent or progressive malignant glioma. *J. Clin. Oncol.* **1999**, *17*, 1516–1525.

170 Jiang, J.F., Lui, W.J., Ding, J. Regulation of telomerase activity in camptothecin induced apoptosis of leukemia HL-60 cells. *Acta Pharmacol. Sin.* **2000**, *21*, 759–764.

171 Van Uden, W., Pras, N., Malingre, T.M. The accumulation of podophyllotoxin-β-D-glucoside by cell suspension cultures derived from the conifer *Callitris drummondii*. *Plant Cell Rep.* **1990**, *9*, 257–260.

172 Empt, U., Alfermann, A.W., Pras, N., Peterson, M. The use of plant cell cultures for the production of podo-phyllotoxin and related lignans. *J. Appl. Bot.* **2000**, *74*, 145–150.

173 Van Uden, W., Pras, N., Vossebeld, E.M., Mol, J.N.M., Malingre, T.M. Production of 5-methoxypodophyllotoxin in cell suspension cultures of *Linum flavum* L. *Plant Cell Tissue Organ Cult.* **1990**, *20*, 81–87.

174 Konuklugil, B., Schimdt, T.J., Alfermann, A.W. Accumulation of aryltetralin lactone lignans in cell suspension cultures of *Linum nodiflorum*. *Planta Med.* **1999**, *65*, 587–588.

175 Kutney, J.P., Arimoto, M., Hewitt, G.M., Jarvis, T.C., Sakata, K. Studies with plant cell cultures of *Podophyllum peltatum* L. I. Production of podophyllotoxin, deoxy-podophyllotoxin, podophyllotoxone, and 4'-demethylpodophyllotoxin. *Heterocycle* **1991**, *32*, 2305–2309.

176 Dong, H.D., Zhong, J.J. Significant improvement of taxane production in suspension cultures of *Taxus chinensis* by combining elicitation with sucrose feed. *Biochem. Eng. J.* **2001**, *8*, 145–150.

177 Hermann, K.M. The shikimate pathway: Early steps in biosynthesis of aromatic compounds. *Plant Cell* **1995**, *7*, 907–910.

178 Hahlbrock, K., Scheel, D. Physiology and molecular biology of phenylpropanoid metabolism. *Annu. Rev. Plant Physiol. Plant Mol. Biol.* **1989**, *40*, 347–369.

179 Seidel, V., Windhövel, J., Eaton, G., Alfermann, A.W., Arroo, R.R.J., Medarde, M., Petersen, M., Wolley, J.G. Biosynthesis of podophyllotoxin in *Linum album* cell cultures. *Planta* **2002**, *215*, 1013–1039.

180 Lauvergeat, V., Lacomme, C., Lacombe, E., Lasserre, E., Roby, D., Grima-Pettenati, J. Two cinnamoyl-CoA reductase (CCR) genes from *Arabidopsis thaliana* are differentially expressed during development and in response to infection with pathogenic bacteria. *Phytochemistry* **2001**, *57*, 1187–1195.

181 Henges, A. *Biosynthese und Kompartimentierung von Lignanen in Zellkulturen von* Linum album. Dissertation, University of Dusseldorf, **1999**.

182 Molog, M.G., Empt, U., Petersen, M., Van Uden, W., Pras, N., Alfermann, A.W. Deoxypodophyllotoxin 6-hydroxy-lase, a cytochrome P450 monooxygenase from cell cultures of *Linum flavum* involved in biosynthesis of cytotoxic lignans. *Planta* **2001**, *214*, 288–294.

183 Kuhlmann, S., Kranz, K., Lücking, B., Alfermann, A.W., Petersen, M. Aspects of cytotoxic lignan biosynthesis in suspension cultures of *Linum nodiflorum*. *Phytochem. Rev.* **2002**, *1*, 37–43.

184 Hezari, M., Croteau, R. Taxol biosynthesis: an update. *Planta Med.* **1997**, *63*, 291–295.

185 Collu, G., Unver, N., Peltenburg-Looman, A.M.G., Van der Heijden, R., Verpoorte, R., Memelink, J. Geraniol 10-hydroxylase, a cytochrome P450 enzyme involved in terpenoid indole alkaloid biosynthesis. *FEBS Lett.* **2001**, *508*, 215–220.

Part 2
Genetic Modifications, Transgenic Plants and Potential
of Medicinal Plants in Genetechnology and Biotechnology

Medicinal Plant Biotechnology. From Basic Research to Industrial Applications
Edited by Oliver Kayser and Wim J. Quax
Copyright © 2007 WILEY-VCH Verlag GmbH & Co. KGaA, Weinheim
ISBN 978-3-527-31443-0

8

In-Vitro Culturing Techniques of Medicinal Plants

Wolfgang Kreis

8.1
Introduction

As early as 1934, it was demonstrated that plant cells and tissues could be cultivated for long periods of time on appropriate media (Gautheret, 1934; White, 1934). By including indole-3-acetic acid and vitamins in his media, Gautheret (1934) extended the culture period of *Salix* callus to 18 months, and was able to subculture the tissue aseptically. Later, White (1939) reported the "Potentially Unlimited Growth of Excised Plant Callus in an Artificial Nutrient", and found that these callus cultures basically showed no evidence of differentiation or polarity. Being undifferentiated, yet capable of unlimited growth, they appeared to satisfy the two main requirements for a true "tissue culture". Meanwhile, the initiation of plant cell and tissue cultures and cultivation of plant cells and tissues *in vitro* has become a relatively simple task (see, for example, Murashige, 1974; Allan, 1991). During the 1970s and 1980s, plant cell suspension culture was regarded as the most appropriate means of studying plant secondary metabolism, and as the most suitable form of plant tissue culture with a view to producing plant secondary metabolites on a commercial scale. As a consequence, these decades witnessed the establishment of cell suspension cultures from hundreds of plant species, including many medicinal plants, as well as the development of bioreactor environments suitable for the cultivation of plant cells and plant organs on a large scale.

Provided that an appropriate phytohormone regime is chosen, plant cells cultivated *in vitro* may undergo coordinated division and development resulting in the formation of complex structures such as roots, shoots, somatic embryoids, and finally intact plants. Several textbooks, proceedings, and review articles have described this topic more comprehensively than is possible in this chapter (Murashige, 1974; Mantell et al., 1985; Day et al., 1985; Stafford and Warren, 1991; Payne et al., 1992). Issues that will be detailed in this chapter are summarized in Table 8.1 and Figure 8.1.

Medicinal Plant Biotechnology. From Basic Research to Industrial Applications
Edited by Oliver Kayser and Wim J. Quax
Copyright © 2007 WILEY-VCH Verlag GmbH & Co. KGaA, Weinheim
ISBN 978-3-527-31443-0

Table 8.1 *In-vitro* culturing techniques of medicinal plants.

Basic methods and techniques	
Seed germination *in vitro*	Multiple protocorm formation
Embryo culture	Embryo rescue
Callus culture	Initiation, maintenance
Organogenesis	Adventitious shoots, adventitious roots
Haploid technology	Anther culture, microspore culture
Somatic embryogenesis	Somatic embryoids, artificial seeds
Protoplast technology	
Protoplasts	Isolation, cultivation, plant regeneration
Somatic hybridization	Protoplast fusion, plant regeneration
Special techniques	
Gene transfer	*Agrobacterium*-mediated transformation
Germplasm storage	Cryopreservation
Permanent in-vitro *cultures*	
Suspension culture	Initiation, subculture, scale-up
Root culture	Initiation, subculture
Shoot culture	Initiation, subculture, rooting, plant regeneration
Bioreactors	Cell suspension cultures, root cultures, shoot cultures, bioreactor design, operation mode
Methods and techniques related to secondary metabolism	
Inducing variability	Mutation, somaclonal variation
Selection	Cell-aggregate cloning, protoplast cloning
Biotransformation	Bioconversion
Elicitation	Biotic elicitors, abiotic elicitors
Immobilization and permeation	Techniques, chemical and physical methods

8.2
Basic Methods and Techniques

8.2.1
Seed Germination In Vitro

Seed germination under axenic conditions is the simplest *in-vitro* culturing technique used. Surface-sterilized seeds of almost any plant species can be germinated on agar-solidified media on which they develop into small sterile plantlets. These in turn can be used to establish plant tissue cultures, such as callus or organ cultures. Since aggressive chemicals (e.g., hypochlorite or hydrogen peroxide) are generally used for decontamination, it may be of advantage to treat the rather robust seeds instead of the much softer explants, such as leaf or stem pieces.

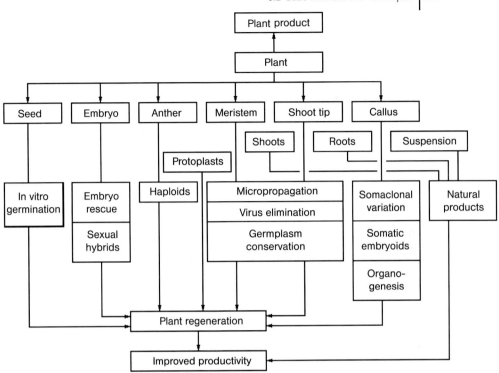

Fig. 8.1 *In-vitro* culture techniques exploited in plant breeding, preservation and propagation, as well as techniques used in the production of secondary metabolites.

The method has a broad application in the commercial propagation of orchids (Sato et al., 1978). Orchids produce tiny seeds in which only a few reserve compounds are stored. In Nature, orchids will not germinate unless infected by a fungus, which feeds the young plants until they are tall enough to produce their own food. Once the seed has germinated the embryo develops into a spherical protocorm that will continue to grow for weeks, months or years, until large enough to differentiate into an orchid plant. When germinated under sterile conditions (i.e., without the fungus) on suitable media, the nutrients provided will substitute for the fungus and allow the seeds to germinate and grow without it. Sterile seeds are usually isolated from surface-sterilized immature capsules; this procedure has the advantage that the seeds themselves have not to be sterilized. Another advantage of using unripe seeds is that they may germinate more quickly than mature seeds, since seed dormancy can be overcome. However, protocorm formation may also be significantly lower in immature seeds than in mature seeds (e.g., Kitsaki et al. 2004). When germinated *in vitro*, the developing protocorm will form additional protocorm-like bodies (multiple protocorm formation). The tendency toward the formation of such structures can be influenced by suitable additions to the nursing medium. Each protocorm-like structure will develop into an orchid plantlet. With *in-vitro* germination large numbers of seedlings can be raised in a relatively short time. As a rule, the young plants transferred to the greenhouse are infected without any further assistance with "their" fungus so that the endomycorrhiza necessary for further plant development will be established. Seed germination *in vitro* is not only used for the propagation of known orchids but can also be employed for the breeding of hybrid plants and their subsequent preservation and propagation.

8.2.2
Embryo Culture

Embryos can be isolated from seeds and cultivated under axenic conditions *in vitro* (Sharma et al., 1996). Grain fruits contain relatively large embryos which can be isolated quite easily, but if the embryos are too tiny to be prepared safely, the complete ovule can be isolated instead. The composition of culture media on which isolated embryos will develop is a problematic issue, and requires much experimental effort; however, if the respective demands are met, then regeneration to intact plants is easily achieved.

By applying the embryo culture technique, it is possible to shorten generation times by breaking seed dormancy. Moreover, embryos present in seeds obtained from sexual crossings that are not able to germinate can be "rescued" and stimulated to develop. Embryos can even be prepared from one seed and then transplanted into the endosperm of a "nurse" seed which, together with the nutrients added to the culture medium, supports further development of the embryo. Actually, the development of various *Triticale* cultivars, the first "synthetic" cereals, would not have been possible without the embryo rescue technique. Embryo rescue thus holds great promise, especially for obtaining plants from inherently weak embryos and for shortening the breeding cycle.

8.2.3
Callus Culture

The cell material described in early reports on plant cell and tissue culture is known as callus, and is initiated by removing explants (e.g., pieces of stem, leaf or root) from the whole plant or plants already kept *in vitro*. Except for those cases where axenic material is used, the respective explants are surface-sterilized and placed on agar-solidified nutrient media. The "wound callus" formed at sites of tissue injury can easily be removed from the initial explant, and its further development can be controlled by exogenous phytohormones or synthetic growth regulators (Fig. 8.2). During prolonged culture *in vitro* cells may achieve hormone autonomy, and often be distinguished as either cytokinin or auxin autotrophic, or completely hormone autotrophic. This phenomenon is called "hormone habituation" (White, 1951), and has been reported in cultures of several species. Hormone

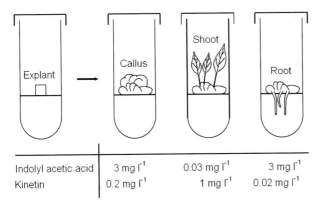

| Indolyl acetic acid | 3 mg l^{-1} | 0.03 mg l^{-1} | 3 mg l^{-1} |
| Kinetin | 0.2 mg l^{-1} | 1 mg l^{-1} | 0.02 mg l^{-1} |

Fig. 8.2 Initiation of callus cultures and induction of organogenesis. Morphogenesis can be triggered by changing the phytohormone regime. (Modified after Kreis, 2003.)

habituation is reversible in some cases, and hence epigenetic rather than genetic changes may be responsible.

Plant cell culture media are composed of many different compounds, including inorganic macronutrients (salts of N, P, S, K, Na, Ca, Mg) and micronutrients (salts of e.g., I, Ni, Fe, Cu, B, Mn, Co), as well as vitamins and a carbohydrate source (usually 2–5% sucrose). The most common media for plant cell culture are those devised by Murashige and Skoog (1962), Gamborg et al. (1968), Schenk and Hildebrandt (1972), and White (1934).

Plant cell culture media support the growth of microbial cells, and therefore any contamination must be avoided in order to provide an aseptic environment. The standard sterilization techniques that are used in microbiology can be applied. Usually, plant cells and tissues are handled under laminar flow in suitable cabinets. Culture media are autoclaved (121 °C, 15 min), while thermolabile compounds (e.g., phytohormones) can be filter-sterilized and then added to the autoclaved media. Various types of vessels are used for cell line preservation and subcultivation, for example disposable, sterile Petri dishes, tubes or Erlenmeyer flasks. Glass flasks and tubes may be sealed with foam or cotton bungs, while metal caps, aluminum foil or sterile clingwrap are used to cover the vessels to avoid them becoming contaminated.

Once established, individual callus isolates can be subcultivated over decades; typically, subculture periods range from three to six weeks. The callus must not be supposed to be homogeneous and composed of identical cells, and therefore subculture is selective. Selection may be either deliberate or unwittingly, but in general rapidly growing cells are favored. Callus becomes more homogeneous with time, and heterogeneity can be reduced by using strict subculture protocols. Callus provides a reliable, self-contained and quite uniform material, and is therefore suited to the study of biosynthetic problems. However, it cannot be advised as a system for the production of secondary metabolites because of the slow growth rates and the lack of suitable bioreactor systems to grow callus on a large scale.

Callus contains meristematic plant cells that are considered as being totipotent – that is, they may be differentiated and finally regenerated to intact plants. Callus can therefore be used as a starting material for organ and plant regeneration.

Plant cell cultures are collected and delivered on request. For example, the DSMZ (German Collection of Microorganisms and Cell Cultures) maintains more than 700 different plant cell lines from more than 80 different plant families. The collection contains a large number of cultures derived from plant species containing secondary metabolites of pharmaceutical importance (http://www.dsmz.de/plant_cell_lines).

8.2.4
Organogenesis

So-called adventitious roots or shoots are formed when organ development is induced in non-meristematic areas of a given plant tissue. In this process, specialized cells (e.g., epidermis cells) may turn meristematic, usually passing through a short period of callus formation (Fig. 8.3). Those plants which can be propagated vegetatively are especially susceptible to adventitious organ formation *in vitro*. Plant regeneration through multiple adventitious shoot differentiation has been established in many species (Bajaj, 1991, 1992a,b,c, 1997a,b). Most often, leaf ex-

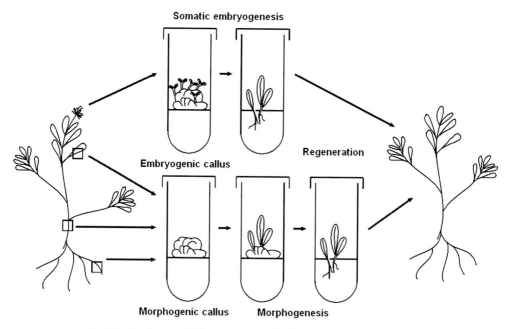

Fig. 8.3 Morphogenesis from root, stem or leaf explants via adventitious organ formation and somatic embryogenesis from embryogenic callus via somatic embryos.

plants are used and shoot formation is then induced by a suitable phytohormone treatment. Explants of younger plants usually respond better than older ones, and herbaceous plants as a rule respond better than woody plants. Adventitious root or shoot formation can also be induced in other plant organs or in callus. Depending on the starting material, growth regulators such as benzyladenine, 4-dichlorophenoxyacetic acid, indole-3-butyric acid, α-naphthalene acetic acid, 2-isopentenyladenine or thidiazuron are added to the basal growth media to induce shoot formation. Shoots are then propagated and rooted as described for shoot cultures (see below). Changes of light conditions, photoperiod, temperature, and/or reduction of the nutrients in the culture medium may help to achieve high rooting efficiency. With some ornamental plants (e.g., geraniums and petunias) adventitious organ formation is the method of choice for clonal propagation.

Although micropropagation by adventitious shoot proliferation or through meristem culture (see below) may be equally successful, it must be considered that adventitious propagation systems yield plants that are genetically more variable compared to those obtained via meristem culture. Adventitious organ formation may also occur in meristematic shoot cultures. A degree of expertise is required to decide whether newly formed shoots arise from true meristematic zones, or whether they were formed adventitiously (i.e., not at the vegetation tips or zones).

8.2.5
Haploid Technology

Haploid technology is an important tool for plant breeding, and involves the cultivation of plant tissue *in vitro* (Hu and Zeng, 1984; Morrisson and Evans, 1988; Bajaj, 1990a; Atanassov et al., 1995). The technique allows for a significant time reduction in the achievement of homozygous breeding lines in crop improvement. As early as 1953 it was found that anthers of ginkgo developed haploid callus when cultivated *in vitro* (Tulecke, 1953), but a further 10 years passed until the first haploid plants (*Datura innoxia*) were successfully regenerated via anther culture (Guha and Maheshwari, 1964). Meanwhile, many plants – including medicinal plants – have been regenerated via anther culture (Jain et al., 1997; Bajaj, 1990a).

Haploids are sporophytes with the chromosome number of gametophytes, and haploid cells are formed during meiosis. Haploid plant cells are present in the embryo sack (megagametophyte) or the pollen (microgametophyte), and consequently haploids can be obtained via gynogenesis from cells of the embryo sack, or by androgenesis from pollen. In this way, monohaploid and dihaploid plants can be obtained from diploid and tetraploid plants, respectively. Colchicine can be used to induce chromosome doubling by inhibiting chromosome segregation during meiosis, and in this way homozygous cells can be obtained. These may form a regenerable callus that can be stimulated to develop into homozygous ("dihaploid") plants, while haploid callus can be produced via gynogenesis or androgenesis, the latter method being the preferred one in practice. In androgenesis, two basic techniques can be distinguished, namely anther culture and microspore culture.

One disadvantage of anther culture is that callus may develop from diploid cells of the anther wall. Thus, the plants may result from somatic embryogenesis (see below) or adventitious shoot formation (see above). These then compete with the haploid plants derived via androgenesis, and must be sorted out by cytogenetic analysis, cultivating microspores alone, i.e. without adhering diploid anther tissue will solve this problem. A number of methods have been developed for the preparation and cultivation of microspores and protocols for the regeneration of haploid plants thereof. Genetic variation, termed "gametoclonal variation" (Evans et al., 1984), may occur during the tissue culture phase.

8.2.6
Somatic Embryogenesis

When roots and shoots develop simultaneously and from a common origin, a differentiation process is seen that leads to shoot and root formation in a coordinated manner. This process resembles the development of zygotic embryos, and is termed "somatic embryogenesis". The structures formed from callus or suspension cells are therefore called somatic embryos or, more precisely, somatic embryoids (embryo-like structures). Globular somatic embryoids, which can be maintained in culture over long periods of time, can differentiate into heart-shaped and later torpedo-shaped embryoids (Merkle et al., 1995). The regeneration of properly developed somatic embryoids to intact plants is quite simple (see Fig. 8.3). For carrot – which serves as a model system for studying somatic embryogenesis – it was calculated that about 50 000 somatic embryoids per liter medium are formed each day (Ducos et al., 1993). Hence, somatic embryogenesis is regarded as a system of choice for mass propagation, but for this purpose the process must be synchronized (Osuga and Komamine, 1994). It must also be considered that embryogenic cell cultures may lose this quality during prolonged cultivation. The advantages of somatic embryoids include high multiplication rates and the potential for scale-up in bioreactors (Eeva et al., 2003).

Somatic embryoids can be encapsulated in alginate as single-embryo beads to produce so-called "artificial seeds" (Redenbaugh et al., 1988; Senaratna, 1992; Bajaj, 1995a; Cervelli and Senaratna, 1995). The quality of artificial seeds depends on the supply of growth regulators and nutrients. Artificial seeds can be desiccated in this way, thereby facilitating year-round production, storage, and distribution. The production of artificial seeds has been reported in many species (Gupta et al., 1993), with germination rates as high as 30% and 65% being found in alfalfa and celery, respectively. Somatic embryoids have also been used for the production of secondary plant metabolites. For example, the torpedo stages of somatic celery embryoids (*Apium graveolens*) will develop the celery flavor (Al-Abta et al., 1979), while embryoids of foxglove (*Digitalis lanata*) are capable of producing cardenolides (Greidziac et al., 1990), albeit in only small quantities.

8.3
Protoplast Technology

8.3.1
Protoplasts

Plant cells possess a thick and quite rigid cell wall which is composed of cellulose and other polysaccharides, such as pectin. In hypertonic solutions, the plasma membranes of cells contract from their walls, with removal of the wall material releasing large populations of spherical, osmotically fragile structures, termed "protoplasts". The main objectives in using protoplast isolation and culture techniques are to:

- regenerate an intact plant from a single cell for proving the concept of totipotency;
- fuse protoplasts of different origin with a view to producing a hybrid cell which subsequently regenerates into a hybrid plant that cannot be obtained by sexual crossing (somatic hybridization); and
- obtain cells accessible for genetic manipulation.

Indeed, many economically important species – including medicinal plants – have been regenerated via protoplasts (Potrykus et al., 1983; Eriksson, 1985; Davey et al., 2005).

During recent years the isolation of protoplasts has become routine since, if chopped plant tissue is treated with pectinase and cellulase, it will release protoplasts (Cocking, 1960). Today, mechanical procedures are rarely employed, though they can be used to isolate the protoplasts of very large cells, or to isolate protoplasts used in patch-clamp experiments in order to avoid unwanted changes in physiological and electrical properties that might result from enzymatic digestion of the cell wall (Binder et al., 2003). In any case, plasmolysed tissue must be used. A slightly hypertonic medium during the isolation and cultivation process is mandatory, as otherwise the protoplasts will burst. When provided with the correct chemical and physical stimuli, each protoplast is capable (in theory) of regenerating a new wall and undergoing repeated mitotic division to produce a callus from which fertile plants may be regenerated.

Isolated protoplasts begin cell wall regeneration almost immediately after their isolation. To achieve this they require osmotic protection until their new cell walls can withstand the normal cell turgor, and this is generally provided by the addition of non-metabolizable sugar alcohols (e.g., mannitol or sorbitol). Protoplasts from different species, or from different tissues of the same species, may vary in their nutritional requirements, but guidance is available from the monographs of the *Biotechnology in Agriculture and Forestry* series, edited by Y.P.S. Bajaj (e.g., Bajaj, 1994a). There is no universal protocol with regard to medium composition and physical parameters, however, and many of the protocols developed have been based on media introduced for plant cell culture. For the culture of protoplasts at low density, coconut milk (Kao and Michayluk, 1975) may be added to the culture medium, and auxins and cytokinins are normally essential for sustained protoplast growth. Isolated protoplasts may be cultured in liquid medium over semi-solid medium containing so-called "nurse" or "feeder" cells. Nurse cells are located in the same vessel as the protoplast, separated only by a semi-permeable membrane or any other device that supports the separation of protoplasts from feeders. Nurse cells may be from either the same or from different species, and can help overcome the problem of cell densities that are too low to allow proliferation. The nurse cells support protoplast growth and division by unknown factors, or via nutrients emitted by the feeder cells and absorbed by the protoplasts. Alternatively, toxic compounds generated during protoplast culture may be adsorbed or sequestered by the nurse cells.

8.3.2
Somatic Hybridization

Somatic hybrids obtained through the fusion of plant protoplasts have widened the genetic variability of plants (Bajaj, 1994b; Nagata and Bajaj, 2001). Several groups have reported the generation of hybrid plants through protoplast fusion, and some improvements have been described in fusion technology since the early days when plant protoplasts were fused using the polyethylene glycol (PEG) method (Kao and Michayluk, 1974). This chemically induced fusion was achieved at high pH (10.5) in the presence of calcium (10 mM) and PEG (10–50%). However in 1979 it was found that protoplasts, when held in an appropriate electric field, would fuse together (Senda et al., 1979), and today electrofusion is the preferred method, mainly because membrane damage is reduced and the entire process is much cleaner and more controllable than PEG-induced fusion (Davey et al., 2005).

Electrofusion is a two-step procedure. In the first step, a non-uniform, alternating electric field is applied which causes the protoplasts to line up perpendicular to the electrodes. The protoplasts chain length is influenced by the distance between the electrodes and the cell density, but ideally the protoplasts agglutinate pair-wise. In the second step, fusion is induced by a single high-voltage DC pulse. It is thought that the charge difference between the outside and inside of the plasma membrane finally crushes the membrane. This causes the electric potential to break down, whereupon the membrane structure can be reformed, including the mixing of membranes of agglutinated protoplasts; the result is cell fusion. A number of recent reports have described the generation of unique plants through somatic hybridization by protoplast fusion (Davey et al., 2000, 2005). As with medicinal plants, quite extensive hybridization studies have been performed in the Solanaceae (*Nicotiana*, *Solanum*) and in *Mentha*. In the latter genus, somatic hybridization was aimed either at modifying the composition of the oil (Sato et al., 1996), or at combining essential oil quality with disease resistance (Krasnyanski et al., 1998).

8.4
Special Techniques

8.4.1
Gene Transfer

The key tool for genetic transformation with a view towards the design and creation of engineered plants or plant tissue cultures, is the Ti plasmid of the plant
pathogen *Agrobacterium tumefaciens*. This bacterium is responsible for crown gall
disease in higher plants, and its Ti plasmid can be used as a natural vector for genetic transformation. This plasmid bears a region termed T-DNA which can be
transferred into the host genome (Chilton et al., 1977). Additional genetic material
can be tailored into the T-DNA, whereas the "disease genes" – which are responsible for the tumor growth of the crown galls – can be either destroyed or removed.

Today, the introduction of genes into plant cells has become routine, and the
available methods include not only the biological approach using *A. tumefaciens*
but also the physical methods of biolistics (which uses microprojectiles coated with
DNA) or electroporation (which uses a high-voltage electric pulse) to achieve trans-
fection (Hoykaas, 2000; Leech et al., 2000). In the biological approach, only one or
a few copies of T-DNA containing the gene(s) of interest are inserted; in contrast,
the physical methods are generally less tidy but offer certain advantages over conventional *Agrobacterium*-mediated gene transfer. One such advantage is the possibility of bypassing *Agrobacterium* host specificity, which allows a broadening of
the range of transformable plants. In many cases either plantlets cultured *in vitro*
or isolated protoplasts are used as the targets for genetic transformation.

In-vitro culture usually forms part of the overall transformation protocol, as the
transformed cells must be separated from those non-transformed and regenerated
to viable plants or permanent tissue cultures. In cases where transformed cells
can be regenerated into whole and fertile plants, cultivars with new traits can be
created.

8.4.2
Germplasm Storage

The maintenance and analysis of hundreds of cell culture lines or organ cultures is
a labor-intensive task, especially when many cell cultures grow quite rapidly and
require transfer to fresh medium almost weekly. In addition, the risk of losing material because of microbial contamination, technical or human errors is always
present. Therefore, efforts have been undertaken to develop methods for storing
plant cells in a convenient way, at minimal personnel expense. One means of limiting growth is to reduce the cultivation temperature. Alternatively (or in addition),
growth-retardant chemicals such as phytohormones (abscisic acid) or osmotics
(mannitol) may be added to the culture medium to achieve reduced growth rates.

The deposition of plant tissue at cryogenic temperatures with retention of viability (cryopreservation) is another option for long-term storage. Initially, cryopreser-

vation was driven by the concern for loss of diversity of crops. Highly sophisticated protocols, including two-stage freezing, vitrification and encapsulation-dehydration, have been developed (e.g., Withers, 1983; Bajaj, 1995b; Towill and Bajaj, 2002). The main drawback for the broad utilization of cryopreservation in the storage of plant germplasm is the lack of standardized protocols. In particular, cell suspension cultures with a high content of secondary metabolites are still recalcitrant to cryopreservation, because of the possible autotoxicity of the secondary compounds that accumulate in these cells. During freezing and thawing membranes may become permeable, such that compounds usually stored in a safe intracellular location may leak and consequently inhibit metabolic processes. With regard to genetic stability, it has been shown that variety-specific characteristics have not been changed by storage in liquid nitrogen (Mix-Wagner et al., 2003).

8.5
Permanent *In-Vitro* Cultures

8.5.1
Cell Suspension Culture

Callus can be submerged in liquid media, where it usually disintegrates into small cell aggregates; suspension cultures can be established in this way. Real single-cell suspensions have been reported, but in most cases cell aggregates are eventually formed. Suspension-cultured plant cells can be maintained in a dedifferentiated state with rather uncoordinated cell division over long periods of time. With regard to homogeneity and genetic stability, the same limitations as described for callus apply, though cells in suspension proliferate more rapidly than callus cells. Growth can be followed simply by assessing various parameters that are associated either directly or indirectly with growth, such as cell number, fresh and dry mass, mitotic index, medium components (e.g., sugar, phosphate), medium conductivity (Allan, 1991), or – non-invasively – by packed cell volume (Blom et al., 1992).

Since plant cell cultures can be cultivated on a large scale in commercial bioreactors they may serve as an alternative source for the production of plant secondary metabolites (Alfermann and Petersen, 1995; Scragg, 1997; Ramachandra Rao and Ravishankar, 2002; Vanisree et al., 2004). Arguments scored for this assumption are that:

- the production will not dependent on geographical, political, and seasonal factors;
- defined protocols and production systems offer reliable yield and quality; and
- products not known from nature can be produced either *de novo* or by biotransformation using cheap precursors.

It was supposed that, concomitant with their totipotency, cells cultivated *in vitro* should also be capable of producing compounds normally associated with the intact plant. Hence, it was tempting to use plant cells for producing important phy-

Table 8.2 Examples of pharmaceuticals accumulating in plant cell cultures in high concentrations.

Plant species	Product	Yield [% DW]
Salvia officinalis	Rosmarinic acid	36.0
Morinda citrifolia	Anthraquinones	18.0
Lithospermum erythrorhizon	Shikonin	12.4
Thalictrum minus	Berberine	10.6
Berberis wilsonae	Jatrorrhizin	10.0
Perilla frutescens	Anthocyanins	8.9
Dioscorea deltoidea	Diosgenin	3.8
Papaver somniferum	Sanguinarine	2.5
Catharanthus roseus	Serpentine	2.2

DW, dry weight.

tochemicals with masses of highly productive cells that could be cultivated in the same way as microorganisms in large bioreactors. However, secondary product formation may be an integral part of a differentiation program, and a result of differential expression of genetic information. It was not too surprising to find that some important target compounds could not be produced by suspension-cultured cells. However, a number of plant cell cultures are able to accumulate larger quantities of secondary metabolites than the intact plant (Table 8.2).

In higher plants, certain biochemical traits are only fully developed in specific organs, or during specific developmental stages, and secondary product formation is often associated with structural differentiation. Product accumulation may be associated with: (i) the presence of certain cell types; (ii) the presence of certain organelles; and (iii) the expression and regulation of biosynthetic or catabolic genes. Therefore, with a view to producing plant secondary metabolites *in vitro*, it is important to consider not only suspensions consisting of very small cell aggregates but also organ cultures, such as root or shoot cultures that can be cultivated on a large scale.

8.5.2
Root Culture

Root cultures can be established using different approaches. One method is based on the infection of suitable tissues with *Agrobacterium rhizogenes*, which recognizes signal molecules (e.g., acetosyringone, α-hydroxy-actosyringone) exuded by susceptible wounded plant cells and attaches to them. The infection of plants with *A. rhizogenes* results in the development of hairy roots at the site of infection. These developing roots can be removed and transferred several times to fresh antibiotic-containing medium before a stable axenic hairy root culture can be established. Usually, hairy roots are fast-growing and require no external supply of growth hormones (Giri and Narasu, 2000). In 1990, Tepfer reported that stable hairy root cultures had been obtained from 116 plant species belonging to 30 dicotyledonous families, and today root cultures of this type are used in many laboratories. Root-derived plant products not produced by cell suspension culture are currently being reinvestigated for production using the hairy root culture technology (Table 8.3).

Adventitious root formation can be induced in callus by changing the phytohormone balance (see Fig. 8.2), whereafter the roots can be excised and cultivated further. Root formation can be triggered back and forth by changing the phytohormone regime. With regard to secondary metabolite formation and accumulation, it has been shown that some metabolites are produced only at the root culture stage (Flores et al., 1987). The growth and productivity of transformed and non-transformed root cultures have been compared in several reports, but convincing evidence to support the assumption that transformed roots generally have faster growth rates or higher production rates than their non-transformed counterparts is still lacking.

Table 8.3 Examples of secondary metabolites produced in hairy root cultures of medicinal plants.

Plant	Product
Artemisia absinthum	Essential oils
Artemisia annua	Artemisinin
Atropa belladonna	Atropine
Cassia obtusifolia	Anthraquinones
Catharanthus roseus	Indole alkaloids
Centranthus ruber	Valepotriates
Cinchona ledgeriana	Quinine
Coleus forskohlii	Forskolin
Datura stramonium	Tropane alkaloids
Digitalis purpurea	Cardenolides
Duboisia myoporoides	Scopolamine
Echinacea purpurea	Alkamides
Fagopyrum esculentum	Flavanol
Glycyrrhiza glabra	Flavonoids
Hyoscyamus muticus	Tropane alkaloids
Hyoscyamus niger	Hyoscyamine
Linum flavum	Lignans
Lithospermum erythrorhizon	Shikonin, benzoquinone
Nicotiana tabacum	Nicotine, anatabine
Panax ginseng	Saponins
Papaver somniferum	Codeine
Rauwolfia serpentina	Reserpine
Rubia tinctorum	Anthroquinone
Sesamum indicum	Naphthoquinone
Solanum laciniatum	Steroidal alkaloids
Tanacetum parthenium	Sesquiterpene coumarin ether
Trigonella foenum graecum	Diosgenin
Valeriana officinalis	Valepotriates

Root cultures show a greater genetic stability than suspension cultures. The selection of well-growing cells during subculture and scale-up is unlikely, and hence productive stability is more reliable than with cell suspension cultures. Root cultures can be cultivated in large vessels and may be regarded as biocatalysts immobilized in their own organic matrix; therefore, the engineering advantages for immobilized cells may apply.

8.5.3
Shoot Culture

Shoot cultures are free of intervening roots and undifferentiated callus. As in the case of root cultures, shoot cultures may produce secondary metabolites that are not seen in non-differentiated cultures (Table 8.4). Although anticipated, shoot cultures will not produce all compounds seen in the leaves of intact plants. If the site of synthesis of a given compound is the root, then it will not appear in cultured shoots even if the leaves are the sink for these metabolites. Even if the compounds of interest are produced in a given shoot culture (see Table 8.4), product patterns and concentrations may differ from those seen in the intact plant.

Shoot cultures can be initiated in several ways (Payne et al., 1992). Typically, they are initiated from sterile germinated seedlings, but they can also be obtained from dissected shoot apical meristems (Fig. 8.4) or from stem sections, as well as from callus. Shoot cultures can be propagated on agar-solidified media or as amphibian cultures in liquid media. These two types of shoot culture may differ considerably with respect to their morphology; in particular, the leaves may remain rather small and rudimentary in amphibian cultures. Since shoot differentiation enables the expression of several biosynthetic pathways, amphibian cultures may accumulate lesser quantities of secondary metabolites than shoots cultured on solid media. Cytokinins stimulate shoot growth and are thus added to the culture media, which often contain reduced levels of some macronutrients (e.g., nitrogen-containing salts) as compared to cell suspension cultures. Although shoot cultures are photosynthetically active, sugars are usually added to the culture media to boost growth.

Table 8.4 Examples of secondary metabolites in shoot cultures of medicinal plants.

Plant species	Product
Artemisia annua	Artemesinin
Atropa belladonna	Atropine
Catharanthus roseus	Vindoline
Cinchona spp.	Vinblastine, quinine
Digitalis lanata	Cardenolides
Digitalis purpurea	Cardenolides
Picrorrhiza kurroa	Kutkin
Stevia rebaudiana	Steviosides
Withania somniferum	Withanolides
Dicentra pergrina	Alkaloids

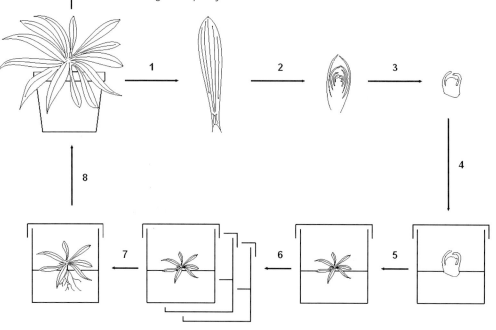

Fig. 8.4 *In-vitro* propagation and regeneration via meristem culture.
1–3: Preparation of apical meristem from axial buds. 4–6: Cultivation and
multiple shoot formation. 7: Rooting *in vitro*. 8: Regeneration of intact plants.

Shoot cultures are regarded as being genetically more stable than non-differentiated callus or suspension cultures; this is an important point, as most shoot cultures have been initiated with a view towards micropropagation (Bajaj, 1991, 1992a,b,c, 1997a,b). Rooting can be achieved when shoots are transferred to agar-solidified media containing auxins only, with roots generally developing within two weeks.

8.5.4
Bioreactors

The initial studies on the production of a particular natural compound by plant tissue culture are carried out with small tissue samples grown in shake flasks or other small vessels. Most of the physiological and biochemical experiments designed to elucidate metabolic pathways can be performed with cells cultured under these conditions. Since mass cultivation of plant cells has been proposed as an alternative way for supplying important phytochemicals, it was necessary to develop systems that would allow plant cells to be cultivated on a large scale (Kreis and Reinhard, 1989; Payne et al., 1992; Ramachandra Rao and Ravishankar, 2002). However scale-up – that is, the reproduction on a larger scale of a production process developed on a smaller scale – is not feasible with this type of vessel. Thus, if cultivation in larger volumes is planned, then bioreactors must be used (Fig. 8.5). One advantage of bioreactors is that they are scalable – that is, it is possible to reproduce

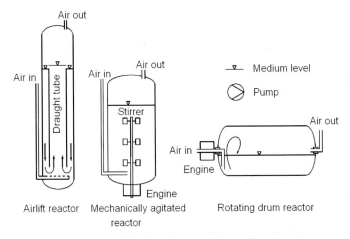

Fig. 8.5 Bioreactors used for the cultivation of plant cells with a view to producing secondary metabolites.

on a large scale those conditions which were found conducive to optimal production on a smaller scale. With regard to scale-up, the maintenance of constant environmental conditions at the various scales of operation is simple in terms of the soluble components (nutrients), but more difficult with respect to the physical environment (shear, mixing, gas transfer).

The most common type of reactor for culturing plant cells is the mechanically agitated vessel, also called a stirred-tank bioreactor. This type of reactor uses impellers of various types for gas–liquid transfer and mixing. Because of the shear sensitivity of plant cells, low agitation speeds of about 100 to 300 rpm are appropriate. Unfortunately, however, these low speeds are generally insufficient to break the incoming gas stream into small bubbles, and in order to obtain sufficient oxygen transfer the incoming gas stream is dispersed as fine bubbles using a perforated ring or sintered glass or metal.

The second type of bioreactors used in plant cell cultivation is the pneumatically agitated bioreactor, most often referred to as the air-lift fermentor. Here, agitation is coupled to aeration, and consequently it may be necessary to use higher gas flow rates to provide mixing than would be required for oxygen transfer. More recently, the trend has been towards the use of mechanically agitated bioreactors, though at present no guidelines can be given concerning the best stirrer-type. Since metabolic activity is a function of the surface area of an organism, it is evident that growth and production rates are much lower in the case of plant cell cultures than microbial cells. In microbes, secondary metabolite production is generally not associated with biomass production, and generally the products are released into the bathing medium. Production rates are generally high, and production phases short. In the case of plant cell cultures, growth is quite slow, metabolite production is low, and usually the products are stored in the cells. Most plant cell culture processes have productivities between 0.025 and 1.0 g L^{-1} per day (see Table 8.5). Thus, the goal must be to make less-productive cells more productive, or to place as many cells as possible into a bioreactor (Matsubara et al. 1989).

Depending on the characteristics of product formation, fermentation can be operated in different modes. The operating mode refers to how the nutrient and product streams are supplied or removed with respect to time, with the appropriate operating mode depending on the timing of product synthesis and growth.

- When all nutrients the culture requires are supplied initially this operating mode is termed *batch operation*, which may be regarded as the standard operation mode:

- In *fed-batch operation* the nutrients (including elicitors and precursors) are fed either at intervals or continuously. A fed-batch mode is justified if it were detrimental to the cells to supply all of the nutrients in one operation.

- In *repeated fed-batch operation*, a substantial portion of the spent medium or the cell suspension is withdrawn and replaced with fresh culture medium, and in this way the culture can be reactivated. Overall productivity may be improved, mainly because the bioreactor can be used for several consecutive production runs without the need of cleaning and sterilization.

- A *two-stage batch operation* is appropriate for non-growth-associated products, and also offers the possibility of medium exchange, where the first medium supports rapid growth and the second medium supports product synthesis.

- The term *chemostat* refers to the operation in which fresh medium is continuously fed to the bioreactor and a stream of suspension is continuously withdrawn. In plant cell culture, a chemostat mode is mainly used as an experimental tool (van Gulik et al., 1993) rather than for secondary product synthesis.

- In a *perfusion operation*, fresh medium is added to the vessel continuously and spent medium (but not cells, as in the chemostat) are withdrawn continuously. In general, the perfusion reactor is appropriate for non-growth-associated product synthesis and immobilized systems.

The demands for bioreactors devised for use with root cultures and shoot cultures are different, and thus considered separately. It was mentioned earlier that shoot or root differentiation sometimes enables the expression of biosynthetic pathways that are expressed at very low levels or not at all in cell suspension cultures. The scale-up of shoot cultures in bioreactors poses several challenges, largely stemming from the cells' unique morphological characteristics, their susceptibility to mechanical damage, possible vitrification, and light requirements (Payne et al., 1992). Production processes with cultured shoots are unlikely to be substituted for field-grown plants, though they may provide excellent systems to study biosynthetic sequences not operative in suspension cultures.

Root cultures, on the other hand, may serve as good candidates for the production of root-specific metabolites. In order to enhance the productivity of hairy root cultures, various methods have been used, including the selection of high-producing clones and elicitation. Bioreactor set-ups have been designed for the cultivation of roots on a large scale (Wysokinska and Chmiel, 1997; Giri and Narasu, 2000), though in general cultured roots form loosely entangled mats and it is not necessary (it may even be deleterious) to use stirrers in order to provide mixing. The use of mist bioreactors also appears promising (Wysokinska and Chmiel, 1997). Roots may be regarded as a form of an immobilized system, and consequently techniques developed for immobilized cells can be employed. The bioreactor system described by Wilson et al. (1990) contains an immobilization matrix, comprised of a series of stainless steel barbs, to which roots may attach. The vessel can be run in either submerged mode or mist-phase mode. Root cultures may also serve as good candidates for processes using perfusion operation.

8.6
Methods and Techniques Related to Secondary Metabolism

8.6.1
Inducing Variability

In-vitro culture techniques can be used to select variant cells from which improved cell lines or plants may be established. Variation can be enhanced by the use of induced physical or chemical mutagenesis (Ahloowahlia, 1998; Donini and Sonnino, 1998). Even without mutagenic treatment plant tissues show genetic instability when cultured *in vitro* (Stafford, 1991). Variability in embryogenic and regenerative potential has frequently been reported, and permanent plant cell cultures may exhibit variation in product yield over successive subcultures. Variability may to some extent also reflect the fact that cell cultures are usually not of clonal origin, and that fast-growing cells may be selected during successive subculture. This involves the fact that rapidly proliferating cells may be more susceptible to mutations, and that the culture conditions themselves may therefore be regarded as mutagenic. This may be problematic for the preservation and micropropagation of elite plants, but for the generation of variant plants this phenomenon offers possibilities (Evans and Sharp, 1986; Bajaj, 1990b). Variation occurring in plants regenerated from tissue culture is termed "somaclonal variation" (Larkin and Scowcroft, 1981). The regenerated "somaclones" may be stable, but this may not be taken as a rule since somaclonal variation is most likely a result of both mutations and reversible epigenetic changes.

Related to somaclonal variation is the question of the stability of plant cell suspension cultures with regard to their biosynthetic potential. As cell cultures are neither homogeneous nor synchronous with regard to their mitotic cycle, rapidly dividing but non-producing cells may eventually overgrow the producing cells, and this would result in productivity loss. Indeed, this has been reported for many plant cell suspension cultures, including those of important medicinal plants (Stafford, 1991). This phenomenon constitutes a clear problem for the commercialization of any plant cell culture process if a given elite cell culture might lose its high-yield character, even during production of the cell mass required for a large culture vessel. The "apparent" stability (meta-stability) of a given cell culture may be the effect of a strict subculture regime (Ulbrich et al., 1985; Fowler, 1988) rather than the result of the clonal origin of the cells.

8.6.2
Selection

Elite cell lines are obtained by selection following a variety of strategies that include macroscopic, microscopic and chemical inspections. Cell-aggregates or protoplasts may be used as the starting material for selection (for a comprehensive description, see Dix, 1990). Selection can be easily achieved if the product of interest is a pigment (Fujita et al., 1984). The screening and selection of high-producing cell lines has not been successful in all cases, however. For example, cell cultures of opium poppy (*Papaver somniferum*) were initiated in many laboratories worldwide with a view to producing codeine in large quantities. As yet, unfortunately, all approaches have failed as poppy cell cultures simply do not produce considerable amounts of morphinanes (Table 8.5).

Table 8.5 Examples of productivity of plant cell cultures.

Plant species	Product	Productivity $[mg\ L^{-1}\ day^{-1}]$
Castanospermum australe	Castanospermine	0.05
Catharanthus roseus	Ajmalicin	4
Coptis japonica	Berberin	600
Dioscorea deltoidea	Diosgenin	10
Lithospermum erythrorhizon	Shikonin	150
Marchantia polymorpha	Arachidonic acid	4.4
Papaver somniferum	Codeine	0.25
	Sanguinarin	34
Taxus canadensis	Paclitaxel	10

8.6.3
Biotransformation

Although many plant cell suspension cultures fail to produce the compounds seen in the plants from which they have been established, these cells may be used in biotransformation processes where exogenous organic compounds are modified by living cells. For any organism, biotransformation represents a means of coping with lipophilic xenobiotics that may easily cross membranes and thus accumulate in cells and tissues, where they may act as toxins. Biotransformation studies in cell suspension cultures have been carried out with a view to: (i) producing new chemicals; (ii) producing known chemicals more economically; (iii) investigating the metabolic fate of xenobiotics; and (iv) elucidating metabolic pathways (Franssen and Walton, 1999; Giri et al., 2001).

The supply of a suitable precursor may result in the formation of a compound known from the intact plant. This would indicate that part of the biosynthetic sequence is still operating in the cultured cells. The demonstration of a biotransformation reaction *in vivo* may be a first step towards elucidating an enzyme-catalyzed conversion, or isolating biosynthetic enzymes or genes.

During the 1970s and 1980s, *Digitalis lanata* cell cultures, which had a large capacity to convert digitoxin derivatives to the respective digoxin derivatives, were selected by cell aggregate cloning. Finally, a cell culture process was developed in which metildigoxin could be prepared with good yields and almost no side reactions from 4‴-*O*-methyldigitoxin (Reinhard et al., 1989). More recently, alternative approaches using *D. lanata* cells to produce digoxin-type cardenolides have been attempted, with special emphasis being placed on the use of digitoxin as the substrate for biotransformation (Kreis and Reinhard, 1992). With regard to the biotechnological application of a biotransformation process using plant cells, it must be borne in mind that plant cell cultures cannot compete with microbial systems in terms of the production rates attainable. Therefore, only those reactions that are restricted to plant cells and which yield products of high economic value can be of commercial interest.

8.6.4
Elicitation

Several "non-invasive" strategies were undertaken with a view to improving productivity in plant cell culture. Some of the major roles of plant secondary metabolites are to protect plants from attack by herbivores and pathogens, or to aid in surviving other biotic and abiotic stresses. Indeed, some strategies for culture production of the metabolites based on this principle have been developed to improve the yield of such plant secondary metabolites. These include treatment with various elicitors, signal compounds, and abiotic stresses (Yukimune et al., 1996; Zhao et al., 2000, 2001, 2005).

Elicitors are compounds that are isolated from microorganisms and known capable of triggering secondary metabolite formation in plant cell cultures (Barz, 1988; Zhao et al., 2005). In whole plants, elicitors stimulate the formation of so-called phytoalexins as a part of pathogen defense. Jasmonates are transducers of elicitor signals for the production of plant secondary metabolites, and their exogenous application to plant cell cultures may stimulate the biosynthesis of various secondary metabolites (Gundlach et al., 1992; Farmer et al., 2003).

Many such treatments effectively promote the production of a wide range of plant secondary metabolites, but to date only the production of shikonin by *Lithospermum erythrorhizon* cell cultures and of taxol by *Taxus* cell cultures have been successfully industrialized. The elicitor approach was not successful in all cases: codeine biosynthesis, for example, has not yet been achieved. However, compounds sharing the same precursors and intermediates, such as the antimicrobial alkaloid sanguinarine, may accumulate in quite high amounts. Using cell cultures of *Papaver somniferum*, the production of sanguinarine was shown to be "elicited" by preparations from fungal mycelia (Eilert and Constabel, 1986; Park et al., 1992).

In the above-mentioned example, the fungal preparation simulated the pathogen attack, whereupon sanguinarine production was increased in the presence of the elicitor compared to the control without fungal mycelia.

8.6.5
Immobilization and Permeation

Immobilization is a technique which traps catalytically active cells or enzymes in a matrix, thus preventing them from entering the mobile phase. The main types of immobilization are adsorption (e.g., to polystyrene or glass), covalent attachment (e.g., by glutaraldehyde or carbodiimide coupling) and entrapment (e.g., alginate or agarose) (Yeoman, 1987; Scragg, 1991). Immobilization has distinct advantages: (i) the biomass can be retained and utilized continuously; (ii) high biomass levels can be achieved; and (iii) cells can easily be separated from the spent medium. However, it is essential that product formation is not strictly associated with growth, and that cell growth can be suppressed in order to prevent disintegration of the immobilization matrix. Assuming that immobilized cells maintain pro-longed viability and biosynthetic capacity, the product should leach out of the cells and beads into the medium. Although Brodelius et al. (1979) were the first to im-mobilize plant cells, this approach has subsequently been used for the *de-novo* bio-synthesis of secondary metabolites as well as for biotransformations (Payne et al., 1992; Ramachandra Rao and Ravishankar, 2002). A special application of an im-mobilization technique may be seen in the entrapment of somatic embryos with a view towards the production of artificial seeds (see Section 8.2.6).

One very important point for the success of an immobilized system is that of product release. In microbial processes, the products are generally released into the bathing medium from which they can be extracted with ease. In plant cells, however, natural products are usually stored in the vacuole, and in order for them to be released the two membranes, namely the tonoplast and plasma membrane, must be penetrated. Several methods have been attempted to release intracellular products without affecting cell viability (Brodelius and Nilsson, 1983; Knorr et al., 1985; Parr et al., 1987). Dimethylsulfoxide is widely used as a permeabilizing agent (e.g., Parr et al., 1984), while ultrasonication, electroporation and ultra-high pres-sure have also been used to recover secondary metabolites (Dornenburg and Knorr, 1993; Brodelius et al., 1988).

8.7
Conclusions

The *in-vitro* cultivation of plant cells and organs has become a relatively simple task. Some of the methods described are commercially used worldwide in plant breeding, micropropagation and the generation of disease-free stocks. The ability to regenerate fertile plants from explants and selected or engineered cells is the most important aspect of medicinal plant biotechnology. Although permanent cell and organ cultures can produce a wide range of compounds used in medicine, there are few success stories of production at the commercial scale. However, *in-vitro* cultures provide useful systems to study the biosynthetic pathways leading to plant secondary metabolites at the enzyme level. In this respect, plant tissue cultures have turned out to be a "a pot of gold" for those seeking to identify and isolate the enzymes and genes of secondary metabolism (Zenk, 1991). *In-vitro* culturing techniques are also exploited in genetic engineering of plants; for example, transgenic *Atropa belladonna* plants producing scopolamine instead of hyoscyamine (Yun et al., 1992) provided the first example of how pharmaceutically important plants could be engineered successfully.

References

Ahloowalia, B.S. In vitro techniques and mutagenesis for the improvement of vegetatively propagated plants. In: Jain, S.M., Brar, D.S., Ahloowalia, B.S. (Eds.), *Somaclonal Variation and Induced Mutations in Crop Improvement*. Kluwer Academic Publishers, Dordrecht, **1998**, pp. 293–310.

Al-Abta, S., Galpin, I.J., Collin, H.A. Flavor compounds in tissue cultures of celery. *Plant Sci. Lett.* **1979**, *16*, 129–134.

Alfermann, A.W., Petersen, M. Natural product formation by plant cell biotechnology. Results and perspectives. *Plant Cell, Tissue Organ Culture* **1995**, *43*, 199–205.

Allan, E. Plant cell culture. In: Stafford, A., Warren, G. (Eds.), *Plant Cell and Tissue Culture*. Open University Press, Milton Keynes, **1991**, pp. 1–24.

Atanassov, A., Zagorska, N., Boyadjiev, P., Djilianov, D. In vitro production of haploid plants. *World J. Microbiol. Biotechnol.* **1995**, *11*, 400–408.

Bajaj, Y.P.S. (Ed.), *Biotechnology in Agriculture and Forestry, Vol. 12, Haploids in Crop Improvement I*. Springer, Berlin, Heidelberg, New York, **1990a**.

Bajaj, Y.P.S. (Ed.), *Biotechnology in Agriculture and Forestry, Vol. 11, Somaclonal Variation in Crop Improvement I*. Springer, Berlin, Heidelberg, New York, **1990b**.

Bajaj, Y.P.S. (Ed.), *Biotechnology in Agriculture and Forestry, Vol. 17, High-Tech and Micropropagation I*. Springer, Berlin, Heidelberg, New York, **1991**.

Bajaj, Y.P.S. (Ed.), *Biotechnology in Agriculture and Forestry, Vol. 18, High-Tech and Micropropagation II*. Springer, Berlin, Heidelberg, New York, **1992a**.

Bajaj, Y.P.S. (Ed.), *Biotechnology in Agriculture and Forestry, Vol. 19, High-Tech and Micropropagation III*. Springer, Berlin, Heidelberg, New York, **1992b**.

Bajaj, Y.P.S. (Ed.), *Biotechnology in Agriculture and Forestry, Vol. 20, High-Tech and Micropropagation IV*. Springer, Berlin, Heidelberg, New York, **1992c**.

Bajaj, Y.P.S. (Ed.), *Biotechnology in Agriculture and Forestry, Vol. 29, Plant Protoplasts and Genetic Engineering V*. Springer, Berlin, Heidelberg, New York, **1994a**.

Bajaj, Y.P.S. (Ed.), *Biotechnology in Agriculture and Forestry, Vol. 27, Somatic Hybridization in Crop Improvement I*. Springer, Berlin, Heidelberg, New York, **1994b**.

Bajaj, Y.P.S. (Ed.), *Biotechnology in Agriculture and Forestry, Vol. 30, Somatic Embryogenesis and Synthetic Seed I*. Springer, Berlin, Heidelberg, New York, **1995a**.

Bajaj, Y.P.S. (Ed.), *Biotechnology in Agriculture and Forestry, Vol. 32, Cryopreservation of Plant Germplasm I*. Springer, Berlin, Heidelberg, New York, **1995b**.

Bajaj, Y.P.S. (Ed.), *Biotechnology in Agriculture and Forestry, Vol. 40, High-Tech and Micropropagation VI.* Springer, Berlin, Heidelberg, New York, **1997a**.

Bajaj, Y.P.S. (Ed.), *Biotechnology in Agriculture and Forestry, Vol. 39, High-Tech and Micropropagation V.* Springer, Berlin, Heidelberg, New York, **1997b**.

Barz, W. Elicitation and metabolism of photoalexins in plant cell cultures. In: *Applications of Plant Cell and Tissue Culture.* Ciba Foundation Symposium 137, John Wiley & Sons, Chichester, **1988**, pp. 239–253.

Binder, K.A., Wegner, L.H., Heidecker, M., Zimmermann, U. Gating of Cl⁻ currents in protoplasts from the marine alga *Valonia utricularis* depends on the trans-membrane Cl⁻ gradient and is affected by enzymatic cell wall degradation. *J. Membr. Biol.* **2003**, *191*, 165–178.

Blom, T.J.M., Kreis, W., Van Iren, F., Libbenga, K.R. A non-invasive method for the routine-estimation of fresh weight of cells grown in batch suspension cultures. *Plant Cell Rep.* **1992**, *11*, 146–149.

Brodelius, P., Nilsson, K. Permeabilization of immobilized plant cells resulting in release of intracellularly stored products with preserved cell viability. *Eur. J. Appl. Microbiol. Biotechnol.* **1983**, *17*, 275–280.

Brodelius, P., Deus, B., Mosbach, K., Zenk, M.H. Immobilized plant cells for the production of natural products. *FEBS Lett.* **1979**, *103*, 93–97.

Brodelius, P., Funk, C., Shillito, R.D. Permeabilization of cultured plant cells by electroporation for the release of intracellularly stored secondary products. *Plant Cell Rep.* **1988**, *7*, 186–188.

Cervelli, R., Senaratna, T. Economic aspects of somatic embryogenesis. In: Aitken-Christie, J., Kozai, T., Smith, L. (Eds.), *Automation and Environmental Control in Plant Tissue Culture.* Kluwer Academic Publishers, Dordrecht, **1995**, pp. 29–64.

Chilton, M.D., Drummond, M.H., Merio, D.J., Sciaky, D., Montoya, A.L., Gordon, M.P., Nester, E.W. Stable incorporation of plasmid DNA into higher plant cells: the molecular basis of crown gall tumorigenesis. *Cell* **1977**, *11*, 263–271.

Cocking, E.C. A method for the isolation of plant protoplasts and vacuoles. *Nature* **1960**, *187*, 927–929.

Davey, M.R., Lowe, K.C., Power, J.B. Protoplast fusion for the generation of unique plants. In: Spier, R.E. (Ed.), *Encyclopaedia of Cell Technology.* John Wiley & Sons, New York, **2000**, pp. 1090–1096.

Davey, M.R., Anthony, P., Powera, J.B., Lowe, K.C. Plant protoplasts: status and biotechnological perspectives. *Biotechnology Adv.* **2005**, *23*, 131–171.

Day, P., Zaitlin, M., Hollaender, A. *Biotechnology in Plant Science.* Academic Press, Florida, **1985**.

Dix, P. *Plant Cell Line Selection: Procedures and Applications.* VCH, Weinheim, **1990**.

Donini, P., Sonnino A. Induced mutation in plant breeding: current status and future outlook. In: Jain, S.M., Brar, D.S., Ahloowalia, B.S. (Eds.), *Somaclonal Variation and Induced Mutations in Crop Improvement.* Kluwer Academic Publishers, Dordrecht, **1998**, pp. 255–292.

Dornenburg, H., Knorr, D. Cellular permeabilization of cultured tissues by high electric field pulses or ultra high pressure for the recovery of secondary metabolites. *Food Biotechnol.* **1993**, *7*, 35–28.

Ducos, J.-P., Bollon, H., Pettard, V. Production of carrot somatic embryos in a bioreactor. *Appl. Microbiol. Biotechnol.* **1993**, *39*, 465–470.

Eeva, M., Ojala, T., Tammela, P., Galambosi, B., Vuorela, H., Hiltunen, R., Fagerstedt, K., Vuorela, P. Propagation of *Angelica archangelica* plants in an air-sparged bioreactor from a novel embryogenic cell line, and their production of coumarins. *Biol. Plant.* **2003**, *46*, 321–480.

Eilert, U., Constabel, F. Elicitation of sanguinarine accumulation in *Papaver somniferum* cells by fungal homogenates: An induction process. *J. Plant Physiol.* **1986**, *125*, 167–172.

Eriksson, T.R. Protoplast isolation and culture. In: Fowke, L.C., Constabel, F. (Eds.), *Plant Protoplasts.* CRC Press, **1985**.

Evans, D.A., Sharp, W.R. Applications of somaclonal variation. *Biotechnology* **1986**, *4*, 528–532.

Evans, D.A., Sharp, W.R., Medina-Filho, H.P. Somaclonal and gametoclonal variation. *Am. J. Bot.* **1984**, *71*, 759–774.

Farmer, E.E., Alméras, E., Krishnamurthy, V. Jasmonates and related oxylipins in plant responses to pathogenesis and herbivory.

Curr. Opin. Plant Biol. **2003**, *6*, 372–378.

Flores, H.E., Hoy, M.W., Puckard, J.J. Secondary metabolites from root cultures. *Trends Biotechnol.* **1987**, *5*, 64–69.

Fowler, F.M. Problems in commercial exploitation of plant cell cultures. In: *Applications of Plant Cell and Tissue Culture.* Ciba Foundation Symposium 137, John Wiley & Sons, Chichester, **1988**, pp. 239–253.

Franssen, M.C.R., Walton, M.J. Biotrans-formations. In: Walton, M.J., Brown, D.E. (Eds.), *Chemicals from Plants, Perspectives on Plant Secondary Products.* Imperial College Press, London, **1999**, pp. 277–325.

Fujita, Y., Takahashi, S., Yamada, Y. Selection of cell lines with high productivity of shikonin derivatives through protoplasts of *Lithospermum erythrorhizon. Proceedings, 3rd European Congress on Biotechnology* **1984**, Volume 1, pp. 161–166.

Gamborg, O.L., Miller, R.A., Ojima, K. Nutrient requirements of suspension cultures of soybean root cells. *Exp. Cell Res.* **1968**, *50*, 151–158.

Gautheret, R.J. Culture du tissus cambial. *C. R. Acad. Sci.* **1934**, *198*, 2195–2196.

Giri, A., Narasu, M.L. Transgenic hairy roots: recent trends and applications. *Biotechnol. Adv.* **2000**, *18*, 1–22.

Giri, A., Dhingra, V., Giri C.C., Singh, A., Ward, O.P., Narasu, M. L. Biotransforma-tions using plant cells, organ cultures and enzyme systems: current trends and future prospects. *Biotechnology Adv.* **2001**, *19*, 175–199.

Greidziac, N., Dietrich, B., Luckner, M. Batch culture of somatic embryos of *Digitalis lanata* in gaslift fermentors. Development and cardenolide accumulation. *Planta Med.* **1990**, *56*, 175–178.

Guha, S., Maheshwari, S.C. *In vitro* produc-tion of embryos from anthers of *Datura. Nature* **1964**, *204*, 497.

Gundlach, H., Müller, M.J., Kutchan, T.M., Zenk, M.H. Jasmonic acid is a signal transducer in elicitor-induced plant cell cultures. *Proc. Natl. Acad. Sci. USA* **1992**, *89*, 2389–2393.

Gupta, P.K., Pullman, G., Timmis, R., Kreitinger, M., Carlson, W.C., Grob, J., Welty, E. Forestry in the 21st Century. *Biotechnology* **1993**, *11*, 454–459.

Hoykaas, P.J.J. *Agrobacterium*, a natural metabolic engineer of plants. In: Verpoorte, R., Alfermann, A.W. (Eds.), *Metabolic Engineering of Plant Secondary Metabolism.* Kluwer Academic Publishers, Dordrecht, Boston, London, **2000**, pp. 51–67.

Hu, H., Zeng, J.Z. Development of new varieties via anther culture. In: Ammirato, P.V., Evans, D.A., Sharp, W.R., Yamada, Y. (Eds.), *Handbook of Plant Oil Culture, Vol. 3.* Macmillan, New York, **1984**, pp. 65–90.

Jain, S.M., Sopory, S.K., Veilleux, R.E.E. (Eds.) In Vitro *Haploid Production in Higher Plants.* Kluwer Academic Publishers, Dordrecht, **1997**.

Kao, K.N., Michayluk, M.R. A method for high-frequency intergeneric fusion of plant protoplast. *Planta* **1975**, *115*, 355–367.

Kitsaki, C.K., Zygouraki, S., Ziobora, M., Kintzios, S. In vitro germination, proto-corm formation and plantlet development of mature versus immature seeds from several *Ophrys* species (Orchidaceae). *Plant Cell Rep.* **2004**, *23*, 284–290.

Knorr, D., Miazga, S.M., Teutonico, R.A. Immobilization and permeabilization of cultured plant cells. *Food Technol.* **1985**, *39*, 139–142.

Krasnyanski, S., Ball, T.M., Sink, K.C. Somatic hybridization in mint: identifica-tion and characterization of *Mentha piperita* (+) M. spicata hybrid plants. *Theoret. Appl. Genet.* **1998**, *96*, 683–687.

Kreis, W. Plant cell, tissue and organ cultures as tools for the elucidation of biosynthetic pathways. In: Sprecher, E., Caesar, W. (Eds.), *Gesellschaft für Arzneipflanzen-forschung: 50 Years 1953–2003: A Jubilee Edition.* Wissenschaftliche Verlagsgesell-schaft, Stuttgart, **2003**, pp. 123–136.

Kreis, W., Reinhard, E. The production of secondary metabolites by plant cells cultivated in bioreactors. *Planta Med.* **1989**, *55*, 409–416.

Kreis, W., Reinhard, E. 12β-Hydroxylation of digitoxin by suspension-cultured *Digitalis lanata* cells. Production of digoxin in 20-litre and 300-litre airlift bioreactors. *J. Biotechnol.* **1992**, *26*, 257–273.

Larkin, P.J., Scowcroft, W.R. Somaclonal variation – a novel source of variability from cell cultures. *Theoret. Appl. Genet.* **1981**, *67*, 197–201.

Leech, M.J., Burtin, D., Hallard, D., Hilliou, F., Kemp, B., Palacios, N., Rocha, P., O'Callaghan, D., Verpoorte, R., Christou, P. Particle gun methodology as a tool in metabolic engineering. In: Verpoorte, R., Alfermann, A.W. (Eds.), *Metabolic Engineering of Plant Secondary Metabolism.* Kluwer Academic Publishers, Dordrecht, Boston, London, **2000**, pp. 69–86.

Mantell, S.H., Matthews, J.A., McKee R.A. *Principles of Plant Biotechnology. An Introduction to Genetic Engineering in Plants.* Blackwell Scientific Publishers, **1985**.

Matsubara, K., Yamada, Y., Kitani, S., Yoshioka, T., Morimoto, T., Fujita, T. High density culture *Coptis japonica* cells increases berberine production. *J. Chem. Technol. Biotechnol.* **1989**, *46*, 61–69.

Merkle, S.A., Parrot, W.A., Flinn, B.S. Morphogenic aspects of somatic embryogenesis. In: Thorpe, T.A. (Ed.), *In Vitro Embryogenesis in Plants.* Kluwer Academic Publishers, Dordrecht **1995**, pp. 155–203.

Mix-Wagner, G., Schumacher, H.M., Cross, R.J. Recovery of potato apices after several years of storage in liquid nitrogen. *Cryoletters* **2003**, *24*, 33–41.

Morrison, R.A., Evans, D.A. Haploid plants from tissue culture: new plant varieties in a shortened time frame. *Biotechnology* **1988**, *6*, 684–690.

Murashige, T. Plant propagation through tissue cultures. *Annu. Rev. Plant Physiol.* **1974**, *25*, 135–166.

Murashige, T., Skoog, F. A revised medium for rapid growth and bioassays with tobacco tissue cultures. *Physiol. Planta* **1962**, *15*, 473–496.

Nagata, T., Bajaj, Y.P.S. (Eds.), *Biotechnology in Agriculture and Forestry, Vol. 49, Somatic Hybridization in Crop Improvement II.* Springer, Berlin, Heidelberg, New York, **2001**.

Osuga, K., Komamine, A. Synchronization of somatic embryogenesis from carrot cells at high frequency as a basis for the mass production of embryos. *Plant Cell Tissue Organ Cult.* **1994**, *39*, 125–135.

Park, J.M., Yoon, S.Y., Giles, K.L., Songstad, D.D., Eppstein, D., Novakovski, D., Roewer, I. Production of sanguinarine by suspension culture of *Papaver somniferum* in bioreactor. *J. Ferm. Bioeng.* **1992**, *74*, 292–296.

Parr, A.J., Robins, R.J., Rhodes, M.J.C. Permeabilization of *Cinchona ledgeriana* cells by dimethyl sulfoxide. Effect on alkaloid release and long term membrane integrity. *Plant Cell Rep.* **1984**, *3*, 262–265.

Parr, A.J., Robins, R.J., Rhodes, M.J.C. Release of secondary metabolites by plant cell cultures. In: Webb, C., Mavituna, F. (Eds.), *Plant and Animal Cells: Process Possibilities.* Ellis-Horwood, Chichester, **1987**, pp. 229–237.

Payne, G.F., Bringi, V., Prince, C., Shuler, M.L. *Plant cell and tissue culture in liquid systems.* Hanser Publishers, Munich, Vienna, New York, Barcelona, **1992**.

Potrykus, I., Harms, C.T., Hinnen, A., Hütter, R., King, P.J., Shillito, R.D. *Protoplasts 1983.* Lecture Proceedings of 6th International Protoplast Symposium, Basel, August 12–16, 1983. Experientia Supplement, Vol. 46. Birkhäuser Verlag, Basel, **1983**.

Ramachanda Rao, S., Ravishankar, G.A. Plant cell cultures: Chemical factories of secondary metabolites. *Biotechnology Adv.* **2002**, *20*, 101–153.

Redenbaugh, K., Fugii, J.A., Slade, D. Encapsulated plant embryos. In: Mizrahi, A. (Ed.), *Biotechnology in Agriculture.* Liss, New York, **1988**, pp. 225–248.

Reinhard, E., Kreis, W., Barthlen U., Helmbold, U. Semicontinuous cultivation of *Digitalis lanata* cells. Production of β-methyldigoxin in a 300-L airlift bioreactor. *Biotechnol. Bioeng.* **1989**, *34*, 502–508.

Sato, D.M., Barnes, D., Murashige, T. Bibliography on clonal multiplication of orchids through tissue culture. *Methods Cell Sci.* **1978**, *4*, 783–786.

Sato, H., Yamada, K., Mii, M., Hosomi, K., Okuyama, S., Uzawa, M., Ishikawa, H., Ito, Y. Production of an interspecific somatic hybrid between peppermint and gingermint. *Plant Sci.* **1996**, *115*, 101–107.

Schenk, R.U., Hildebrandt, A.C. Medium and techniques for induction and growth of monocotyledonous and dicotyledonous plant cell cultures. *Can. J. Bot.* **1972**, *50*, 199–204.

Scragg, A. Plant cell bioreactors. In: Stafford, A., Warren, G. (Eds.), *Plant Cell and Tissue Culture.* Open University Press, Buckingham, **1991**, pp. 221–239.

Scragg, A.H. The production of aromas by plant cell cultures. In: Schepier, T. (Ed.), Advances in Biochemical Engineering/Biotechnology, Volume 55. Springer-Verlag, Berlin, **1997**, pp. 239–263.

Senaratna, T. Artificial seeds. *Biotechnology Adv.* **1992**, *10*, 379–392.

Senda, M., Takeda, J., Abe, S., Nakamura, T. Induction of cell fusion of plant protoplasts by electrical stimulation. *Plant Cell Physiol.* **1979**, *20*, 1441–1443.

Sharma, D.R., Kaur, R., Kumar, K. Embryo rescue in plants, a review. *Euphytica* **1996**, *89*, 325–337.

Stafford, A. Genetics of cultured plant cells. In: Stafford, A., Warren, G. (Eds.), *Plant Cell and Tissue Culture*. Open University Press, Milton Keynes, **1991**, pp. 25–47.

Tepfer, D. Genetic transformation using *Agrobacterium rhizogenes*. *Physiol. Planta* **1990**, *79*, 14–16.

Towill, L.E., Bajaj, Y.P.S. (Eds.), *Biotechnology in Agriculture and Forestry, Vol. 50, Cryopreservation of Plant Germplasm II*. Springer, Berlin, Heidelberg, New York, **2002**.

Tulecke, W. A tissue derived from the pollen of *Ginkgo biloba*. *Science* **1953**, *117*, 599–600.

Ulbrich, B., Wiesner, W., Arens, H. Large-scale production of rosmarinic acid from plant cell cultures of *Coleus blumei* Benth. In: Neumann, K.H., Barz, W., Reinhard, E. (Eds.), *Primary and Secondary Metabolism of Plant Cell Cultures*. Springer-Verlag, Berlin, Heidelberg, New York, Tokyo, **1985**, pp. 293–303.

van Gulik, W.M., ten Hoopen, H.J.G., Heijnen, J.J. A structured model describing carbon and phosphate limited growth of *Catharanthus roseus* plant cell suspensions in batch and chemostat culture. *Biotechnol. Bioeng.* **1993**, *41*, 771–780.

Vanisree, M., Lee, C.-Y., Lo, S.-F., Nalawade, S.M., Lin, C.Y., Tsay, H.-S. Studies on the production of some important secondary metabolites from medicinal plants by plant tissue cultures. *Bot. Bull. Acad. Sinica* **2004**, *45*, 1–22.

White, P.R. Potentially unlimited growth of excised tomato root tips in a liquid medium. *Plant. Physiol.* **1934**, *9*, 585–600.

White, P.R. Potentially unlimited growth of excised plant callus in an artificial nutrient. *Am. J. Bot.* **1939**, *26*, 59–64.

White, P.R. Neoplastic growth in plants. *Q. Rev. Biol.* **1951**, *26*, 1–16.

Wilson, P.D.G., Hilton, M.G., Meehan, P.T.H., Waspe, C.R., Rhodes, M.J.C. The cultivation of transformed roots, from laboratory to pilot plant. In: Nijkamp, H.J.J., Van der Plaas. L.H.W., Van Aartrijk. J. (Eds.), *Progress in Plant Cellular and Molecular Biology*. Kluwer Academic Publishers, Dordrecht, **1990**, pp. 700–705.

Withers, L.A. Germplasm maintenance and storage in plant biotechnology. In: Mantell, S.H., Smith, H. (Eds.), *Plant Biotechnology. Society for Experimental Biology Seminar Series 18*. Cambridge University Press, Cambridge **1983**, pp. 187–218.

Wysokinska, H., Chmiel, A. Transformed root cultures for biotechnology. *Acta Biotechnol.* **1997**, *17*, 131–159.

Yeoman, M.M. Techniques, characteristics, properties, and commercial potential of immobilized plant cells. In: Constabel, F., Vasil, I. (Eds.), *Cell Culture and Somatic Cell Genetics of Plants, Vol. 4*. Academic Press, San Diego, **1987**, pp. 197–215.

Yukimune, Y., Tabata. H., Higashi, Y., Hara, Y. Methyl jasmonate-induced overproduction of paclitaxel and baccatin III in *Taxus* cell suspension cultures. *Nature Biotechnol.* **1996**, *14*, 1129–1132.

Yun, D.-J., Hashimoto, H., Yamada, Y. Metabolic engineering of medicinal plants: transgenic *Atropa belladonna* with an improved alkaloid composition. *Proc. Natl. Acad. Sci. USA* **1992**, *89*, 11799–11803.

Zenk, M.H. Chasing the enzymes of secondary metabolism: plant cell cultures as a pot of gold. *Phytochemistry* **1991**, *30*, 3861–3863.

Zhao, J., Zhu, W.H., Hu, Q., He, X.W. Improved indole alkaloid production in *Catharanthus roseus* suspension cell cultures by various chemicals. *Biotechnol. Lett.* **2000**, *22*, 1221–1226.

Zhao, J., Fujita, K., Yamada, J., Sakai, K. Improved beta-thujaplicin production in *Cupressus lusitanica* suspension cultures by fungal elicitor and methyl jasmonate. *Appl. Microbiol. Biotechnol.* **2001**, *55*, 301–305.

Zhao, J., Davis, L.C., Verpoorte, R. Elicitor signal transduction leading to production of plant secondary metabolites. *Biotechnol. Adv.* **2005**, *23*, 283–333.

9
Biotechnological Methods for Selection of High-Yielding Cell Lines and Production of Secondary Metabolites in Medicinal Plants

Donald P. Briskin

9.1
Introduction

There is widespread interest in the application of plant tissue culture methods and biotechnological approaches to the production of medicinal plants and isolation of medicinal secondary products. When compared to traditional agricultural growth, plant tissue culture production of medicinal plants offers a number of unique advantages, including the possibility of year-round, continuous production of plant medicinal compounds under highly controlled conditions. As the *in-vivo* production of secondary metabolites by plants can be highly influenced by plant growth environment factors such as climactic and soil conditions, pathogen attack and herbivory [1, 2], *in-vitro* tissue culture growth of medicinal plants can provide a route for consistent medicinal chemical isolation from plant materials [3]. Indeed, the lack of consistency in the levels of active chemicals in herbal medicines has been a continual issue [4] and variation in secondary metabolite production in agriculturally grown plant material can certainly contribute to this problem. Plant tissue culture growth of medicinal plants can also be scaled up using continuous-culture systems such as "bioreactors", and this could allow for automated, high-level isolation of medicinal secondary products [5, 6]. This would be particularly advantageous for secondary metabolite production from medicinal plants that are slow-growing *in vivo*, and could eliminate concerns regarding over-harvesting of medicinal plants which are either rare or endangered [3, 5, 6].

In order to be a viable approach for the commercial production of herbal medicines or pharmaceutical chemical precursors, plant tissue culture growth must result in a high-yield recovery of secondary metabolites. Enhanced production of medicinal plant secondary metabolites in cell culture has been achieved in some cases via the optimization of culture conditions, immobilization and physical shock, and the use of host-pathogen "elicitor" compounds [7, 8]. The ability to transform *in-vitro* plant cell cultures with *Agrobacterium rhizogenes* to produce "hairy root cultures" has had a major impact on this field as this approach leads to fast-growing tissue cultures that are easy to maintain and which produce elevated levels of sec-

Medicinal Plant Biotechnology. From Basic Research to Industrial Applications
Edited by Oliver Kayser and Wim J. Quax
Copyright © 2007 WILEY-VCH Verlag GmbH & Co. KGaA, Weinheim
ISBN 978-3-527-31443-0

ondary metabolites [9, 10]. In addition, recent advances in molecular methods have allowed the modification ("bioengineering") of metabolic pathways in cell cultures, which as resulted in either enhanced secondary metabolite production or the generation of unique metabolites not produced in the original plant material [11, 12].

This chapter will focus on methods utilized in plant tissue culture growth for the recovery of phytomedicinal chemicals. A major emphasis will be placed on describing approaches for the enhancement of secondary metabolite production in cell cultures and the development of systems for the large-scale recovery of phytomedicinal plant secondary products.

9.2
Medicinal Plant Tissue Cultures and the In-Vitro Production of Phytomedicinal Secondary Metabolites

A number of medicinal plants have been successfully introduced into tissue culture (see Table 9.1). Typically, these tissue cultures involve either callus cells growing on a semi-solid media, or liquid suspension cultures. Both systems, involving the unorganized growth of plant cells, have the advantage of allowing straightforward, continual propagation of cultures and in the case of liquid suspension cells, cell production can be scaled up to high levels using bioreactors [7, 8]. In this respect, suspension cell cultures have the advantage of higher growth rates than callus cultures [7]. Moreover, secondary metabolites can be more easily extracted from liquid suspension culture than organized growth systems. Hence, if adequate levels of secondary metabolites can be produced by such a system, this would be ad-

Table 9.1 Examples of medicinal plant growth and production of secondary metabolites in cell cultures.

Plant	Culture type	Phytomedicinal compound(s)	Function	Reference
Catharanthus roseus	Suspension culture	Vinblastine/vincristine	Antitumor	13
Capsicum frutescens	Suspension culture	Capsaicin	Diaphoretic	14
Cinchona sp.	Suspension culture	Alkaloids	Antimalarial	15
Coptis japonica	Suspension culture	Berberine	Antimicrobial	16
Digitalis purpurea	Suspension culture	Cardiac glycosides	Cardiac function	17
Ginkgo biloba	Suspension culture	Ginkgolide A	Antioxidant	18
Glycyrrhiza glabra	Callus culture	Triterpenes	Anti-inflammatory	19
Hypericum perforatum	Suspension culture	Hypericins	Antidepressant	20
Panax ginseng	Callus culture	Ginsenosides	Tonic	21
Panax notoginseng	Suspension culture	Ginsenosides	Tonic	22
Papaver somniferum	Suspension culture	Opiates	Anesthetic	23
Piper methysticum	Callus culture	Kavapyrones	Sedative	24
Podophyllum hexandrum	Suspension culture	Podophyllotoxin	Antitumor	25
Rauwolfia serpentina	Suspension culture	Reserpine	Antihypertensive	26
Taxus sp.	Suspension culture	Taxol	Antitumor	27

vantageous for the commercial production of these compounds. An recent extensive survey by Vanisree et al. [8] included over 90 reports of successful culture of medicinal plants, and in most cases the work involved production of either callus or liquid suspension cultures. These authors also provide details regarding the culture media required for the successful culture of each medicinal plant species.

As noted in Table 9.1, analysis of the cell cultures revealed the production of medicinal secondary products as found for the *in-vivo* plant. One major problem, however, is that often the level of secondary metabolite production in such unorganized plant cell systems may be low, and below levels that would make industrial production feasible [3, 7]. As pointed out by Walton et al. [7], unorganized cell systems, such as callus and suspension cells, may lack the biochemical control mechanisms that specify secondary metabolite production, and secondary metabolite production may be restored or elevated with the induction of organized cell growth in organ cultures. Moreover, herbal medicines often represent a complex mixture of secondary metabolites, and cell cultures may not produce the appropriate spectrum and relative levels of secondary metabolites necessary for an effective phytomedicinal preparation [7]. Indeed, studies conducted with Kava callus cultures by our laboratory group demonstrated levels of kavapyrones quite different from those isolated from the intact roots typically utilized in production of the sedative herbal medicine [24]. Nevertheless, it has been possible to increase secondary metabolite production in some medicinal plant callus or suspension cultures, and this was the case for some of the examples included in Table 9.1. The achievement of elevated levels of secondary metabolite production in cell culture can require modification of the growth media components, including levels of plant hormones, the addition of fungal elicitors, or the addition of metabolic precursors for the secondary metabolite [7, 8]. A more detailed discussion of the strategies and factors utilized to enhance secondary metabolite production in callus or liquid suspension cultures is presented in the next section.

9.3
Factors Leading to Elevated Production of Secondary Metabolites in Medicinal Plant Cell Cultures

Invariably, the approach to improve secondary metabolite production in medicinal plant cell cultures is empirical. Details of some factors that have led to increases in secondary metabolite production are discussed in the following sections.

9.3.1
Cell Culture Media Components

Several research groups have found that modification of the basic components of the nutrient medium can elevate secondary metabolite production. With respect to mineral nutrients, this can involve elevation of specific nutrients or subjecting the cultures to low nutrient stress. For example, in Ginseng (*Panax ginseng*), the production of ginsenosides – the active medicinal compounds – can be increased by optimization of levels of media nitrogen, potassium, and phosphate [30–32]. Production of the antimicrobial compound berberine in *Coptis japonica* cell cultures [16, 33] and the antitumor agent paclitaxel in *Taxus cuspidata* cell cultures [34] was also achieved by empirical optimization of mineral nutrients present in the growth medium. In contrast, the production of capsaicin – the alkaloid compound responsible for the pungent effects in pepper – was elevated by subjecting *Capsicum annum* cells to low nutrient stress [14]. It should be noted that low nutrient stress is also an environmental factor which can lead to elevated secondary product biosynthesis in intact plants [35].

The level and form of the carbon supply in the culture medium can have a strong effect upon secondary metabolite production in cell cultures. In several different studies, elevation of sucrose was found to promote paclitaxel biosynthesis in *Taxus* sp. grown in culture [36, 37]. Paclitaxel biosynthesis by cell cultures was also further enhanced by the inclusion of fructose [36]. Likewise, elevated sucrose in the growth media has been shown to increase alkaloid biosynthesis in *Catharanthus roseus* cultures [38], anthocyanin biosynthesis in *Perilla frutesens* cell suspensions [22], and steroidal alkaloid production in *Solanum aviculare* cultures [39]. For cell cultures of ginseng, the initial presence of glucose and sucrose followed by sucrose alone during the later phase of culture growth promoted the production of ginsenosides [40]. This effect of sugar supply on secondary product biosynthesis could result directly from these sugars as a specific carbon input, or it could reflect a shift in metabolism due to changes in general carbon/nitrogen balance. Indeed, carbon/nitrogen balance can have a strong effect on the levels of secondary product biosynthesis in intact plants [35]. This influence of carbon/nitrogen balance was also observed to be critical for promoting secondary metabolite biosynthesis in a large-scale culture system for *Catharanthus roseus* [41]. Here, secondary metabolism was elevated during the later phase of culture growth and correlated with a depletion of nitrate in the growth medium. Tal et al. [42] also demonstrated an influence of carbon versus nitrogen levels in determining diosgenin biosynthesis in cultures of *Dioscorea deltoidea*.

In addition to levels of mineral nutrients and carbon source/supply levels, the forms of plant hormones, as well as media components such as amino acids, can further affect secondary metabolite production in cell cultures. Levels of plant hormones present in the growth medium were shown to influence ginsenoside production in ginseng cell cultures [40], and hormone removal in a large-scale culture system of *Papaver somniferum* resulted in about a threefold elevation in production of the medicinal alkaloids, codeine and morphine [23]. The effect of amino acids on paclitaxel production in *Taxus cuspidata* cell cultures was examined by Fett-Neto et al. [34], and the presence of phenylalanine in the growth medium was found to promote maximum levels of paclitaxel biosynthesis.

9.3.2
Elicitors and Jasmonates

As many secondary metabolites have a key role in plant defense responses against pathogens, it is not surprising that the addition of molecules involved in the signaling of pathogen attack would increase secondary metabolite production. In this respect, there are a number of reports where the addition of an elicitor to the growth medium increases secondary product biosynthesis [43]. It should be noted that in this discussion the term "elicitor" refers to carbohydrate oligomers either of plant or fungal origin (excluding chitosan) involved in early pathogen recognition events leading to a plant defense response [44]. Once an elicitor is recognized by an appropriate plant receptor, the activation of signal transduction pathways results in a response which can involve an array of components including protein kinases, jasmonate, salicylic acid, and reactive oxygen species [43]. However, in the plant tissue culture literature, the term "elicitor" has been used in a broader context and can refer to any chemical substance (biological or inorganic) or treatment resulting in increased secondary product metabolism [8].

This relationship between the role of an elicitor in a pathogen defense response and the potential for induction of secondary product metabolism was clearly shown in early studies by Funk et al. [45]. Here, a fungal elicitor fraction which induced glyceollin biosynthesis in soybean (*Glycine max*) could also increase berberine biosynthesis fourfold in cell suspension cultures of *Thalictrum regosum*. Production of the phytoalexin, glyceollin, is a well-characterized early defense response in soybean to fungal pathogen attack which is initiated by elicitor recognition [44].

Several studies have demonstrated the value of fungal elicitors in promoting biosynthesis of the antitumor agent, paclitaxel, by *Taxus* sp. cell cultures, and these elicitors have been derived from a variety of microbial sources. For example, Ciddi et al. [46] utilized elicitor fractions from extracts of *Penicillium minioluteum*, *Botrytis cinerea*, *Verticillium dahliae* and *Gilocladium deliquescens*, and found that all elicitor fractions increased the biosynthesis of paclitaxel in cell cultures. Paclitaxel biosynthesis in *Taxus chinensis* cell cultures was also increased by the addition of an elicitor fraction generated from *Aspergillus niger*, an endophytic fungus which infects the inner bark of this plant *in vivo* [47].

Fungal elicitors have also been used to elevate ginsenoside production in cell suspension cultures of ginseng. In studies conducted by Lu et al. [48], a yeast extract was found to serve as an elicitor, resulting in an over 20-fold stimulation of ginsenoside production. Likewise, Archambault et al. [49] were able to obtain a twofold stimulation in the production of the antimicrobial compound, sanguinarine, through the use of a chitosan elicitor in *Papaver somniferum* cell cultures.

Jasmonic acid and its derivative forms such as methyl jasmonate (collectively referred to as "jasmonates") represent important plant signal transduction molecules involved in the response to pathogen attack [50]. Not surprisingly, these molecules have been shown to induce secondary product metabolism in cell cultures. Indeed, Gundlach et al. [51] showed that jasmonates are involved in the elicitor-based elevation in secondary metabolite production for a wide range of cell cultures. This general role of jasmonate in linking elicitor-recognition to secondary product biosynthesis was also demonstrated in studies measuring phytoalexin biosynthesis in cell cultures of rice [52]. Here, jasmonate levels were shown to correlate with phytoalexin biosynthesis, while the inhibition of jasmonate production using ibuprofen reduced both jasmonate and phytoalexin levels. Moreover, this inhibition of phytoalexin biosynthesis could be reversed by the addition of exogenous jasmonate. As such, the effects of jasmonic acid and its methyl ester (methyl jasmonate) have been extensively tested for their effects in stimulating secondary metabolite biosynthesis in a variety of medicinal plant cell cultures.

As with the effects of fungal elicitors, several studies have shown jasmonate to be effective for stimulating paclitaxel biosynthesis in cell cultures of *Taxus* sp. [8]. Yukimune et al. [53] showed that the addition of methyl jasmonate to cell cultures of *Taxus media* resulted in a substantial increase in production, with a preferential elevation in paclitaxel relative to other taxoids. A similar result was observed by Ketchum et al. [54] for cultures of *Taxus candensis* and *Taxus cuspidata*, although taxoids in addition to paclitaxel were increased by this treatment. Quite interestingly, Mirjalili and Linden [55] found that combining a methyl jasmonate treatment with ethylene exposure led to an even further increase in paclitaxel biosynthesis by up to 19-fold. In contrast, these authors found that the addition of methyl jasmonate alone resulted in only about a threefold elevation in paclitaxel biosynthesis with this system. This apparent synergistic response may reflect an interaction among multiple signal transduction pathways involved in this pathogen response [43].

Jasmonates have also been effective for increasing the production of ginsenosides by cell cultures of ginseng. Treatment of *Panax ginseng* cell cultures with methyl jasmonate has been shown to elevate ginsenoside production substantially, without much effect on cell proliferation [48]. These authors also found an antagonistic interaction between methyl jasmonate and 2, 4-dichlorophenoxyacetic acid (2, 4-D), so that optimal effects were obtained when this growth regulator was removed from the medium. With *Panax notoginseng*, methyl jasmonate was found not only to increase the level of ginsenoside production (over ninefold) but also (apparently) to alter the relative levels of different ginsenoside forms in the extract [56]. This could prove to be a problem for the use of such a system in the generation of a phytomedicinal, as its effectiveness may depend upon a specific relative amounts of each ginsenoside component that is present.

Jasmonic acid has been shown to increase the production of hypericins and to promote cell proliferation in cell cultures of St. John's wort (*Hypericum perforatum*) [57]. The hypericins are naphthodianthrones which contribute to the antidepressant activity of St. John's wort extracts [58]. Quite interestingly, fungal elicitor fractions did not stimulate hypericin production in this system.

9.3.3
Exogenous Substances

A variety of inorganic substances have been shown to increase secondary metabolite biosynthesis, and could prove useful for stimulating phytomedicinal compound production by cell cultures. For example, the addition of forms of vanadate to cell cultures have been shown to increase secondary product biosynthesis. The addition of sodium orthovandate to cultures of *Eschscholtzia californica* elevated production of benzophenanthride alkaloids to levels observed with fungal elicitors [59]. Likewise, the addition of vanadyl sulfate to cultures of *Catharanthus roseus* cell cultures elevated production of the antitumor agent, vinblastine [13]. Vanadate is an inhibitor of the plasma membrane proton pump and other phosphohydrolases [60], and these elevations in secondary metabolites might represent a cell stress response.

Other substances utilized in this way have included silver ions (Ag^+) and lanthanum (La^{3+}). For example, the addition of silver ions was shown to increase biosynthesis of the diterpenoid tanshinones in *Salvia miltiorrhiz* cultures, and this process demonstrated many similarities to elicitor-based stimulation of secondary product biosynthesis [61]. Likewise, the biosynthesis of paclitaxel in suspension cell cultures of *Taxus yunnanensis* was elevated by treatment with lanthanum, and this also appeared similar to elicitor-based effects [62]

9.3.4
Immobilization

The immobilization of cultured cells to form aggregates has been observed to enhance secondary metabolite production [63]. Typically, culture cell immobilization has been achieved through the use of alginate, polyurethane foam cubes and growing cells as aggregates [64–66]. As noted by Verpoorte et al. [67], the basis for this effect may be cell-to-cell contact and some degree of cell differentiation. As an approach to achieve large-scale production of secondary metabolites, immobilization may be problematic since such systems are difficult to maintain on a large-scale basis, and secondary metabolites may not be extensively released to the growth media [67, 68].

9.3.5
Physical Stress

The exposure of plant cell cultures to physical stresses that affect membrane permeability has been shown to enhance secondary metabolite production. Moreover, any increases in membrane permeability can also lead to a greater release of the secondary metabolite to the growth medium for recovery. In suspension cultures of *Lithospermum erythrorhizon*, Lin and Wu [69] found that exposure to low-energy ultrasound resulted in about a 60–70% increase in shikonin production, and an additional increase in the recovery of this metabolite due to enhanced membrane permeability. The combined effect resulted in about a threefold increase in recovery of the metabolite. An enhancement of taxsuyannanin C biosynthesis in *Taxus chinensis* was similarly achieved by exposure of suspension cultures to a pulsed electrical field [70]. The proposal was that such exposure induced a defense-type response that resulted in enhanced secondary product biosynthesis.

9.4

Enhancement of Secondary Metabolite Production through
***Agrobacterium rhizogenes* Transformation**

An alternative biotechnological approach for obtaining improved secondary metabolite biosynthesis in medicinal plant tissue cultures can involve the transformation of plant cells with the soil bacterium, *Agrobacterium rhizogenes*. This results in the development of a fast-growing root culture which can, for many plants, also involve elevated secondary metabolite production [9, 10]. These root cultures are called "hairy roots" due to the large number of root hairs typically associated with the highly branched root structures [7].

Hairy root cultures are usually generated by *Agrobacterium rhizogenes* infection of sterile plant explants such as leaf and leaf petioles [7, 9]. Transformation involves bacterial plasmid transfer and the incorporation of key sections of the plasmid DNA into the host plant genome. In particular, the *Agrobacterium rhizogenes* plasmid TL section with its three genes, *Rol* A, B, and C, is important for root induction and growth [71]. Once the roots have grown for a sufficient period of time, they can be excised from the explant tissue and then cultured in a growth medium containing an antibiotic, ultimately to free the cultures of residual *Agrobacterium*. These root cultures grow very quickly and generate a highly branched network structure with rapid rates of linear extension [7]. Overall, the resulting hairy root cultures are fast-growing and can be maintained on a very simple culture medium, without plant hormones [9, 10].

With respect to secondary product metabolism, one particular advantage of hairy root cultures is that they exhibit biosynthetic levels equal to or even greater than those observed for the same plant grown *in vivo* [72–74]. Examples of medicinal plants which have been introduced into hairy root culture are listed in Table 9.2. The data in the table also indicate the relevant phytomedicinal chemicals produced by the cultures and their medicinal functions.

Table 9.2 Examples of medicinal plants and phytomedicinal compounds generated with growth in hairy root culture.

Plant species	Phytomedicinal compound	Function	Reference
Artemisia annua	Artemisinin	Antimalarial	75
Catharanthus roseus	Indole alkaloids	Antihypertensive	76
Coleus forskohlii	Forskolin	Antihypertensive	77
Datura candida × *D. aureas*	Tropane alkaloids	Anticholinergic	78
Glycyrrhiza glabra	Isoprenylated flavonoids	Antimicrobial	88
Paulownia tomentosa	Verbascoside	Antibacterial, antiviral	79
Scutelleria baicalensis	Flavonoid glycosides	Antibacterial, antioxidant	80
Solanum aviculare	Solasodine	Steroid drug precursor	81
Valeriana wallichi	Valeopotriates	Sedative	82

The pattern of secondary metabolite production by hairy root cultures may be similar to that which is observed for roots of the intact plant, or there may an accumulation of secondary metabolites in the root cultures that would normally be present in other plant regions [7, 9, 10]. For example, hairy root cultures of ginseng accumulate high levels of ginsenosides which are normally produced in the roots [56]. On the other hand, hairy root cultures of *Artemisia annua* accumulate the sesquiterpenoid artemisinin [83–85]. This compound is normally accumulated in the aerial (oil-containing) parts of the *in-vivo* plant [86]. To some extent this may be due to the fact that, in the root cultures, long-distance transport processes are not present and the metabolites would remain and hence, accumulate in the roots [7, 10].

There have been some instances where hairy root cultures produced novel secondary metabolites not associated with the native plant species. For example, hairy root cultures of *Panax ginseng* were found to produce four novel polyacetetylene compounds not observed for the native plant [87]. Likewise, two novel isoprenylated flavonoids with antimicrobial activity were produced by hairy root cultures of *Glycyrrhiza glabra* [88].

Walton et al. [7] point out that an important aspect of the capacity of hairy root cultures to generate elevated levels of secondary metabolites relates to their root morphology. These authors cite the example of *Nicotiana rustica* roots, where mechanical damage to the root morphology in culture led to a decrease in the capacity for nicotine production.

9.5
Metabolite Engineering of Medicinal Plants in Culture

In contrast to "inducing" native enzyme pathways in plant cell cultures for increased secondary metabolite production, metabolic engineering can serve to directly modify pathways, the result being an increased production of secondary metabolites or the production of metabolites not normally produced by a given plant species. While a major limitation to this approach is the requirement for a thorough knowledge of the pathway steps involved, the application of bioinformatic approaches coupled with biochemical studies promises to provide the necessary information to make metabolic engineering a reality. With respect to strategies to enhance secondary metabolite production, useful approaches can involve the over-expression of an enzyme (or enzymes) at a key limiting step(s) in a pathway, decreasing the expression of enzymes in competitive pathway branches, increasing an entire pathway through expression of regulatory factors (excluding transcription factors), and the introduction of novel enzymes utilizing a pathway intermediate for the generation of a different secondary metabolite [11, 12, 68]. At present, a limited number of examples are available where these approaches have been utilized in cell cultures resulting in the enhanced production of secondary products. However, these studies demonstrate the potential for application for production of phytomedicinal compounds from medicinal plant cells in culture.

In cell cultures of *Coptis japonica*, isoquinoline alkaloid metabolism leads to the production of berberine, as well as to other alkaloid components [89]. By overexpressing an enzyme located at a key branch point in this pathway (scoulerine 9-*O*-methyltransferase), Sato et al. [90] enhanced berberine (and columbamine) production over that of an alternative metabolite, coptisine. Overexpression of this enzyme resulted in a ca. 20% increase in enzyme activity and an elevation of the amount of berberine and columbamine levels ranging from 79% to 91% of the total alkaloid level. Likewise, in hairy root cultures of *Hyoscyamus muticans*, overexpression of the enzyme hyosciamine 6β-hydroxylase was utilized to shift the relative production of the tropane alkaloids, hyosciamine and scopolamine, towards increased scopolamine biosynthesis [91]. This compound has a number of important pharmaceutical uses, including the reduction of motion sickness [2]. While hairy root cultures produced hyosciamine as the major tropane alkaloid, scopolamine production was increased 100-fold following overexpression of this enzyme. This elevation in scopolamine production was achieved without a concomitant reduction in the level of hyosciamine production.

In *Catharanthus roseus*, one portion of the metabolic pathway for terpenoid indole alkaloid biosynthesis is under the control of ORCA3, a transcription factor which is involved in the jasmonate-dependent induction of this pathway [11]. When this transcription factor was overexpressed in *C. roseus* culture cells, enhanced expression of several genes in the biosynthetic pathway was observed [92]. However, in order to obtain enhanced alkaloid production in the cells it was necessary to include the compound loganin, which is an early metabolic intermediate. Since it was found that ORCA3 did not control expression of a cytochrome P450 involved in an early step in the pathway, this addition was necessary to bypass this region of the pathway [12, 92]. With the inclusion of loganin, alkaloid biosynthesis was increased threefold in the *C. roseus* culture cells [92]. In maize kernels, two transcription factors, R and C1, appear to control anthocyanin biosynthesis [12], and overexpression of these transcription factors in maize cell cultures resulted in induction of the entire flavonoid pathway [93].

The introduction of genes into plant cells in culture conferring heterologous expression of novel enzymes for that species could allow for the generation of new secondary metabolites. For example, Fecker et al. [94] transformed *Nicotiana* hairy root cultures with a bacterial lysine decarboxylase gene, thereby allowing the generation of the piperidine alkaloid anabasine. Similarly, Lodhi et al. [95] introduced a bacterial gene encoding isochorismate synthase into *Rubia peregrina* hairy root cultures, which resulted in an increase in the production of anthraquinones.

While the metabolic engineering of pathways will have tremendous potential for use in *in-vitro* phytomedicinal production, there are some limitations to this approach. Although the overexpression of individual enzymes in a pathway has shown some success in increasing or modifying secondary metabolite biosynthesis, it should be recognized that, according to *metabolic control theory*, overall pathway flux is limited by the net effect of all enzymes present in the pathway [96]. As such, overexpression of a single enzyme might have little effect, as the limitation ("control") exerted by other enzymes in the pathway may overshadow any enhanced activity from a single enzyme. Recent studies have also supported the role of "metabolons" in many pathways, where an entire pathway or portion of a pathway is mediated by a multienzyme complex [96, 97]. In this case, the enzymes in a metabolon may be coordinately expressed, and intermediary metabolites are passed directly from one enzyme to the next via metabolite channeling, and without free metabolite pools. As such, the effect of overexpression of a single enzyme would be limited by the levels of the other enzymes present in the metabolon for assembly, and the lack of metabolic pools due to metabolite channeling could limit the effectiveness of novel enzyme introduction to generate new metabolites. Finally, as pointed out by Oksman-Caldentey and Inze [11], the autotoxicity of secondary metabolites could pose a problem when their levels of synthesis are elevated by metabolic engineering. With *in-vivo* plants, many toxic secondary metabolites are accumulated in the vacuole, and transport processes could prove to be a limiting factor [11]. Alternatively, secondary metabolites might be glycosylated as a means of preventing toxicity. This latter possibility was observed when a phenolic metabolism was modified by the introduction of a bacterial gene for chorismate-pyruvate lyase into *Nicotiana* cell cultures [98, 99]. While this enzyme would lead to the production of 4-hydroxybenzoate, the glycosylated forms of this compound were found.

9.6
Large-Scale Production of Medicinal Secondary Metabolites in Bioreactor Systems

The scaling-up of plant culture systems to allow the industrial production of secondary metabolites has been achieved for a limited number of medicinal plant species. These production systems, or "bioreactors", can allow for continuous culture growth and possibly, continuous recovery of secondary metabolites. A number of different types of bioreactor have been developed for plant culture systems which differ in terms of how the cultures are mixed, how the medium is supplied to the culture, and how gas exchange is controlled [63, 100, 101]. Some examples of medicinal plant species, together with the active secondary metabolites produced in bioreactors, are listed in Table 9.3. These data indicate that bioreactors have been

Table 9.3 Some examples of medicinal plants grown and phytomedicinal compounds recovered in bioreactor systems.

Plant species	Culture type	Phytomedicinal compound	Function	Reference
Artemisia annua	Hairy roots	Artemisinin	Antimalarial	101
Atropa belladona	Hairy roots	Atropine	Anticholinergic	102
Coleuis blumei	Suspension	Rosmarinic acid	Antioxidant	103
Coptis japonica	Suspension	Berberine	Antimicrobial	104
Datura stramonium	Hairy roots	Hyoscyamine	Anticholinergic	105
Papaver somniferum	Suspension	Sanguinarine	Antimicrobial	106
Taxus baccata	Suspension	Taxol	Antitumor	107

used with both suspension cultures and hairy root cultures. The scaling-up of a plant culture with a bioreactor requires careful consideration of several growth parameters, including oxygen supply, nutrient medium composition, light supply, and factors required to induce secondary product synthesis [63, 100]. With such an approach, only the ginsenosides from ginseng and the antimicrobial compound berberine have been produced commercially, on a large-scale basis [104, 108]. The production of taxol in bioreactors has provided very favorable results, and a clear potential for commercial use [11]. For other compounds to be produced at commercial levels, the bioreactor design may need further optimization, and the methods used to recover the secondary metabolites may need to be improved.

9.7
Summary and Perspective

While plant cell culture systems can, potentially, have a number of advantages in the production of phytomedicinal compounds, there are a number of technical challenges that would need to be resolved. Whilst it is generally possible to introduce most plants into tissue culture, the production of adequate levels of particular secondary metabolites may be problematic. Identifying growth conditions to induce secondary product production, either through growth medium optimization or by the addition of elicitors or jasmonates, can lead to increased production of secondary metabolites. However, this process is both empirical and time-consuming. On the other hand, the transformation of medicinal plants using *Agrobacterium rhizogenes* to form hairy root cultures has the potential benefits of fast growth and rates of secondary metabolite production equal to or greater than that found for the intact plant. Moreover, hairy root cultures can be scaled-up for bioreactor production to allow for the large-scale recovery of medicinal products. Finally, advances in molecular methods and in knowledge relating to secondary metabolite pathways can lead to the use of metabolic engineering as a means of directly modifying pathways for increased phytomedicinal product biosynthesis.

References

1 Sudha, G., Ravishankar, G.A. *Plant Cell Tissue Organ Culture* **2002**, *71*, 181–212.

2 Wink, M., Schimmer, O. In: Wink, M. (Ed.), *Functions of Plant Secondary Metabolites and Their Exploitation in Biotechnology*. CRC Press, USA, **1999**, pp. 17–133.

3 Wawrosch, C. In: *Handbook of Medicinal Plants*. CRC Press, USA, **2005**, pp. 261–278.

4 Wills, R.B.H., Bone, K., Morgan, M. *Nutrit. Res. Rev.* **2000**, *13*, 47–77.

5 Kim, Y., Wyslouzil, B.E., Weathers, P.J. *In Vitro Cell. Dev. Biol.* **2002**, *38*, 1–10.

6 Gupta, R.. In: Wijesekora, R.O.B. (Ed.), *The Medicinal Plant Industry*. CRC Press, **1991**, pp. 43–57.

7 Walton, N.J., Alfermann, A.W., Rhodes, M.J.C. In: Wink, M. (Ed.), *Functions of Plant Secondary Metabolites and Their Exploitation in Biotechnology*. CRC Press, USA, 1999, pp. 311–345.

8 Vanisree, M., Lee, C.-Y., Lo, S.-F., Nalawade, M.S, Lin, C.Y., Tsay, T.-S. *Bot. Bull. Acad. Sing.* **2004**, *45*, 1–22.

9 Shanks, J.V., Morgan, J. *Curr. Opin. Biotechnol.* **1999**, *10*, 151–155.

10 Giru, A., Narasu, M.L. *Biotechnol. Adv.* **2000**, *18*, 1–22.

11 Oksman-Caldentey, K.-M., Inze, D. *Trends Plant Sci.* **2004**, *9*, 433–440.

12 Verpoorte, R., Memelink, J. *Curr. Opin. Biotechnol.* **2002**, *13*, 181–187.

13 Smith, J.I., Smart, N.J., Misawa, W.G.W., Kurtz, S.G. *Plant Cell Rep.* **1987**, *58*, 142–145.

14 Lindsey, K., Yeoman, M.M. *Annuum. Planta* **1984**, *162*, 495–501.

15 Koblitz, H., Koblitz, D., Schmauder, H.P., Groger, D. *Plant Cell Rep.* **1983**, *2*, 122–125.

16 Sato, F., Yamada, Y. *Phytochemistry* **1984**, *23*, 281–285.

17 Hagimori, M., Matsumoto, T., Obi, Y. *Plant Cell Physiol.* **1983**, *23*, 1205–1211.

18 Carrier, D., Chauret, N., Mancini, M., Coulombe, P., Neufeld, R., Weber, M., Archambault, J. *Plant Cell Rep.* **1991**, *10*, 256–259.

19 Ayabe, S., Takano, H., Fujita, T., Hirota, H., Takahashi, T. *Plant Cell Rep.* **1990**, *9*, 181–184.

20 Bais, H.P., Walker, T.S., McGrew, J.J., Vivanco, J.M. *In Vitro Cell. Dev. Biol.* **2002**, *38*, 58–65.

21 Furuya, T., Kojima, H., Syono, K., Ishi, T., Uotani, K., Nishio, M. *Chem. Pharm. Bull.* **1973**, *21*, 98–101.

22 Zhong, J.J., Zhu, Q.X. *Appl. Biochem. Biotechnol.* **1995**, *55*, 241–246.

23 Siah, C.L., Doran, P.M. *Plant Cell Rep.* **1991**, *10*, 349–353.

24 Briskin, D.P., Kobayashi, H., Lila, M.A., Gawienowski, M. In: Dyer, L.A., Palmer, A.D.N. (Eds.), *Piper. A model genus for studies of phytochemistry, ecology and evolution*. Kluwer Academic Press, New York, USA, **2004**, pp. 140–155.

25 Uden, W., Pras, N., Vossebeld, E.M., Mol, J.N.M., Malingre, T.M. *Plant Cell Tiss. Cult.* **1990**, *20*, 81–87.

26 Yamamoto, O., Yamada, Y. *Plant Cell Rep.* **1986**, *5*, 50–53.

27 Wu, J., Wang, C., Mei, X. *J. Biotechnol.* **2001**, *85*, 67–73.

28 Bourgaud, F., Gravot, A., Milesi, S., Gontier, E. *Plant Sci.* **2001**, *161*, 839–851.

29 Ravishankar, G.A., Sharma, K.S., Venkataraman, L.V., Kadyan, A.K. *Curr. Sci.* **1988**, *57*, 381–383.

30 Franklin, C.I., Dixon, R.A. In: Dixon, R.A., Gonzales, R.A. (Eds.), *Plant Cell Culture – A Practical Approach*. 2nd edn. IRL Press, UK, **1994**, pp. 1–25.

31 Zhang, Y.H., Zhong, J.J. *Enzyme Microb. Technol.* **1997**, *21*, 59–63.

32 Liu, S., Zhong, J.J. *J. Biotechnol.* **1996**, *52*, 121–126.

33 Morimoto, T., Hara, Y., Kato, Y., Hiratsuka, J., Yoshioka, T., Fujita, Y., Yamada, Y. *Agric. Biol. Chem.* **1988**, *52*, 1835–1836.

34 Fett-Neto, A.G., Stewart, J.M., Nicholson, S.A., Pennington, J.J., DiCosmo, F. *Biotechnol. Bioeng.* **1994**, *44*, 967–971.

35 Wink, M.. In: Wink, M. (Ed.), *Biochemistry of Plant Secondary Product Metabolism*. CRC Press, USA, 1999, pp. 1–16.

36 Hirasuna, T.J., Pestchanker, L.J., Srinivasan, V., Shuler, M.L. *Plant Cell Tissue Org. Cult.* **1996**, *44*, 95–102.

37 Ellis, D.D., Zeldin, E.L., Brodhagen, M., Russin, W.A., McCowan, B.H. *J. Nat. Prod.* **1996**, *59*, 246–250.

38 Vazquez-Flota, F., Moreno-Valenzuela, O., Miranda-Hamm, M.L., Coello-Coello, J., Loyola-Vargas, V.M. *Plant Cell Org. Cult.* **1994**, *38*, 273–279.

39 Yu, S., Kwok, K.H., Doran, P.M. *Enzyme Microb. Technol.* **1996**, *18*, 238–243.

40 Furuya, T., Yoshikawa, T., Orihara, Y., Oda, H. *J. Nat. Prod.* **1984**, *47*, 70–75.

41 Schlatmann, J.E., Moreno, P.R.H., Selles, M., Vinke, J.L., Ten Hoopen, H.U.G., Verpoorte, R., Heijnen, J.J. *Biotechnol. Bioeng.* **1995**, *47*, 53–59.

42 Tal, B., Rokem, J.S., Goldberg, I. *Plant Cell Rep.* **1983**, *2*, 219–222.

43 Zhao, J., Davis, L.C., Verpoorte, R. *Biotech. Adv.* **2005**, *23*, 283–333.

44 Benhamou, N. *Trends Plant Sci.* **1996**, *1*, 233–240.

45 Funk, C., Gügler, K., Brodelius, P. *Phytochemistry* **1987**, *26*, 401–405.

46 Ciddi, V., Srinivasan, V., Shuler, M.L. *Biotechnol. Lett.* **1995**, *17*, 1343–1346.

47 Wang, C.G., Wu, J.Y., Mei, X.G. *Appl. Microbiol. Biotech.* **2001**, *55*, 404–410.

48 Lu, M.B., Wong, H.L., Teng, W.L. *Plant Cell Rep.* **2001**, *20*, 674–677.

49 Archambault, J., Williams, R.D., Bedard, C., Chavarie, C. *J. Biotech.* **1996**, *46*, 95–106.

50 Pozo, M.J., Van Loon, L.C. *J. Plant Growth Reg.* **2004**, *23*, 211–222.

51 Gundlach, H., Muller, M.J., Kuchan, T.M., Jenk, M.H. *Proc. Natl. Acad. Sci. USA* **1992**, *89*, 2389–2393.

52 Nojiri, H., Sugimori, M., Yamanhe, H., Nishimura, Y., Yamada, A. *Plant Physiol.* **1996**, *110*, 387–392.

53 Yukimune, Y., Tabata, H., Higashi, Y., Hara, Y. *Nature Biotech.* **1996**, *14*, 1129–1132.

54 Ketchum, R.E.B., Gibson, D.M., Croteau, R.B., Shuler, M.L. *Biotechnol. Bioeng.* **1999**, *62*, 97–105.

55 Mirjalili, N., Linden, J.C. *Biotech. Prog.* **1996**, *12*, 110–118.

56 Wang, W., Zhong, J.J. *J. Biosci. Bioeng.* **2002**, *93*, 48–53.

57 Walker, T.S., Bais, H.P., Vivanco, J.M.. *Phytochemistry* **2002**, *60*, 289–293.

58 Briskin, D.P. *Plant Physiol.* **2000**, *124*, 507–514.

59 Villegas, M., Sommarin, M., Brodelius, P.E. *Plant Physiol. Biochem.* **2000**, *38*, 233–241.

60 Briskin, D.P., Hanson, J.B. *J. Exp. Bot.* **1992**, *43*, 269–289.

61 Zhang, C.H., Yan, Q., Cheukl, W.K., Wu, J.Y. *Planta Med.* **2004**, *70*, 147–151.

62 Wu, J.Y., Wang, C.G., Mei, X.G. *J. Biotech.* **2001**, *85*, 67–73.

63 Roberts, S.C., Shuler, M.L. *Curr. Opin. Biotechnol.* **1997**, *8*, 154–159.

64 Hulst, A.C., Tramper, J.J. *Enzyme Microbiol. Technol.* **1989**, *11*, 564–568.

65 Doernenburg, H., Knorr, D. *J. Biotech.* **1995**, *50*, 55–62.

66 Gontier, E., Sangwan, B.S., Barbotin, J.N. *Plant Cell Rep.* **1994**, *13*, 533–536.

67 Verpoorte, R., van der Heijden, R., ten Hoopen, H.J.G., Memelink, J. *Biotechnol. Lett.* **1999**, *21*, 467–479.

68 Verpoorte, R., van der Heijden, R., Memelink, J. *Transgenic Res.* **2000**, *9*, 323–343.

69 Lin, L.D., Wu, J.Y. *Biotechnol. Bioeng.* **2002**, *78*, 81–88.

70 Ye, H., Huang, L.L., Chen, S.D., Zhong, J.J. *Biotechnol. Bioeng.* **2004**, *89*, 788–795.

71 Micheal, A., Spena, A. *Methods Mol. Biol.* **1995**, *44*, 207–222.

72 Charlwood, B.V., Charlwood, K.A., Molina-Torres, J. In: Charlwood, B.V., Rhoades, M.J.C. (Eds.), *Secondary Products from Plant Tissue Culture.* Clarendon Press, Oxford, UK, **1990**, pp. 201–226.

73 Flores, H.E., Vivanco, J.M., Loyola-Vargas, V.M. *Trends Plant Sci.* **1999**, *4*, 220–226.

74 Kim, Y., Wyuslouzil, B.E., Weathers, P.J. *In Vitro Cell. Dev. Biol.* **2002**, *38*, 1–10.

75 Weathers, P.J., Hemmavanh, D.D., Walcerz, D.B., Cheetham, R.D. *In Vitro Cell. Dev. Biol. Plant* **1997**, *33*, 306–312.

76 Rijhwani, S., Shanks, J.V. *Biotech. Prog.* **1998**, *14*, 442.

77 Sasaki, K., Udagawa, A., Ishimaru, H., Hayashi, T., Alfermann, A.W., Nakanishi, F., Shimomura, K. *Plant Cell Rep.* **1998**, *17*, 457–459.

78 Nussbaumer, P., Kapetanidis, I., Christan, P. *Plant Cell Rep.* **1998**, *17*, 405–409.

79 Wysokinska, H., Rozga, M. *J. Plant Physiol.* **1998**, *152*, 78–83.

80 Nishikawaa, K., Ishimaru, K. *J. Plant Physiol.* **1997**, *151*, 633–636.

81 Kittipongpatana, N., Hock, R.S., Porter, J.R. *Plant Cell Tissue Org. Cult.* **1998**, *52*, 133–143.

82 Banerjee, S., Rahman, L., Uniyal, G.C., Ahuja, P.S. *Plant Sci.* **1998**, *131*, 203–208.

83 Weathers, P.J., Cheetham, R.D., Follansbee, E., Teoh, T. *Biotechnol. Lett.* **1994**, *16*, 1281–1286.

84 Jaziri, M., Shimomura, K., Yoshimatsu, K., Fauconnier, M.-L., Homes, J. *J. Plant Physiol.* **1995**, *145*, 175–177.

85 Liu, C.Z., Wang, Y.C., Zhao, B., Guop, C., Ouyang, F., Ye, H.C., Li, G.F. *In Vitro Cell Dev. Biol.* **1999**, *35*, 271–274.

86 Wallaart, T.E., Pras, N., Quax, W.J. *J. Nat. Prod.* **1999**, *62*, 1160–1162.

87 Kwon, B.M., Ro, S.H., Kim, M.K., Nam, J.Y., Jung, H.J., Lee, I.R., Kim, Y.K., Bok, S.H. *Planta Med.* **1997**, *63*, 552–553.

88 Asada, Y., Li, W., Yoshikawa, T. *Phytochemistry* **1998**, *47*, 389–392.

89 Roberts, M.F., Strack, D. In: Wink, M. (Ed.), *Biochemistry of Plant Secondary Product Metabolism.* CRC Press, USA, 1999, pp. 17–78.

90 Sata, F., Hashimoto, T., Haciya, A., Tamura, K., Choi, K.-B., Morishige, T., Fujimoto, H., Yamada, Y. *Proc. Natl. Acad. Sci. USA* **2001**, *98*, 367–372.

91 Jouhikainen, J., Lindgren, L., Jokelainen, T., Hiltunen, R., Teeri, T.H., Oksman-Caldentey, K.-M. *Planta* **1999**, *208*, 545–551.

92 van der Frits, L., Memelink, J. *Science* **2000**, *289*, 295–297.

93 Grotewold, E., Chamberlin, M., Snook, M., Siame, B., Butler, L., Swenson, J., Maddock, S., St. Clair, G., Bowen, B. *Plant Cell* **1998**, *10*, 721–740.

94 Fecker, L.H., Hildebrandt, S., Rügenhagen, C., Herminghaus, S., Landsmann, J., Berlin, J. *Biotechnol. Lett.* **1992**, *14*, 1035–1040.

95 Lodhi, A.H., Bongaerts, R.J.M., Verpoorte, R., Coomber, S.A., Charlwood, B.V. *Plant Cell Rep.* **1996**, *16*, 54–57.

96 Winkel, B.S.J. *Annu. Rev. Plant Biol.* **2004**, *55*, 85–107.

97 Spivey, H.O., Ovadi, J. *Methods* **1999**, *19*, 306–321.

98 Li, S.-M., Wang, Z.-X., Wemako, E., Heide, L. *Plant Cell Physiol.* **1997**, *38*, 844–850.

99 Siebert, S., Sommer, S., Li, S., Wang, Z., Severin, K., Heide, L. *Plant Physiol.* **1996**, *112*, 811–819.

100 Sajc, L., Grubisic, D., Vunjak-Novakoivic, G. *Biochem. Eng. J.* **2000**, *4*, 89–99.

101 Kim, Y., Wyslouzil, B.E., Weathers, P.J. *Plant Cell Rep.* **2001**, *20*, 451–455.

102 Lee, K., Suzuki, T., Yamakawa, T., Kodama, T., Igarashi, Y., Shimomura, K. *Plant Cell Rep.* **1999**, *18*, 567–571.

103 Ulbrich, B., et al. In: Neuman, K.H. (Ed.), *Primary and Secondary Metabolism of Plant Cell Cultures.* Springer-Verlag, Berlin, **1985**, pp. 293–303.

104 Fujita, Y., Tabatha, M. In: Green, C.E., Somers, D.A., Hacket, W.P., Biesboer, D.D. (Eds.), *Plant Tissue and Cell Culture.* Alan R. Liss, New York, USA, **1987**, pp. 169–185.

105 Hilton, M.G., Rhoades, M.J.C. *Appl. Microbiol. Biotech.* **1990**, *33*, 132–138.

106 Eilert, U., et al. *J. Plant Physiol.* **1985**, *119*, 65–76.

107 Srinivasan, V., Pestchanker, L., Moser, S., Hirasuna, T.J., Taticek, R.A., Shuler, M.L. *Biotech. Bioeng.* **1995**, *47*, 666–676.

108 Hara, Y. In: DiCosmo, F., Misawa, M. (Eds.), *Plant Cell Culture Second Metabolism – Towards Industrial Application.* CRC Press, Boca Raton, FL, USA, **1996**, pp. 187–202.

10

Impact of Whole-Genome and Expressed Sequence Tag Databases on the Study of Plant Secondary Metabolism

Jillian M. Hagel, Jonathan E. Page, and Peter J. Facchini

10.1
Introduction

Plants produce a wide diversity of low-molecular-weight natural products via a network of typically complex secondary metabolic pathways. To date, more than 100 000 natural products, such as terpenoids, phenylopropanoids and alkaloids, have been identified [1]. Despite the widespread application and economic importance of these compounds as dyes, pigments, flavors, aromas, medicines and poisons, the physiological relevance of most natural products has not been determined. However, secondary metabolites – deemed *secondary* only because they are not required for normal growth and development – generally play physiological roles ranging from plant defense against herbivores to the attraction of pollinators.

The past few decades have seen the elucidation of major secondary metabolic pathways and the discovery of many novel biosynthetic enzymes. The early development of radioactive tracing techniques, the later use of plant tissue culture methods, and the more recent application of molecular biological approaches have each represented a *revolution* in the efficiency and efficacy of research on plant secondary metabolism. The most recent – and perhaps the most profound – revolution involves the widespread use of genomics, which can be defined as the use of current knowledge of whole or partial genome sequences to answer broad biological questions [2]. Research involving plant secondary metabolism has taken advantage of the genomics revolution by using new tools to isolate novel genes, to elucidate evolutionary relationships between plant species, and to understand intricate signaling and regulatory networks. In this chapter, we review the application of plant genome and related sequence resources to the discovery of novel biosynthetic genes involved in terpenoid, phenylpropanoid, and alkaloid metabolism in order to demonstrate the potential of genomics to advance research in the field of plant secondary metabolism.

Medicinal Plant Biotechnology. From Basic Research to Industrial Applications
Edited by Oliver Kayser and Wim J. Quax
Copyright © 2007 WILEY-VCH Verlag GmbH & Co. KGaA, Weinheim
ISBN 978-3-527-31443-0

10.2
Whole-Genome Sequences

Plant biology was formally ushered into a new era in December 2000, with the completion of the *Arabidopsis thaliana* genome sequence. Representing 115 million base pairs (Mb) of euchromatin out of the estimated 125-Mb genome, recent estimates total the number of genes at 30 700 (version 5 annotation; The Arabidopsis Information Resource, TAIR; http://www.arabidopsis.org). The International Rice Genome Sequencing Project declared the rice genome sequence complete in December 2004, while large-scale efforts are being made toward the sequencing of tomato (*Lycopersicon esculentum*), lotus (*Lotus corniculatus*), barrel medic (*Medicago truncatula*), and black cottonwood (*Populus trichocarpa*) genomes. Certainly, the sequencing of an entire plant genome is a daunting task; nuclear genomes of plants are often notoriously large, requiring immense resources, investment and coordination. However, the real challenge of this new era lies not with the amassment of sequence data, but with the application of genomic information to biological problems. The high quality of the *A. thaliana* annotation is expected to strengthen broad comparisons involving proteome content, transcriptional patterns, and epigenetic state, with other plants and distantly related model organisms [3]. Comparative genomics may be used to identify functional elements, which are more likely to be conserved through time while neutral mutations accumulate. When multiple gene sequences are aligned, conserved regulatory regions, noncoding genes, and protein coding genes become evident [4, 5]. Ultimately, information garnered from genome analyses may be applied to functional genomics in plants for which little or no sequence data are available, rendering whole-genome studies of model organisms invaluable to medicinal plant biology.

Computers, software, and the World Wide Web are integral components to the analysis of, and access to, genome sequence information. Reiser et al. [6] provided a survey of plant genome data resources, including those containing information for single or multiple plant species. The types of data available through public databases include DNA and protein sequences, precomputed phylogenetic profiles of completed genomes, sequence analysis software, maps, clones and seed stocks. The National Center for Biotechnology Information (NCBI; http://www.ncbi.nih. gov) provides a data-rich platform in support of genomic research by integrating data from more than 20 biological databases, using a flexible search and retrieval system named Entrez [7]. Entrez-Nucleotide, a core database in Entrez, includes GenBank [8], which is a primary database of nucleotide sequences synchronized daily with the DNA Databank of Japan [9] and the European Molecular Biological Laboratory [10]. The Entrez system also covers a suite of more specialized databases of particular relevance to genomics, such as Entrez Genomes, Unigene, Entrez Gene, and HomoloGene. Entrez Genome contains genomic sequence and annotations for over 1000 organisms, including 39 complete sequences for plant chromosomes, plastids, and mitochondria. Taking advantage of the same database technology as Entrez, and supporting text queries with Boolean logic, the NCBI Map Viewer (http://www.ncbi.nih.gov/mapview/) is used to display genomic maps for many plants and animal genomes. Performing a BLAST search using an accession, GI, or sequence in FASTA format leads to Map Viewer displays in which the genomic context of the hits can be seen. This function is useful for identifying non-coding, putative regulatory elements in model organisms, and finding conserved regulatory regions in plants for which genome sequence is unavailable.

Comparative plant genomics resources are also available through PlantGDB (http://www.plantgdb.org/), a database of plant molecular sequences. PlantGDB contains data of plant sequences extracted from public sequence repositories such as GenBank, including assembled expressed sequence tag (EST) and genome survey sequence (GSS) contigs, and the complete genome sequences for *Arabidopsis* and rice [11]. The extracted sequences are sorted by taxonomic classification to provide fast and easy access to sequence subsets limited to individual species or phylogenetic group, and the site is synchronized daily with source public repositories. Currently, PlantGDB contains sequences from over 24 000 plant species, representing more than 6000 genera. Unique to PlantGDB, researchers have access to three online sequence data analysis tools: BLAST@PlantGDB, GeneSeqer@PlantGDB, and PatternSearch@PlantGDB. Although nearly all sequence databases provide an online BLAST server, most restrict researchers to one database at a time, making PlantGDB's multi-source database an attractive alternative. For example, the current NCBI BLAST server requires a selection of predefined database options (e.g., "*nr*", "*est*", "*gss*", etc.), whereas this distinction is not necessary at PlantGDB.

Research progress toward the completion of large-scale genome sequencing projects may be accessed at http://www.ncbi.nih.gov/genomes/PLANTS/PlantList.html. Following the links leads to information regarding species-specific genome characteristics (i.e., size, chromosome number, ploidy) and the research center or consortium currently engaged in genome sequencing. A more comprehensive listing of genome sequencing initiatives is available (http://www.ncbi.nih.gov/genomes/leuks.cgi), although many of the projects listed have not yet started, or are operating on a "small-scale" basis. Plant species for which genome sequencing initiatives are in progress include *Arabidopsis* family members *A. lyrara* and *Capsella rubella*, cultivated grains *Sorghum bicolor* and *Triticum aestivum*, fruits, legumes, trees, and evolutionarily important fern (*Selaginella moellendorffii*) and moss (*Physcomitrella patens*) species. The genomes of several green algae are also being sequenced.

As sequence data accumulate for myriad plant species, the model genome of *Arabidopsis thaliana* will continue to serve as our genetic workhorse. Plant research has placed enormous expectation in this small, weedy "supermodel" plant, prompting the launch of the ambitious *Arabidopsis* 2010 Program, the aim of which is to determine the function of every gene within the span of a decade [12]. The *Arabidopsis* genome contains an impressive array of genes, encoding enzymes involved in primary and secondary metabolism. More than 300 members of the cytochrome P450 gene family are represented, in addition to a large number of transcription factors (~1500), many of which are unique to plants [3]. Despite these features, and although *Arabidopsis* has been used extensively to study aspects of primary metabolism, secondary metabolism is more often studied in "exotic" or medicinal plants [13]. Ironically, once genes that encode enzymes involved in the biosynthesis of a class of secondary metabolites (e.g., alkaloids) are found in such species, sequence comparisons often reveal that the *Arabidopsis* genome contains related sequences [14]. However, sequence homology is often insufficient evidence on which to base predictions of specific enzyme function, let alone the overall biosynthetic capacity of a certain plant. Due to examples of minor, even single, amino acid substitutions conferring distinct substrate specificities in enzymes involved in natural product metabolism, caution is advised when interpreting sequence data [15, 16]. While available genome information cannot be used to replace traditional, empirical enzyme characterization, gene annotations may be used to identify conserved sequence domains suggesting putative enzyme function.

Gene discovery and annotation within sequenced genomes provide tools for the identification and isolation of homologues in non-model plants. Reference plants often produce metabolites known to be involved in the biosynthesis of nutraceutical or pharmaceutical compounds, thereby facilitating the study of these compounds in medicinal plant species. The completion of genome sequences of model organisms also permits broader applications, such as comparative genomics. Comparative mapping has revealed extensive genome co-linearity between species in the same family, and microsyntenic correlations between distantly related species [17]. Syntenic relationships can extend genetic maps established in one species to related species, and provide unambiguous identification of gene orthologues [18]. Comparative sequence analysis of large DNA regions across species has identified long-range, *cis*-regulatory elements that are difficult to find by conventional methods [19–21]. The first comparative genetic maps were created during the late 1980s [22, 23]. As a natural product of the "genome", comparative mapping gained the precision necessary for a more exact delineation of the syntenic relationships between members of the grass and legume families [24–26].

The most comprehensive comparative dataset to date is from the grass family, which contains all of the major cereals [27]. The Poaceae family includes wheat, maize, barley, sorghum, oats, sugarcane, and the model plant, rice. Synteny is fairly conserved across the cereal genomes, and local regions of co-linearity will be of immense use in positional cloning efforts in larger cereal genomes [28]. Availability of genomic sequence from two subspecies of rice (*O. japonica* and *O. indica*) provides a profound resource for adaptation and evolution studies. Phylogenetic approaches have been used to identify conserved, noncoding sequences (CNS) in plant genes in a number of studies [29–31], including a comparative analysis of *phytochrome A* gene promoters from sorghum, maize, and rice, revealing a CNS that spanned known *cis*-regulatory sequences [32]. Buchanan et al. [33] carried out a phylogenetic analysis of 5′-noncoding regions from ABA-responsive *rab16/17* gene family of sorghum, maize and rice, with the goal of determining how to use sequence data to identify *cis*-elements that control gene expression in grass species. Using the FootPrinter software package (http://bio.cs.Washington.edu/software) and the well-characterized maize *rab17* as a model, several conserved 5′-noncoding regions were identified as putative transcription factor binding sites.

Evidently, the concept that conserved genome structure can facilitate transfer of knowledge among related plant species has been exemplified in grasses. Sequencing of *Medicago truncatula* and *Lotus japonicus* has encouraged a similar trend within the legume family [34]. Fabaceaeous plants include the crop species soybean (*Glycine max*), peanut (*Arachis hypogaea*), and alfalfa (*Medicago sativa*), along with numerous beans and peas. Using *M. truncatula* as the central point of comparison, an in-depth analysis of legume macrosynteny within pea, mungbean and alfalfa was reported by Choi et al. [26, 35]. Microsynteny, which refers to conserved gene content and order at the sequence level in a short, physically defined DNA contig, has been estimated between *M. truncatula* and soybean using a hybridization strategy involving bacterial artificial chromosome (BAC) contigs [36]. The demonstrably conserved genome structure between *M. truncatula* and crop legumes has permitted map-based cloning of genes required for nodulation in crop legumes, using *M. truncatula* as a surrogate genome [37, 38]. Beyond comparisons between members of the grass or legume families, genome-wide comparisons of gene families within more divergent species have been reported. Open reading frames (ORFs) encoding P-type ATPase ion pumps [39], *CONSTANS*-like genes [40], cryptochrome ORFs [41] and calcium-sensing gene families [42] that have been examined in both rice and *Arabidopsis* have shed light on conserved genes and pathway components in these two model species.

10.3
Expressed Sequence Tags

In plants for which the complete genome sequence is available, information on both the physical and functional annotation of the genome can be gained through transcriptional genomics [43, 44]. This advantage is especially true for the model plant *Arabidopsis*, whose genome annotation is particularly strong. In plant species for which complete genomes sequences are not available, cDNA libraries, and increasingly, EST databases have been used as a source of DNA sequence information [13]. ESTs differ from cDNAs in that they are generally shorter and/or incomplete copies of mRNA sequence. Generating sequence information from EST fragments serves two purposes: (i) the discovery of new genes; and (ii) the assessment of their expression levels in the source tissue [45]. The approach is based on the premise that the level of an mRNA molecule in a specific tissue is mirrored by the frequency of the occurrence of its corresponding EST in a clone library. In this regard, EST sequencing methods are distinct from ratio-based techniques, such as microarraying, in that they are immediately quantitative [46]. EST-based projects are attractive because they do not rely on existing sequence information from the organism under study. For this reason, the construction of EST databases presents a distinct advantage for non-model plants. A disadvantage of EST sequencing approaches, however, is the expense. Even at a few dollars per sequence, the process can be costly if one desires to progress beyond a cursory screening of abundant transcripts to an in-depth analysis [47]. Beyond the statistical questions raised by sampling small numbers of a large population [48], there are also bias problems involved with cloning and cDNA synthesis. Consequently, EST sequencing on a larger scale is favored as this minimizes such concerns. In theory, expression profiles can be derived for very weakly expressed genes if ESTs are sequenced in sufficient number.

Exhaustive sequencing of ESTs is a common method for gene expression profiling, while the primary objective of EST sequencing is usually to generate genic sequence information [49]. To reduce the amount of sequencing required in achieving a survey of expressed genes, auxiliary techniques such as subtractive hybridization [50], representational difference analysis (RDA) [51] and suppression subtractive hybridization (SSH) [52] can be used. Expressed sequence tag data are generated by bulk, single-pass, partial sequencing of cDNA clones (~500 base pairs). Comparisons of EST frequencies in libraries constructed from different tissues may reveal differential gene expression patterns [53]. Unfortunately, public plant EST libraries are, in general, too small or from too many sources for accurate expression analyses.

As of November 2005, there were 420 789 *Arabidopsis thaliana* ESTs or cDNAs in GenBank (http://www.ncbi.nlm.nih.gov/dbEST). The abundance of ESTs in the GenBank *Arabidopsis* library does not accurately reflect gene expression levels, however, since most of these clones were generated either from a single library derived from multiple tissues or were selected from normalized libraries [49, 54, 55]. Although model plants and crop species are well represented in GenBank EST databases, few and often no ESTs are available for "exotic" and/or medicinal plants. Other public or non-profit sites containing EST sequence information include http://www.tigr.org and http://www.plantgdb.org. Specialized internet-accessible databases exist for individual crop species [11] and plants such as grapevine (*Vitis vinifera*) (http://www.vitigen.com) and various conifer species (http://treenomix.com). In some cases, EST databases amassed by private corporations may be mined for academic purposes [56], usually with proprietary/monetary considerations.

Software developments for use in EST analyses are routinely published [11, 57, 58]. Because EST sequences are typically redundant, assembly of overlapping ESTs into putative contigs constitutes the first step in generating a usable library. EST assembly remains a computational challenge given the large number of ESTs currently available. Computational requirements can be reduced using parallel EST clustering programs such as PaCE [59] (http://bioinformatics.iastate.edu/bioinformatics2go/PaCE/), which is employed by PlantGDB. Researchers working on a particular organism often generate their own species-specific contigs, either independently or in collaboration with groups specializing in genomic analysis. Additionally, PlantGDB makes available data analysis tools such as GeneSeqer@-PlantGDB and PatternSearch@GDB. In particular, GeneSeqer@PlantGDB allows researchers to "thread" EST sequences onto genomic DNA across all plant species, helping identify linking patterns between genes. The PatternSearch tool permits searches for relatively short matches – possibly interspersed with mismatches, insertions, or deletions – against PlantGDB sequences. For a more complete review of these and related EST analysis programs, the reader is referred to Dong et al. [11].

10.4
Terpenoids

The isoprenoids comprise the largest and most diverse family of natural products [60]. More than 30000 individual terpenoid compounds have been identified, of which at least half are synthesized by plants [61]. Although the vast majority of terpenoids are classified as secondary metabolites, a relatively small number of isopentenyl-derived compounds are involved in plant primary metabolism including, for example, the phytol side chain of chlorophyll, the carotenoid pigments, the phytosterols of cellular membranes, and the gibberellin plant hormones. Mixtures of terpenoids form the basis of a range of commercially important products, such as essential oils, turpentines, and resins [62, 63]. The flavors and aromas of many herbs, spices and fruit can be attributed to the presence of volatile isoprenoids. Mints, sages and basils synthesize and store terpenoid cocktails in glandular trichomes, and citrus fruit owes its scent to the sesquiterpenoid valencene [64–66]. For centuries, plant varieties such as rose have provided humankind with a source of natural perfume [67, 68], while many flowers have relied for much longer on the monoterpene constituents of floral scent to attract pollinating insects [69–71]. Constitutive and induced terpenoids are important defense compounds for many plants against potential herbivores and pathogens, as seen in the traumatic resin response of Norway spruce to insect infestation [72, 73]. In terms of our own defense, several terpenoids are of pharmacological significance, including the dietary anticarcinogen limonene [74] (Fig. 10.1), the antimalarial artemisinin [75], and the anticancer drug paclitaxel (Taxol) [76, 77].

All terpenoids are derived from the central precursor isopentenyl diphosphate (IPP). In plants, IPP is synthesized in the cytosol via the classical mevalonate (MVA) pathway [78], by which sesquiterpenes (C_{15}) and triterpenes (C_{30}) are formed, and in the plastids via the non-mevalonate, or methylerythritol phosphate (MEP) pathway [79, 80]. New evidence in snapdragon, however, has suggested that both mono- and sesquiterpenes may be synthesized in the plastid [81]. These two spatially distinct pathways have recently been elucidated in plants and microorganisms, and the genes encoding all the enzymes in both pathways have now been identified [82, 83]. As shown in Figure 10.1, these three acyclic prenyl diphosphates serve as immediate substrates for terpene synthases (TPSs), which generate a vast array of mono-, sesqui-, and diterpenoid compounds. Although the first three TPS genes were isolated from distantly related plants [84–86], sequence comparisons revealed a degree of relatedness [86–89]. Later analysis of the deduced amino acid sequences of 33 TPSs from angiosperms and gymnosperms allowed the identification of six TPS gene subfamilies [90]. Conservation between these subfamilies has eased subsequent efforts to clone new genes based on genome sequence data alone [91].

Prior to the availability of complete genome sequences, the search for isoprenoid biosynthetic genes began in plants well-known for their production of essential oils, resins, or terpenoid volatiles. In the case of *Arabidopsis*, a reversal of this trend occurred. Analysis of the *A. thaliana* genome sequence revealed a family of 40 TPS genes [92] which, when taken together with supporting evidence [93, 94], suggest-

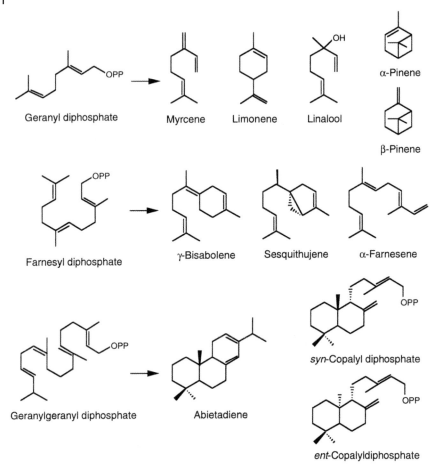

Fig. 10.1 Representative monoterpenes, sesquiterpenes, and diterpenes produced from geranyl diphosphate, farnesyl diphosphate, and geranylgeranyl diphosphate, respectively.

ed the presence of terpenoid metabolism in the model plant. This hypothesis was confirmed by Chen et al. [95] and Aharoni et al. [96], who within the same year published conclusive data concerning the emission of volatile isoprenoids from *Arabidopsis* flowers. The inflorescences were found to emit a complex mixture of more than 20 sesquiterpenes that was primarily dominated by (–)-(E)-β-caryophyllene, in addition to eight monoterpenes [97]. Proton-transfer-reaction mass spectrometry was used to show that *Arabidopsis* roots emit 1,8-cineole [98], a monoterpene formed by the action of a root-specific TPS [99]. The *Arabidopsis* genome sequence has proven to be an invaluable resource, enabling the rapid cloning, characterization, and localization of new genes involved in isoprenoid metabolism [97, 100]. Genomic evidence has suggested that *A. thaliana* also contains higher terpenes, including triterpenes. Fazio et al. [101] used genomic information to clone a predicted oxidosqualine synthase capable of synthesizing tricyclic triterpenoids *in vitro*, adding to previous studies demonstrating the presence of genes encoding similar, higher terpene-synthesizing enzymes in *Arabidopsis* [102–104].

The more recent completion of the rice genome [105] has similarly eased the process of cloning new genes involved in terpenoid metabolism. In particular, there is current interest in the biosynthesis of labdadienyl/copalyl diphosphate (CDP)-derived isoprenoids in rice, since these diterpenoids serve as precursors to important primary metabolites such as gibberellin hormones, as well as to myriad natural products [106–108]. As seen in Figure 10.1, the C_{20} terpenoid GGPP is converted to *ent*-CDP by the terpene synthase $OsCPS_{ent}$, for which more than one isoform exists. Two isoforms of $OsCPS_{ent}$ were reported by Prisic et al. [109], with one of the two being mined from rice genomic sequence. Weeks earlier, a very similar report of two $OsCPS_{ent}$ isoforms had been published [108]. In both papers, the authors concluded that one $OsCPS_{ent}$ isoform was UV-inducible and specific for phytoalexin biosynthesis, whereas the other terpene synthase isoform was not inducible and was specific for gibberellin biosynthesis. Rice genomic sequence was also used to isolate a third TPS, specifically responsible for *syn*-CDP-derived phytoalexin/allelopathic compounds ($OsCPS_{syn}$) [107].

With the exceptions of *Arabidopsis* and rice, plant isoprenoid metabolism research relies most heavily on EST sequence databases. In some cases, combinations of public and private EST databases are mined for appropriate clones. Schnee et al. [56] searched a proprietary maize database assembled by Pioneer Hi-Bred International in addition to public databases [110] for ESTs with sequence homology to known TPSs. The selected ESTs were aligned with each other to form contigs. One of these contigs was used to isolate a full-length cDNA encoding a sesquiterpene synthase, which surprisingly exhibited low overall similarity to other plant TPSs. Transcript levels of the corresponding gene *terpene synthase 1* increased with herbivore damage, and the heterologously expressed enzyme was found to catalyze the formation of (*E*)-β-farnesene, (*E*)-nerolidol, and (*E,E*)-farnesol. Continued study of maize isoprenoid metabolism showed that allelic variation of two TPS genes, *tps4* and *tps5*, caused variability in sesquiterpene emissions between two maize cultivars [111]. Clones of both genes were obtained by searching the Pioneer Hi-Bred EST sequence libraries. Genomics-based selection of triterpene glycosyltransferases from the model legume *Medicago truncatula* was reported recently [112], as was the cytochrome P450 monooxygenase CYP720B1 (abietadienol/abietadienal oxidase; PtAO) from loblolly pine (*Pinus taeda*) [113]. In conifers, P450s are involved in the formation of a suite of diterpene resin acids. PtAO was cloned and identified using phylogenetic cluster analysis of P450-like ESTs from loblolly pine, functional cDNA screening in yeast (*Saccharomyces cerevisiae*) and *in-vitro* enzyme characterization. Conifer TPSs have been a research focus for ecologists and molecular biochemists alike over the past decade [73]. Resin terpenoids are sequestered in specialized anatomical structures, such as resin ducts or resin blisters. Stem-boring insects damage the resin ducts, exuding monoterpenoid-solvated diterpenes such as pinene (Fig. 10.1). The release of terpenoids creates a long-lasting chemical and physical barrier at the site of insect attack. The evolution of gymnosperm TPSs of the *TPS-d* subfamily was examined following the cloning and functional characterization of nine Norway spruce (*Picea abies*) TPS genes [114]. Four clones were isolated by cDNA library filter hybridization, whereas a combination of similarity-based PCR, EST-mining and rapid amplification of cDNA ends (RACE) cloning strategies enabled the isolation of the remaining five TPS genes.

Many other species possess specialized anatomical structures known to biosynthesize and/or store isoprenoid compounds. Plants secrete a diverse array of secondary metabolites within modified epidermal hairs called glandular trichomes. The products that accumulate in or exude from plant glandular trichomes are biosynthesized by secretory cells located at the apex of the trichome [115]. Due to the low level of trichome biomass relative to the organs on which they are located, it is not clear how well trichome-expressed genes are represented in whole-organ cDNA libraries. In response to this problem, specialized EST libraries have been developed from the glandular trichome cells of peppermint (*Mentha×piperita*) [116], sweet basil (*Ocimum basilicum*) [64], alfalfa (*Medicago sativa*) [117] and wild and cultivated tomatoes (http://www.tigr.org/tdb/tgi/). While a significant proportion of ESTs from mint and tomato trichome libraries represented genes involved in terpenoid metabolism, the peltate glands of basil appeared to be richer with clones dedicated to phenylpropene biosynthesis [64]. In the case of alfalfa, no ESTs corresponded to enzymes of cyclized terpenoid biosynthesis [117]. The isolated secretory cells of peppermint oil glands are capable of *de-novo* biosynthesis of monoterpenes from primary carbohydrate precursors [118], and have been shown to be highly enriched in the enzymes of monoterpenoid biosynthesis [119]. The monoterpenoid (*R*)-(+)-menthofuran (Fig. 10.2) is a common component of the essential oil of several *Mentha* species, of which levels in excess of a few percent can decrease the quality of the distilled commercial product. Drawing a candidate cytochrome P450 monooxygenase clone from their EST sequence library [64], Bertea et al. [120] characterized the responsible enzyme menthofuran synthase (MFS) (Fig. 10.2). Exploitation of this highly enriched library has further led to the cloning and functional expression of isopiperitenone reductase (IPR) and pulegone reductase (PR) [121], isopiperitenol dehydrogenase (IPD) [122], and the two menthol reductases (–)-menthone:(–)-(3*R*)-menthol reductase (MMR) and (–)-menthone:(+)-(3*S*)-neomenthol reductase (MNR) [123]. Taken together, these three papers describe the entire complement of cDNAs encoding the redox enzymes of (–)-menthol biosynthesis in peppermint [123]. The biosynthetic pathway and corresponding enzymes are illustrated in Figure 10.2.

Although the EST library of sweet basil was especially rich with clones encoding phenylpropene-related enzymes [64], Iijima et al. [124] reported the isolation of a cDNA encoding geraniol synthase using the sequence database. In a related publication, Iijima et al. [125] examined the biochemical and molecular basis for the divergent patterns in the biosynthesis of terpenes and phenylpropenes in the peltate glands of three cultivars of basil. Interestingly, the total amount of terpenes was correlated with total levels of TPS activities, and negatively correlated with levels of phenylpropanoids and phenylalanine ammonia lyase (PAL) activity. Using an annotated EST database for the three basil cultivars [64, 124], an exhaustive search re-

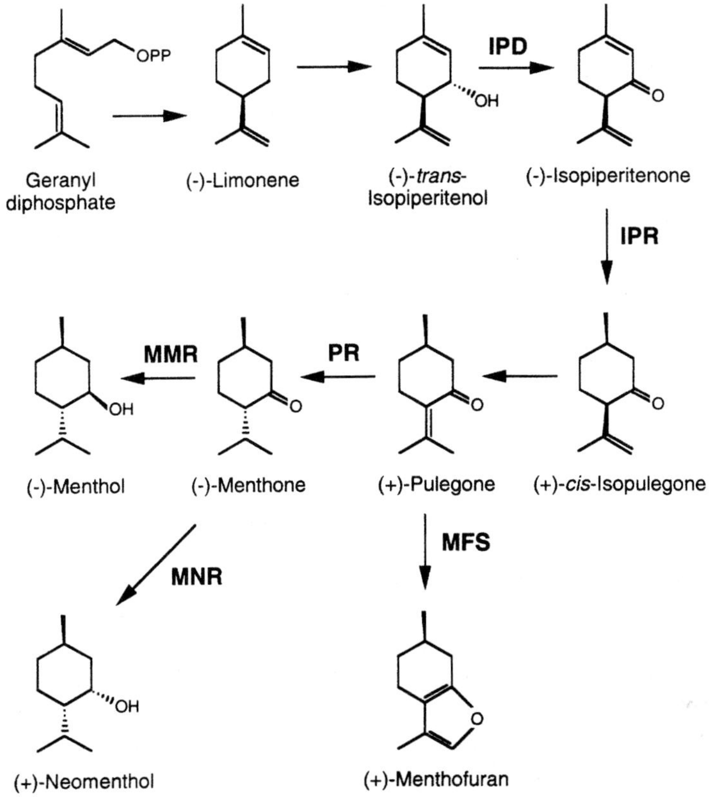

Fig. 10.2 Biosynthesis of (–)-menthol, (+)-neomenthol, and (+)-menthofuran. Enzymes for which corresponding molecular clones have been isolated using a genomics-based approach are shown. Abbreviations: IPD, (–)-*trans*-isopiperitenol dehydrogenase; IPR, (–)-isopiperitenol reductase; PR, (+)-pulegone reductase; MFS, (+)-menthofuran synthase; MMR, (–)-menthone : (–)-(3*R*)-menthol reductase; MNR, (–)-menthone : (+)-(3*S*)-neomenthol reductase.

vealed a total of nine contigs encoding proteins with sequence homology for known TPSs. After obtaining full-length cDNAs for all the contigs, the corresponding protein sequences were aligned and compared, and phylogenetic relationships assessed [125]. TPSs have been similarly isolated and compared in grapevine flowers and berries [126]. Recently, two sesquiterpene synthase cDNAs encoding (+)-valencene synthase and (–)-germacrene D synthase were obtained by *in-silico* screening of a database developed for the grapevine cultivar Gewürztraminer (http://www.vitigen.com) [126, 127]. In snapdragon, three genes of a new TPS subfamily were cloned and characterized using a flower EST library containing 792 sequences [71]. Further work with these genes, which encode two myrcene synthases (Fig. 10.1) and one (*E*)-β-ocimene synthase, led to the discovery that the nonmevalonate pathway supports both monoterpene and sesquiterpene formation in snapdragon flowers [81].

Genomics approaches have become increasingly popular to investigate the metabolism of natural products in less-characterized exotic and/or medicinal plant species. The root of ginseng (*Panax ginseng*) is known to be rich in ginsenosides, which are glycosylated triterpenes (saponins) and considered to be the main active compounds in ginseng root. Despite the commercial interest in ginseng, little is known about the genes and biochemical pathways involved in ginsenoside biosynthesis. To create a genomic resource, Jung et al. [128] sequenced 11 636 ESTs from five different ginseng libraries. Numerous putative biosynthetic enzymes were identified, including four oxidosqualine cyclase candidates putatively involved in modification of the triterpene backbone. Random sequencing of an induced *Taxus* cell cDNA library was carried out to identify clones involved in Taxol biosynthesis [129]. Taxol is a structurally complex taxane diterpenoid (taxoid) with well-established use as a chemotherapeutic agent. First isolated from bark of the Pacific yew (*Taxus brevifolia*), Taxol is now produced semi-synthetically using natural source taxoid intermediates [130]. To circumvent this dependence on biological sources for Taxol manufacture, research has focused on elucidating the entire Taxol biosynthetic pathway. Jennewein et al. [129] reported surprisingly high abundances for transcripts of several previously defined genes, cDNAs encoding two new cytochrome P450 taxoid hydroxylases, and candidate genes for all but one of the remaining uncharacterized steps. Natural rubber biosynthesis was investigated using a genomics-based approach, wherein the latex of the Brazilian rubber tree (*Hevea brasiliensis*) was used to develop a sequence library of over 20 000 cDNA-AFLP-based transcription-derived fragments (TDFs) and 1176 ESTs [131]. Despite the availability of petroleum-based synthetics, natural rubber (*cis*-1,4-polyisoprene) is highly valued because no synthetic substitute has comparable elasticity, resilience and resistance to high temperature [132, 133]. Surprisingly, only seven gene families accounted for more than 51% of the latex transcriptome, with rubber particle proteins REF (rubber elongation factor) and SRPP (small rubber particle protein) comprising 29% of the total ESTs. Several candidate rubber biosynthetic genes were present, albeit at lower levels.

Logically, the next step toward gene discovery included the use of sequenced elements for transcriptional profiling. Several microarrays of ESTs isolated from plants known to produce isoprenoids either constitutively or inducibly have been reported. Recent examples have included microarray analyses of spider mite-infested tomato [134] and cucumber [135] leaves. In both cases, spider mite (*Tetranuchus urticae*) herbivory-induced transcriptional up-regulation of enzymes involved in the biosynthesis of monoterpenes and diterpenes, although a three-day delay between transcriptional up-regulation and emission of volatile terpenoids was noted in tomato. The late increase in volatile production coincided with an increased olfactory preference of predatory mites (*Phytoseiulus persimilis*) for infested plants, leading to the conclusion that tomato activates indirect defenses (volatile production) to complement direct defense responses against spider mites. Transcriptional analysis was used to identify novel scent-related genes in rose petals, using rose flowers from tetraploid scented and nonscented cultivars. DNA chips were prepared from an annotated petal EST database of ~2100 unique genes from both cultivars. Identification of secondary metabolism-related genes whose expression coincided with scent production, combined with detailed analysis of volatile composition in the two rose varieties, led to the discovery of several novel scent-related candidate genes. The biological function of some of these clones, including a germacrene D synthase, was determined using bacteria-expressed recombinant enzymes.

10.5
Phenylpropanoids

Phenylpropanoids are natural products derived from the aromatic amino acid L-phenylalanine. Coumarins, stilbenes, flavonoids/isoflavonoids, lignins and lignans are among the important classes of metabolites that arise from a core phenylpropanoid pathway, which begins with the deamination of phenylalanine via PAL (Fig. 10.3). Although a large proportion of polyphenolic products play defensive and/or structural roles in the plant [136], certain volatile C_6C_1 benzoic acid derivatives, which are generally included in discussions of phenylpropanoids because of their presumed biosynthetic origin via side-chain shortening of hydroxycinnamic acids (HCAs), serve to attract pollinating insects to flowers [137, 138]. Benzoic/benzenoic volatiles are also partly responsible for imparting the unique aromas associated with such herbs as sweet basil (*Osimum basilicum*) [139]. Certainly, more than one metabolic route may exist *in planta* for these compounds. Isochorismate synthase (ICS), for example, was shown to be required for the production of salicylic acid (SA) [140], circumventing PAL-catalyzed phenylpropanoid biosynthesis. The C_6C_3 phenylpropane-based HCAs comprise a central pathway in phenylpropanoid metabolism (Fig. 10.3). Following phenylalanine-derived cinnamic acid formation, cinnamate 4-hydroxylase (C4H) catalyzes the initial aromatic ring oxidation to generate *p*-coumaric acid. The enzyme 4-coumarate:coenzyme A ligase (4CL) then activates *p*-coumaric acid with an addition of a thioester-bonded coenzyme A group. Although the details are uncertain [136], a series of methyl transfers, hydroxylations, and coniferylaldehyde dehydrogenase (CAD)-catalyzed reductions generate the monolignol precursors for lignin biosynthesis. Complex phenylpropanoids such as flavonoids, isoflavonoids, and stilbenes are formed by the condensation of a phenylpropane unit with malonyl-CoA (MCoA)-derived acetate groups.

As with terpenoid research, the availability of the complete genome sequence of *Arabidopsis thaliana* has proven an important tool to the area of phenylpropanoid biochemistry, especially with respect to lignin biosynthesis. In a study by Costa et al. [141], an exhaustive analysis of TIGR and TAIR databases, together with compiled EST sequence data, was carried out to determine the extent to which various metabolic networks from phenylalanine to the monolignols were organized and/or could be predicted. Although some 65 genes in *A. thaliana* had been annotated as encoding putative enzymatic steps in monolignol biosynthesis, many of them had only low homology to monolignol pathway genes of known function in other plant

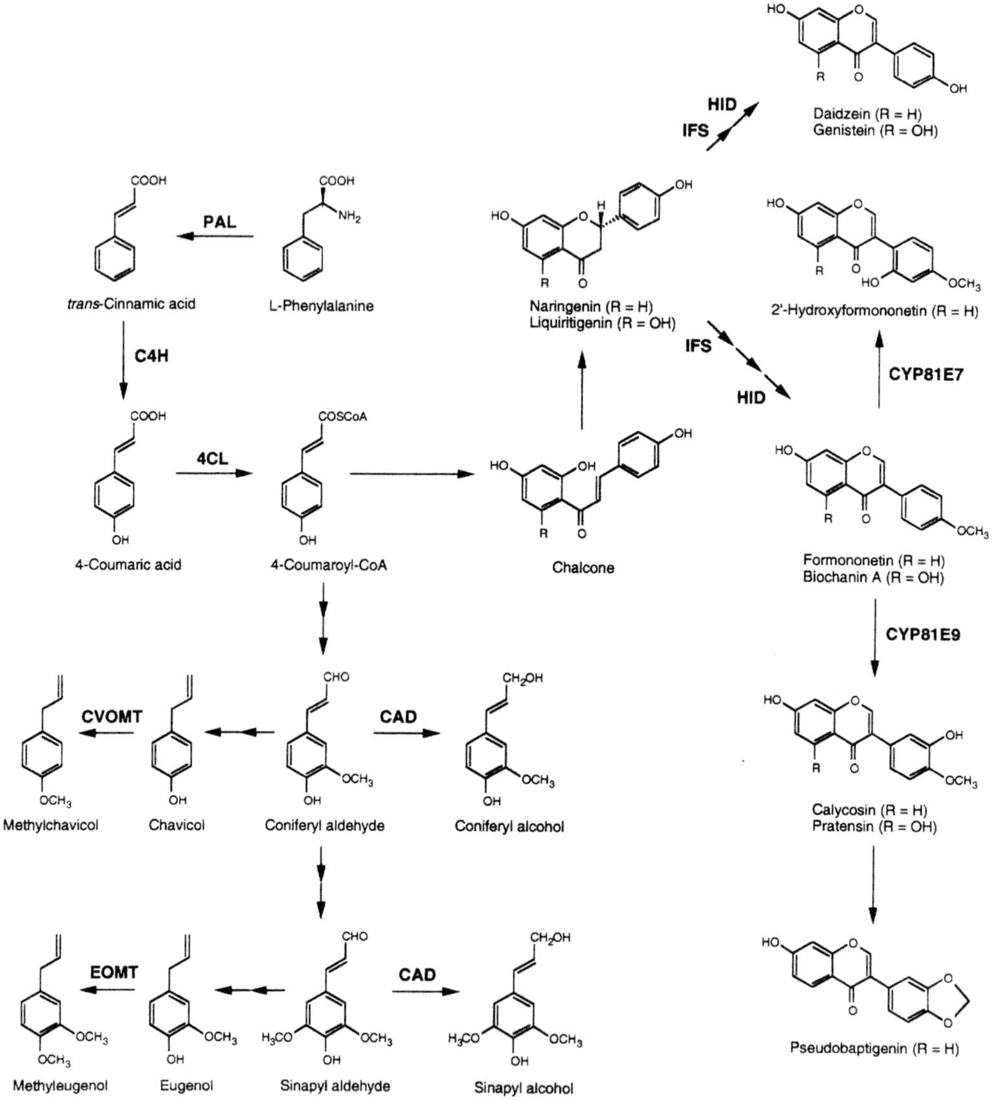

Fig. 10.3 Biosynthesis of phenylpropanoids and isoflavonoids. Enzymes for which corresponding molecular clones have been isolated using a genomics-based approach are shown. Abbreviations: PAL, phenylalanine ammonia lyase; C4H, cinnamate 4-hydroxylase; 4CL, 4-coumaroyl-CoA ligase; CVOMT, chavicol O-methyltransferase; EOMT, eugenol O-methyltransferase; CAD, coniferylaldehyde dehydrogenase; IFS, isoflavone synthase; HID, hydroxyisoflavonone dehydratase; CYP81E7, isoflavonoid 2′-hydroxylase; CYP81E9, isoflavonoid 5′-hydroxylase.

systems. As a result, only 13 genes could be classified as having *bona fide* function in phenylpropanoid metabolism, whereas the remaining 52 genes were resigned to "undetermined" physiological roles. The authors admitted that biochemical enzyme characterization and physiological studies were necessary to complement genomic database mining in order to discover new genes. In a follow-up report, Costa et al. [142] expressed 11 of the 14 undetermined genes originally annotated as 4CL homologues, and assayed empirically for 4CL enzyme activity. It was found that four of the 11 recombinant proteins were catalytically active, confirming that the 4CL gene family in *A. thaliana* had only four members.

The *Arabidopsis* genome sequence has also proven useful for identification of glucosyltransferase genes involved in sinapate metabolism. Sinapic acid, a major phenylpropanoid of the Brassicaceae family, is an intermediary compound in two distinct metabolic pathways, leading to sinapoyl ester formation and lignin, respectively [143–145]. Glucosyltransferases play key roles in the production of these intermediates, either through the formation of sinapoylglucose leading to the production of sinapoylmalate and sinapoylcholine, or through the production of sinapyl alchohol-4-*O*-glucoside, potentially leading to the syringyl units found in lignins. In combining *Arabidopsis* genome sequence information with biochemical data obtained via screening recombinant proteins for appropriate activities, Lim et al. [145] identified five new sinapoyl-glucosyltransferase clones. As well as sinapate metabolism, glycosyltransferases involved in flavonol glycoside biosynthesis in *A. thaliana* have been cloned [146]. Candidate genes were selected based on sequence homology to other known flavonoid glycosyltransferases, and corresponding T-DNA knockout lines were used to identify plants lacking any glycosylated flavonols. *In planta* observations and *in vitro* enzyme studies led to the cloning of UDP-rhamnose:flavonol-3-*O*-rhamnosyltransferase and UDP-glucose:flavonol-3-*O*-glycoside-7-*O*-glucosyltransferase. A genomics-oriented review of glycosyltransferases involved in plant secondary metabolism was recently published [147].

While *Arabidopsis* has provided a practical model system for the study of certain glycosylated phenylpropanoid products, the use of this species to elucidate lignin biosynthesis in conifers has conjured doubt. It came therefore as a surprise that expressed genes from wood-forming tissues of loblolly pine (*Pinus taeda*) displayed substantial homology with *A. thaliana* genes [148]. Initially, analysis of 59 797 pine ESTs revealed that only 50% of the expressed sequences had homologues in *A. thaliana* or any other angiosperm in public databases. However, very different results were found when the pine ESTs were assembled into contigs containing long, high-quality sequence data. For example, in contigs containing more than 1100 bp, over 90% had an apparent *Arabidopsis* homologue. In another study, global transcript profiling of primary stems from *A. thaliana* identified candidate genes for missing links in lignin biosynthesis and transcriptional regulators of fiber differentiation [149]. Using near-full-genome *Arabidopsis*-spotted 70-mer oligonucleotide arrays, discrete sets of candidate genes involved in various aspects of fiber cell differentiation and maturation were identified, including novel candidates for transcriptional regulation, monolignol polymerization, monolignol transport, and phenylalanine biosynthesis.

Notwithstanding the argument for *A. thaliana* as a model plant for the study of lignin biosynthesis in conifers, EST databases have been developed and used successfully in several coniferous plants, including loblolly pine, *Eucalyptus globulus* and white spruce (*Picea glauca*). In fact, transcriptomics in pine predates the completion of the *Arabidopsis* genome project, wherein 1097 ESTs from gravistimulated immature xylem were sequenced [150]. This database was later employed for the molecular cloning and expression of five laccase cDNAs in loblolly pine [151]. Laccases are multicopper oxidases believed to be involved in lignin oxidation and/or degradation [152, 153].

Since the report of Allona et al. [150], the number of xylem-derived pine ESTs has expanded to over 75 000 (http://web.ahc.umn.edu/biodata/nsfpine/) [154]. Investigations involving microarray analyses have also been reported in *Eucalyptus* species. Kirst et al. [155] correlated transcript abundance with trunk diameter variation, revealing coordinated down-regulation of genes encoding lignin biosynthetic enzymes in fast-growing individuals. Additionally, quantitative trait locus (QTL) analysis of transcript levels for lignin-related genes demonstrated that mRNA abundance is regulated by two genetic loci. These two loci were shown to co-localize with QTLs for trunk diameter increases, indicating that the same genetic regions were responsible for regulating trunk growth and lignin biosynthesis. Further microarray-based investigations of QTL involvement in differentiating xylem were reported for a *Eucalyptus* hybrid, along with *E. grandis* and *E. globulus* parents [156]. In this case, the genetic architecture of transcript regulation in different genetic backgrounds was compared. Beyond pine and *Eucalyptus*, transcript profiles of stress-related genes in developing white spruce somatic embryos have been examined [157]. The effects of polyethylene glycol (PEG) on the transcript levels of 512 stress-related genes were analyzed via microarray technique. As well as genes encoding heat shock proteins, glutathione-*S*-transferases and cysteine proteases, several genes involved in lignin biosynthesis were found to be differentially regulated upon PEG treatment.

It is likely that model plant species such as pine, *Eucalyptus* and spruce will be used well into the future for phenylpropanoid research. Although investigations of lignin metabolism have relied heavily on coniferous systems, members of the Leguminosae family have been used extensively to study flavonoids and isoflavonoids, along with myriad other products [158, 159]. While flavonoids are found throughout the plant kingdom, isoflavonoids are more restricted, and are particularly prevalent in the Papilonoideae subfamily of the Leguminosae. Due to the purported health-promoting activities of both flavonoids and isoflavones, in addition to their uses as colorants [160] and phytoestrogenic agents [161], respectively, major efforts have been made to elucidate the biosynthesis of these compounds. Prior to the initiation of the *Medicago* Genome Initiative [162], soybean EST sequence databases were being developed and used for the cloning of P450 enzymes involved in isoflavone metabolism [163, 164]. Isoflavone synthase (IFS) catalyzes the first committed step of isoflavone biosynthesis (Fig. 10.3), and was identified by searching a fungal-infected soybean seed database of 1700 EST sequences [164].

Since these initial reports, most studies have relied on either *Medicago truncatula* or *Lotus japonica* as model legume systems [159, 165]. Searches of public EST databases of both these species, in addition to EST databases for soybean (http://www/tigr.org/tgi/) was necessary to identify 2-hydroxyisoflavonone dehydratase (HID) (Fig. 10.3). Following the IFS-catalyzed biosynthesis of 2-hydroxyiso-flavonone, a water molecule must be shed to convert the flavonone to a flavone. The next steps in forming the more complex isoflavonoid products rely on P450 enzymes, three of which were identified by Liu et al. [166]. Mining of public *M. truncatula* EST databases and screening of a root cDNA library led to identification of the cytochrome P450 81E enzymes CYP81E7, CYP81E8 and CYP81E9; the catalytic role of two of these is indicated in Figure 10.3. These monooxygenases were found to be responsible for regiospecific hydroxylations of various isoflavonoids, although some pathway details remained unclear. In the same year, Xie et al. [167] randomly sequenced a cDNA library from young *M. truncatula* seeds, leading to the identification of a clone encoding anthocyanidin reductase (BANYULS, abbreviated MtBAN). Like isoflavonoids, flavonoids are derived from precursor flavonones, although branching of the flavonoid pathway occurs at leucoanthocyanidin, which can form anthocyanins via ANS (anthocyanin synthase) or condensed tannins via BAN.

In contrast to the aforementioned natural products, little is known about the biosynthesis of phenylpropenes, a group of small phenolic molecules that are key flavoring agents in many important herbs and spices. Phenylpropenes are found in peppercorns, cloves, nutmeg, cinnamon, allspice, pimenta, tarragon and basil [168]; they are also important components of the defensive arsenal of plants or function as pollinator attractants. Eugenol, for example, has antiherbivory and antimicrobial properties [169, 170], while methyl-eugenol attracts pollinating moths and bees to many flower types [65]. Conflicting reports have left unanswered questions regarding how the allyl/propenyl side chain of the phenylpropenes is formed [171–174]. Several recent elaborations on downstream reactions in the pathway have been made, however, for which references are found in Section 10.4. Sweet basil EST sequence databases have been mined for both terpenoid-related genes and phenylpropenoid-related genes, as basil peltate glandular trichomes contain volatiles belonging to both classes of secondary metabolites. Particular attention was paid to phenylpropenes by Gang et al. [175], when a biochemical genomics approach was used to isolate chavicol *O*-methyltransferase (CVOMT) and eugenol *O*-methyltransferase (EOMT). Gang et al. [139] used the same EST sequence database to isolate a cDNA encoding a 3′-hydroxylase (CS 3′H, or CYP98A13).

10.6
Alkaloids

Alkaloids are a large, diverse group of natural products found in approximately 20% of plant species. Generally characterized by the presence of a nitrogen atom within a heterocyclic ring, many of the 12 000 known alkaloids possess pharmacological activities and are widely used in medicine. Alkaloids draw on the products of primary metabolism for their biosynthesis, with amino acids serving as their main precursors [176]. Unlike most other groups of secondary metabolites, the structurally diverse alkaloids have independent biosynthetic origins. Tropane alkaloids such as cocaine, atropine and scopolamine are derived from ornithine, benzylisoquinoline alkaloids such as morphine, codeine and berberine are produced from tyrosine, whereas tryptophan is used for the biosynthesis of the indole alkaloids vinblastine and strychnine. The biosynthesis of benzylisoquinoline alkaloids palmatine and berberine is shown in Figure 10.4, while nicotine/nornicotine biosynthesis is illustrated in Figure 10.5. Typically, multiple catalytic steps are required for the formation of the basic structural nuclei of alkaloids, along with subsequent carbon-ring modifications and "decorative" hydroxylations, methylations, acetylations and glycosylations [177]. Many of the different enzyme classes involved in plant alkaloid metabolism are engaged in primary metabolism, or in the formation of other natural products.

To date, about 40 genes encoding alkaloid biosynthetic enzymes have been cloned [176, 178]. The vast majority of these clones were obtained using traditional biochemical approaches, requiring enzyme purification and protein sequencing. Morishige et al. [179] were the first to use EST sequence data for the molecular cloning of a new gene involved in alkaloid biosynthesis. Using a high-metabolite-producing *Coptis japonica* cell culture, 1014 cDNA clones were isolated and sequenced. Sequences corresponding to three previously reported *C. japonica* O-methyltransferases – norcoclaurine 6-O-methyltransferase (6OMT), 3′-hydroxy-N-methylcoclaurine 4′-O-methyltransferase (4′OMT), and scoulerine 9-O-methyltransferase (SOMT) – were present in the collection (Fig. 10.4). In addition, a novel O-methyltransferase-like cDNA, S-adenosyl-L-methionine:columbamine O-methyltransferase (CoOMT), which catalyzes the conversion of columbamine to palmatine (Fig. 10.4), was also found. The study illustrated the utility of EST analysis for the isolation of new genes involved in benzylisoquinoline alkaloid biosynthesis. Comparative macroarray analysis of opium poppy and various morphine-free *Papaver* species was used to identify a *P. somniferum* O-methyltransferase clone [180]. In this case, *P. somniferum* seedlings were used to develop a cDNA sequence library. Of the 849 sequenced elements, three were shown on a macroarray to be differentially expressed in *P. somniferum* compared to non-morphine-producing species. Whereas two of these cDNAs showed no significant homology to any known protein, one was found to encode a protein identified as S-adenosyl-L-methionine(R, S)-3′-hydroxy-N-methylcoclaurine 4′-OMT (4′OMT). Recently, an EST sequence database was used to obtain the *P. somniferum* clone of (S)-norcoclaurine synthase (NCS; Fig. 10.4), which catalyzes the first committed step in benzylisoquinoline alkaloid metabolism [181].

A similar gene discovery strategy was used by Pilazke-Wunderlich and Nessler [182], in which homologues for cell-wall-degrading enzymes were found during random sequencing of an opium poppy latex cDNA library. Poppy elements putatively encoding pectin methyltransferase (PME), pectin acetylesterase (PAE) and pectate lyase (PL) were shown to express specifically in laticifers. Although the corresponding enzymes were not heterologously expressed, assays of latex serum confirmed the presence of PME, PAE and PL activities. Pectin-degrading enzymes

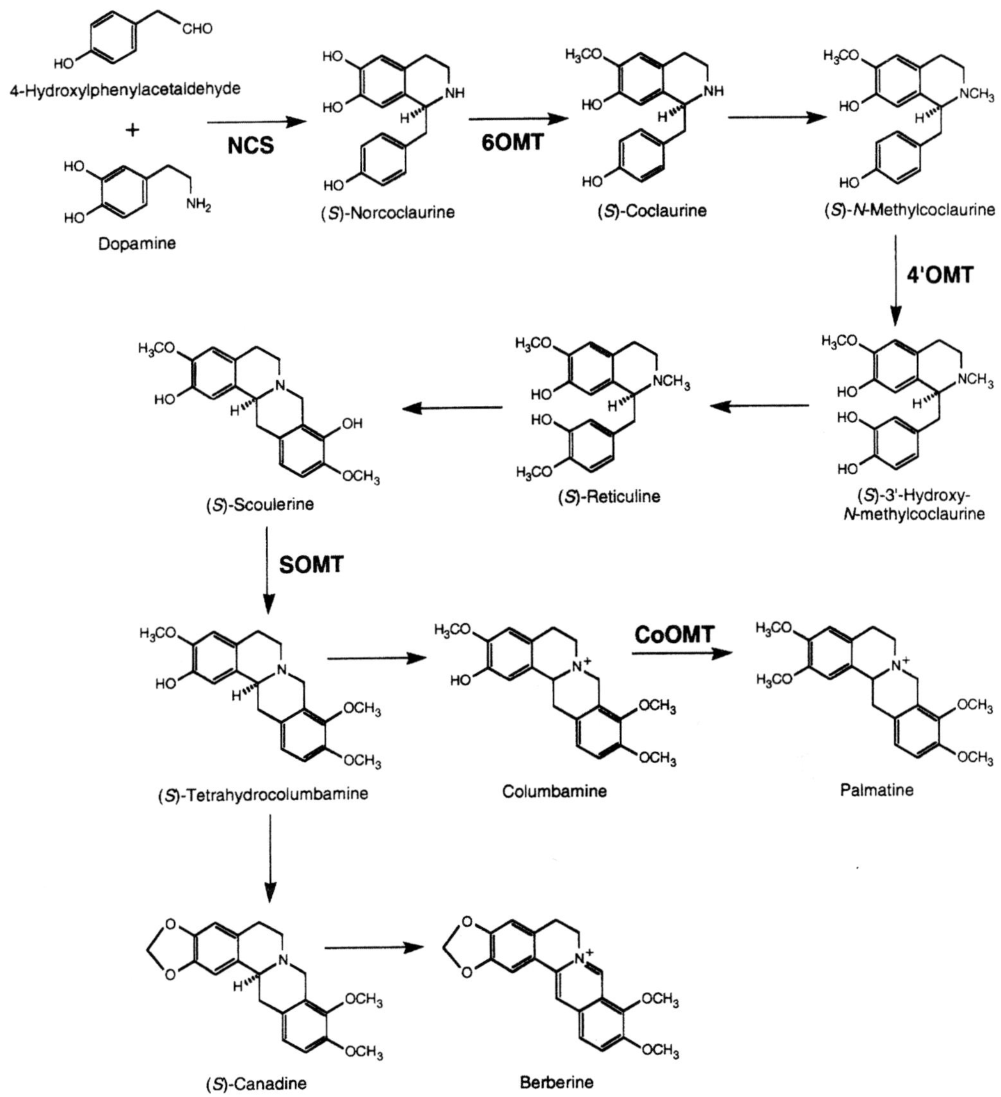

Fig. 10.4 Biosynthesis of the benzylisoquino-line alkaloids berberine and palmatine. Enzymes for which corresponding molecular clones have been isolated using a genomics-based approach are shown. Abbreviations: NCS, norcoclaurine synthase; 6OMT, norcoclaurine 6-O-methyltransferase; 4′OMT, 3′-hydroxy-N-methylcoclaurine 4′-O-methyltransferase; SOMT, scoulerine 9-O-methyltransferase; CoOMT, columbamine O-methyltransferase.

tabolism [197]. An important undercurrent of the present debate is the well-known, yet often overlooked, tenet that sequence homology is an insufficient datum on which to base predictions of specific enzyme function, much less the overall biosynthetic capacity of a plant. The problem is due in part to the protocol used to annotate genomic and EST data. Such annotations are intended only to identify conserved sequence domains and to suggest putative functions as a guide to empirical characterization. A combination of biochemical and molecular phylogenetic approaches was used to investigate the evolution of benzylisoquinoline alkaloid biosynthesis in angiosperms [181]. As part of the study, Consensus Bayesian trees were derived from putative or functionally characterized FAD-dependent proteins homologous to BBE. Gene products functionally characterized as BBE grouped together and were found only distantly related to putative BBE-like proteins from non-alkaloid-producing plants such as *Arabidopsis*. Sequence similarity between BBE-like genes from *A. thaliana* and functionally characterized BBE clones may reflect a shared, monophyletic origin, but the question of whether *Arabidopsis* homologues retain BBE activity has yet to be investigated. Beyond BBE, a phylogenetic analysis of proteins sharing homology with NCS was performed. Functionally characterized NCSs from *P. somniferum* and *T. flavum* clustered together, distant from uncharacterized homologues [181].

Since NCS catalyzes the first committed step in benzylisoquinoline alkaloid biosynthesis, plant species lacking NCS activity likely do not have the capacity for alkaloid production. A parallel conclusion may be drawn for plants lacking functional SS enzyme, which catalyzes the formation of the structural nucleus common to all monoterpenoid indole alkaloids [176]. Homologues for SS also occur in the *Arabidopsis* genome, albeit with low sequence identity with functionally characterized SSs. The limited similarity between NCS, or SS, and their *Arabidopsis* sequence homologues contrasts with the extensive identity between other groups of alkaloid biosynthetic genes and their counterparts in the *Arabidopsis* genome. Sequence identities of 50–70% are not uncommon when comparing the decarboxylases, methyltransferases, and P450-dependent monooxygenases involved in the biosynthesis of berberine with sequence homologues in *Arabidopsis* [14]. Interestingly, compared to the relatively isolated phylogenetic clustering of functional NCS enzymes, the Consensus Bayesian tree for selected *O*-methyltransferases (OMTs) shows a greater dispersal of functionally characterized OMTs [181]. The ability of a plant species to synthesize a specific alkaloid may depend on the presence of a functional, key entry-point enzyme, even if downstream enzymes retain some catalytic activity. As suggested by Facchini et al. [14], the abundance of *Arabidopsis* sequences sharing homology with genes encoding benzylisoquinoline and monoterpenoid indole alkaloid biosynthetic enzymes may reflect an evolutionary phenomenon known as exaptive evolution [198]. Genetic drift of a key enzyme no longer under selective pressure could cause an entire "shutdown" of a biosynthetic pathway, as would hypothetically occur if either NCS or SS activity were to be lost. Perhaps *Arabidopsis* displays a genomic fingerprint for complex alkaloid biosynthesis, but the substrate specificity of some – if not all – of the gene products has changed.

10.7
Conclusion

Plant secondary metabolites can be defined as compounds that have no recognized role in the maintenance of fundamental life processes in the plants that synthesize them. Despite this rather dismissive description, secondary products play important roles in the interaction of a plant with its environment, and many are economically, medically, and traditionally valuable. The question of how these often highly complex compounds are biosynthesized *in planta* has motivated intense research, and required the application of new technologies and approaches. Revolutionary advances have been made in large steps, beginning with the use of radiotracing methods in the mid-twentieth century, followed by a reliance on tissue culture techniques and subsequently, molecular biology. Genomics represents the most recent revolution in biology. The use of whole genome sequences and species-specific EST collections has allowed rapid discovery of new genes involved in plant secondary metabolism. Additionally, genomic tools have provided the means necessary to understand intricate signaling and regulatory pathways, complex phylogenic relationships, and overall genetic architecture. The genomics revolution has already had a major impact in the field of natural products research, and will undoubtedly continue to lead to new discoveries.

Acknowledgments

P.J.F. is privileged to hold the Canada Research Chair in Plant Biotechnology. Research in his laboratory is funded by grants from the Natural Sciences and Engineering Research Council of Canada. J.M.H. is the recipient of an Alberta Ingenuity doctoral scholarship.

References

References marked * are of interest; those marked ** are of considerable interest

1 Wink, M. *Functions of Plant Secondary Metabolites and Their Exploitation in Biotechnology*. Sheffield Academic Press, Sheffield, UK, **1999**.

2 Cullis, C.A. *Plant Genomics and Proteomics*. John Wiley & Sons, Hoboken, New Jersey, **2004**.

3 Borevitz, J.O., Ecker, J.R. Plant genomics: the third wave. *Annu. Rev. Genomics Hum. Genet.* **2004**, *5*, 443–477.

4 Cliften, P., Sudarsanam, P., Desikan, A., Fulton, L., Fulton, B., et al. Finding functional features in *Saccharomyces* genomes by phylogenetic footprinting. *Science* **2003**, *301*, 71–76.

5 * Kellis, M., Patterson, N., Endrizzi, M., Birren, B., Lander, E.S. Sequencing and comparison of yeast species to identify genes and regulatory elements. *Nature* **2003**, *423*, 241–254. [This paper was among the first to use comparative genomics for the discovery of novel genes and regulatory elements. Similar approaches have since been applied to plants, especially in the grass and legume families.]

6 * Reiser, L., Mueller, L.A., Rhee, S.Y. Surviving in a sea of data: a survey of plant genome data resources and issues in building data management systems.

Plant Mol. Biol. **2002**, *48*, 59–74. [This paper discussed resources available in the public domain and describes some of the software for data management systems currently available for plant research.]

7 ** Wheeler, D.L., Smith-White, B., Chetvernin, V., Resenchuk, S., Dombrowski, S.M., Pechous, S.W., Tatusova, T., Ostell, J. Plant genome resources at the National Center for Biotechnology Information. *Plant Physiol.* **2005**, *138*, 1280–1288. [The paper outlines resources available through NCBI for use in nucleotide sequence analysis.]

8 Benson, D.A., Karsch-Mizrachi, I., Lipman, D.J., Ostell, J., Wheeler, D.L. GenBank: update. *Nucleic Acids Res.* **2005**, *33*, D34–D38.

9 Tateno, Y., Saitou, N., Okubo, K., Sugawara, H., Gojobori, T. DDBJ in collaboration with mass-sequencing teams on annotation. *Nucleic Acids Res.* **2004**, *33*, D25–D28.

10 Kanz, C., Aldebert, P., Althorpe, N., Baker, W., Baldwin, A., Bates, K., Browne, P., van den Broek, A., Castro, M., Cochrane, G., Duggan, K., Eberhardt, R., Faruque, N., Gamble, J., Diez, F.G., Harte, N., Kulikova, T., Lin, Q., Lombard, V., Lopez, R., Mancuso, R., McHale, M., Nardone, F., Silventoinen, V., Sobhany, S., Stoehr, P., Tuli, M.A., Tzouvara, K., Vaughan, R., Wu, D., Zhu, W., Apweiler, R. The EMBL nucleotide sequence database. *Nucleic Acids Res.* **2005**, *33*, D29–D33.

11 Dong, Q., Lawrence, C.J., Schlueter, S.D., Wilkerson, M.D., Kurtz, S., Lushbough, C., Brendel, V. Comparative plant genomics resources at PlantGDB. *Plant Physiol.* **2005**, *139*, 610–618.

12 Ausubel, F.M. Summaries of the National Science Foundation-sponsored *Arabidopsis* 2010 projects and National Science Foundation-sponsored plant genome projects that are generating *Arabidopsis* resources for the community. *Plant Physiol.* **2002**, *129*, 394–437.

13 Fridman, E., Pichersky, E. Metabolomics, genomics, proteomics, and the identification of enzymes and their substrates and products. *Curr. Opin. Plant Biol.* **2005**, *8*, 242–248.

14 * Facchini, P.J., Bird, D.A., St. Pierre, B. Can *Arabidopsis* make complex alkaloids? *Trends Plant Sci.* **2004**, *9*, 116–122. [This review critically examines the suggestion that *Arabidopsis* can produce complex alkaloids based on the occurrence of numerous alkaloid biosynthetic gene homologues in the genome sequence of this plant.]

15 Frey, M., Chomet, P., Glawischnig, E., Stettner, C., Grun, S., Winklmair, A., Eisenreich, W., Bacher, A., Meeley, R.B., Briggs, S.P., Simcox, K., Gierl, A. Analysis of a chemical plant defense mechanism in grasses. *Science* **1997**, *277*, 696–699.

16 Frick, S., Kutchan, T.M. Molecular cloning and functional expression of *O*-methyltransferases common to iso-quinoline alkaloid and phenylpropanoid biosynthesis. *Plant J.* **1999**, *17*, 329–339.

17 Schmidt, R. Plant genome evolution: lessons from comparative genomics at the DNA level. *Plant Mol. Biol.* **2002**, *48*, 21–37.

18 Brunner, A.M., Busov, V.B., Strauss, S.H. Poplar genome sequence: functional genomics in an ecologically dominant plant species. *Trends Plant Sci.* **2004**, *9*, 49–56.

19 Vu, T.H., Hoffman, A.R. Comparative genomics sheds light on mechanisms of genomic imprinting. *Genome Res.* **2000**, *10*, 1660–1663.

20 King, G.J. Through a genome, darkly: comparative analysis of plant chromosomal DNA. *Plant Mol. Biol.* **2002**, *48*, 5–20.

21 Van der Peer, Y. 'Horozontal' plant biology on the rise. *Genome Biol.* **2004**, *6*, 302

22 Bonierbale, M.W., Plaisted, R.L., Tanksley, S.D. RFLP maps based on a common set of clones reveal modes of chromosomal evolution in potato and tomato. *Genetics* **1988**, *120*, 1095–1103.

23 Chao, S., Sharp, P.J., Worland, A.J., Warham, E.J., Koebner, R.M.D., Gale, M.D. RFLP-based genetic maps of wheat homologous group 7 chromosomes. *Theoret. Appl. Genet.* **1989**, *78*, 495–504.

24 Devos, K.M., Gale, M.D. Comparative genetics in the grasses. *Plant Mol. Biol.* **1997**, *35*, 3–15.

25 Gale, M.D., Devos, K.M. Plant comparative genetics after 10 years. *Science* **1998**, *282*, 656–659.

26 Choi, H.-K., Mun, J.-H., Kim, D.-J., Zhu, H., Baek, J.-M., Mudge, J., Roe, B., Ellis, N., Doyle, J., Kiss, G.B., Young, N.D., Cook, D.R. Estimating genome conservation between crop and model legume species. *Proc. Natl. Acad. Sci. USA* **2004**, *101*, 15289–15294.

27 Devos, K.M. Updating the crop circle. *Curr. Opin. Plant Biol.* **2005**, *8*, 155–162.

28 Rensink, W.A., Buell, C.R. *Arabidopsis* to rice. Applying knowledge from a weed to enhance our understanding of a crop species. *Plant Physiol.* **2004**, *135*, 622–629.

29 Koch, M.A., Weisshaar, B., Kroymann, J., Haubold, B., Mitchell-Olds, T. Comparative genomics and regulatory evolution: conservation and function of the *Chs* and *Apetala* promoters. *Mol. Biol. Evol.* **2001**, *18*, 1882–1891.

30 Hong, R.L., Hamaguchi, L., Busch, M.A., Weigel, D. Regulatory elements of the floral homeotic gene *AGAMOUS* identified by phylogenetic footprinting and shadowing. *Plant Cell* **2003**, *15*, 1296–1309.

31 Colinas, J., Birnbaum, K., Benfey, P.N. Using cauliflower to find conserved noncoding regions in *Arabidopsis*. *Plant Physiol.* **2002**, *129*, 451–454.

32 Morishige, D.T., Childs, K.L., Moore, L.D., Mullet, J. Targeted analysis of orthologous *phytochrome A* regions of the sorghum, maize, and rice genomes using comparative gene-island sequencing. *Plant Physiol.* **2002**, *130*, 1614–1625.

33 Buchanan, C.D., Klein, P.E., Mullet, J.E. Phylogenetic analysis of 5′-noncoding regions from the ABA-responsive *rab16/17* gene family of sorghum, maize and rice provides insight into the composition, organization and function of *cis*-regulatory modules. *Genetics* **2004**, *168*, 1639–1654.

34 Zhu, H., Choi. H.-K., Cook. D.R., Shoemaker. R.C. Bridging model and crop legumes through comparative genomics. *Plant Physiol.* **2005**, *137*, 1189–1196.

35 Choi, H.-K., Kim, D., Uhm, T., Limpens, E., Lim, H., Mun, J.H. Kalo, P., Penmetsa, R.V., Seres, A., Kulikova, O., Roe, B.A., Bisseling, T., Kiss, G.B., Cook, D.R. A sequence-based genetic map of *Medicago truncatula* and comparison of marker colinearity with *M. sativa*. *Genetics* **2004**, *166*, 1463–1502.

36 Yan, H.H., Mudge, J., Kim, D.-J., Shoemaker, R.C., Cook, D.R., Young, N.D. Estimates of conserved microsynteny among genomes of *Glycine max*, *Medicago truncatula*, and *Arabidopsis thaliana*. *Theoret. Appl. Genet.* **2003**, *106*, 1256–1265.

37 Endre, G., Kereszt, A., Kevei, Z., Mihacea, S., Kalo, P., Kiss, G.B. A receptor kinase gene regulating symbiotic nodule development. *Nature* **2002**, *417*, 962–966.

38 Limpens, E., Franken, C., Smit, P., Willemse, J., Bisseling, T., Geurts, R. LysM domain receptor kinases regulating rhizobial Nod factor-induced infection. *Science* **2003**, *302*, 630–633.

39 Baxter, I., Tchieu, J., Sussman, M.R., Boultry, M., Palmgren, M.G., Gribskov, M., Harper, J.F., Axelsen, K.B. Genomic comparison of P-type ATPase ion pumps in *Arabidopsis* and rice. *Plant Physiol.* **2003**, *132*, 618–628.

40 Griffiths, S., Dunford, R.P., Coupland, G., Laurie, D.A. The evolution of *CONSTANS-like* gene families in barley, rice and *Arabidopsis*. *Plant Physiol.* **2003**, *131*, 1855–1867.

41 Matsumoto, N., Hirano, T., Iwasaka, T., Yamamoto, N. Functional analysis and intracellular localization of rice cryptochromes. *Plant Physiol.* **2003**, *133*, 1494–1503.

42 Kolukisaoglu, U., Weinl, S., Blazevic, D., Batistic, O., Kudla, J. Calcium sensors and their interacting protein kinases: genomics of the *Arabidopsis* and rice CBL-CIPK signaling networks. *Plant Physiol.* **2004**, *134*, 43–58.

43 Hughes, T.R., Mao, M., Jones, A.R., Burchard, J., Marton, M.J., Shannon, K.W., Lefkowitz, S.M., Ziman, M., Schelter, J.M., Meyer, M.R., Kobayashi, S., Davis, C., Dai, H., He, Y.D., Stephaniants, S.B., Cavet, G., Walker, W.L., West, A., Coffey, E., Shoemaker, D.D., Stoughton, R., Blanchard, A.P., Friend, S.H., Linsley, P.S. Expression

profiling using microarrays fabricated by an ink-jet oligonucleotide synthesizer. *Nature Biotechnol.* **2001**, *19*, 342–347.

44 Shoemaker, D.D., Schadt, E.E., Armour, C.D., He, Y.D., Garrett-Engele, P., McDonagh, P.D., Loerch, P.M., Leonardson, A., Lum, P.Y., Cavet, G., Wu, L.F., Altschuler, S.J., Edwards, S., King, J., Tsang, J.S., Schimmack, G., Schelter, J.M., Koch, J., Ziman, M., Marton, M.J., Li, B., Cundiff, P., Ward, T., Castle, J., Krolewski, M., Meyer, M.R., Mao, M., Burchard, J., Kidd, M.J., Dai, H., Phillips, J.W., Linsley, P.S., Stoughton, R., Scherer, S., Boguski, M.S. Experimental annotation of the human genome using microarray technology. *Nature* **2001**, *409*, 922–927.

45 Donson, J., Fang, Y., Espiritu-Santo, G., Xing, W., Salazar, A., Miymoto, S., Armendarez, V., Volkmuth, W. Comprehensive gene expression analysis by transcript profiling. *Plant Mol. Biol.* **2002**, *48*, 75–97.

46 Bohnert, H.J., Ayoubi, P., Borchert, C., Bressan, R.A., Burnap, R.L., Cushman, J.C., Cushman, M.A., Deyholos, M., Fischer, R., Galbraith, D.W., Sawegawa, P.M., Jenks, M., Kawasaki, S., Koiwa, H., Kore-eda, S., Lee, B.H., Michalowski, C.B., Misawa, E., Nomura, M., Ozturk, N., Postier, B., Prade, R., Song, C.P., Tanaka, Y., Wang, H., Zhu, J.K. A genomics approach towards salt stress tolerance. *Plant Physiol. Biochem.* **2001**, *39*, 1–17.

47 Ohlrogge, J., Benning, C. Unraveling plant metabolism by EST analysis. *Curr. Opin. Plant Biol.* **2000**, *3*, 224–228.

48 Audic, S., Claverie, J.M. The significance of digital gene expression profiles. *Genome Res.* **1997**, *7*, 986–995.

49 Meyers, B.C., Galbraith, D.W., Nelson, T., Agrawal, V. Methods for transcriptional profiling in plants. *Plant Physiol.* **2004**, *135*, 637–652.

50 Sargent, T.D. Isolation of differentially expressed genes. *Methods Enzymol.* **1987**, *152*, 423–432.

51 Hubank, M., Schatz, D.G. Identifying differences in mRNA expression by representational difference analysis of cDNA. *Nucleic Acids Res.* **1994**, *25*, 5640–5648.

52 Diatchenko, L., Lau, Y.F., Campbell, A.P., Chenchik, A., Moqadam, F., Huang, B., Lukyanov, S., Lukyanov, K., Gurskaya, N., Sverklov, E.D., Siebert, P.D. Suppression subtractive hybridization: a method for generating differentially regulated or tissue-specific cDNA probes and libraries. *Proc. Natl. Acad. Sci. USA* **1996**, *12*, 6025–6030.

53 Ewing, R.M., Kahla, A.B., Poirot, O., Lopez, F., Audic, S., Claverie, J.M. Large-scale statistical analyses of rice ESTs reveal correlated patterns of gene expression. *Genome Res.* **1999**, *9*, 950–959.

54 Newman, T., de Bruijn, F.J., Green, P., Keegstra, K., Kende, H., McIntosh, I., Ohlrogge, J., Raikhel, N., Somerville, S., Thomashow, M., et al. Genes galore: a summary of methods for accessing results from large-scale partial sequencing of anonymous *Arabidopsis* cDNA clones. *Plant Physiol.* **1994**, *106*, 1241–1255.

55 Delseny, M., Cooke, R., Raynal, M., Grellet, F. The *Arabidopsis thaliana* cDNA sequencing projects. *FEBS Lett.* **1997**, *405*, 129–132.

56 Schnee, C., Köllner, T.G., Gershenzon, J., Degenhardt, J. The maize gene *terpene synthase 1* encodes a sesquiterpene synthase catalyzing the formation of (*E*)-β-farnesene, (*E*)-nerolidol, and (*E, E*)-farnesol after herbivore damage. *Plant Physiol.* **2002**, *130*, 2049–2060.

57 Lee, Y., Tsai, J., Sunkara, S., Karamycheva, S., Pertea, G., Sultana, R., Antonescu, V., Chan, A., Cheung, F., Quackenbush, J. The TIGR Gene Indices: clustering and assembling ESTs and known genes and integration with eukaryotic genomes. *Nucleic Acids Res.* **2005**, *33*, D71–D74.

58 Schoof, H., Spannagl, M., Yang, L., Ernst, R., Gundlach, H., Haase, D., Haberer, G., Mayer, K.F.X. Munich Information Center for protein sequences plant genome resources. A framework for integrative and comparative analyses. *Plant Physiol.* **2005**, *138*, 1301–1309.

59 Kalyanaraman, A., Aluru, S., Kothari, S., Brendel, V. Efficient clustering of large EST data sets on parallel computers. *Nucleic Acids Res.* **2003**, *31*, 2963–2974.

60 Gershenzon, J., Kreis, W. Biochemistry of terpenoids: monoterpenes, sesquiterpenes, diterpenes, sterols, cardiac glycosides, and steroid saponins. In: Wink, M. (Ed.), *Biochemistry of Plant Secondary Metabolism*. CRC Press, Boca Raton, Florida, **1999**, pp. 222–299.

61 Buckingham, J. *Dictionary of Natural Products on CD-ROM, Version 6.1*. Chapman & Hall, London, UK, **1998**.

62 Zinkel, D.F., Russell, J. *Naval Stores: Production, Chemistry, Utilization*. Pulp Chemicals Association, New York, **1989**.

63 Dawson, F.A. The amazing terpenes. *Naval Stores Rev*. **1994**, *104*, 6–12.

64 Gang, D.R., Wang, J.H., Dudareva, N., Nam, K.H., Simon, J.E., Lewinsohn, E., Pichersky, E. An investigation of the storage and biosynthesis of phenylpropenes in sweet basil. *Plant Physiol*. **2001**, *125*, 539–555.

65 Pichersky, E., Gershenzon, J. The formation and function of plant volatiles: perfumes for pollinator attraction and defense. *Curr. Opin. Plant Biol*. **2002**, *5*, 237–243.

66 Sharon-Asa, L., Shalit, M., Frydman, A., Bar, E., Holland, D., Or, E., Lavi, U., Lewinsohn, E., Eyal, Y. Citrus fruit flavor and aroma biosynthesis: isolation, functional characterization, and developmental regulation of *cstps1*, a key gene in the production of the sesquiterpene aroma compound valencene. *Plant J*. **2003**, *36*, 664–674.

67 Knudsen, J.T., Tollesten, L., Bergström, G.L. Floral scent: a checklist of volatile compounds isolated by head-space techniques. *Phytochemistry* **1993**, *33*, 252–280.

68 Guterman, I., Shalit, M., Menda, N., Piestun, D., Dafny-Yelin, M., Shalev, G., Bar, E., Davydov, O., Ovadis, M., Emanuel, M., Wang, J., Adam, Z., Pichersky, E., Lewinsohn, E., Zamir, D., Vainstein, A., Weiss, D. Rose scent: genomics approach to discovering novel floral fragrance-related genes. *Plant Cell* **2002**, *14*, 2325–2338.

69 Dobson, H.E.M. Floral volatiles in insect biology. In: Bernays, E. (Ed.), *Insect–Plant Interactions*. CRC Press, Boca Raton, Florida, **1994**, Volume 5, pp. 47–81.

70 Dudareva, N., Pichersky, E. Biochemical and molecular aspects of floral scents. *Plant Physiol*. **2000**, *122*, 627–634.

71 Dudareva, N., Martin, D., Kish, C.M., Kolosova, N., Gorenstein, N., Fäldt, J., Miller, B., Bohlmann, J. (*E*)-β-Ocimene and myrcene synthase genes of floral scent biosynthesis in snapdragon: function and expression of three terpene synthase genes of a new terpene synthase subfamily. *Plant Cell* **2003**, *15*, 1227–1241.

72 Bohlmann, J., Croteau, R. Diversity and variability of terpenoid defenses in conifers: molecular genetics, biochemistry and evolution of the terpene synthase gene family in grand fir (*Abies grandis*). In: Chadwick, D.J., Goode, J.A. (Eds.), *Insect–Plant Interactions and Induced Plant Defense*. John Wiley & Sons, West Sussex, UK, **1999**, pp. 132–146.

73 Huber, D.P.W., Ralph, S., Bohlmann, J. Genomic hardwiring and phenotypic plasticity of terpenoid-based defenses in conifers. *J. Chem. Ecol*. **2004**, *30*, 2399–2418.

74 Crowell, P.L., Gould, M.N. Chemoprevention and therapy of cancer by *d*-limonene. *Crit. Rev. Oncog*. **1994**, *5*, 1–22.

75 Van Geldre, E., Vergauwe, A., Van den Eekchout, E. State of the art of the production of the antimalarial compound artimisinin in plants. *Plant Mol. Biol*. **1997**, *33*, 199–209.

76 Holmes, F.A., Kudelka, A.P., Kavanagh, J.J., Huber, M.H., Ajani, J.A. Current status of clinical trials with paclitaxel and docetaxel. In: Georg, G.I., Chen, T.T., Ojima, I., Vyas, D.M. (Eds.), *Taxane Anticancer Agents: Basic Science and Current Status*. American Chemical Society Symposium Series 583, Washington DC, **1995**, pp. 31–57.

77 Jennewein, S., Croteau, R. Taxol: biosynthesis, molecular genetics, and biotechnological applications. *Appl. Microbiol. Biotechnol*. **2001**, *57*, 13–19.

78 Newman, J.D., Chappell, J. Isoprenoid biosynthesis in plants: carbon partitioning within the cytoplasmic pathway. *Crit. Rev. Biochem. Mol. Biol*. **1999**, *34*, 95–106.

79 Eisenreich, W., Schwarz, M., Cartayrade, A., Arigoni, D., Zenk, M.H., Bacher, A. The deoxyxylose phosphate pathway of terpenoid biosynthesis in plants and microorganisms. *Chem. Biol.* **1998**, *5*, R221–R233.

80 Lichtenthaler, H.K. The 1-deoxy-D-xylulose-5-phosphate pathway of isoprenoid biosynthesis in plants. *Annu. Rev. Plant Physiol. Plant Mol. Biol.* **1999**, *50*, 47–66.

81 * Dudareva, N., Anderssonn, S., Orlova, I., Gatto, N., Reichelt, M., Rhodes, D., Boland, W., Gershenzon, J. The nonmevalonate pathway supports both monoterpene and sesquiterpene formation in snapdragon flowers. *Proc. Natl. Acad. Sci. USA* **2005**, *102*, 933–938. [This paper shows that only the methylerythritol phosphate (MEP or nonmevalonate) pathway, and not the mevalonate pathway, provides isopentyl diphosphate and dimethylallyl diphosphate for the biosynthesis of volatile terpenes in snapdragon flowers.]

82 Rodriguez-Concepcion, M., Boronat, A. Elucidation of the methylerythritol phosphate pathway for isoprenoid biosynthesis in bacteria and plastids. A metabolic milestone achieved through genomics. *Plant Physiol.* **2002**, *130*, 1079–1089.

83 Lange, B.M., Ghassemian, M. Genome organization in *Arabidopsis thaliana*: a survey for genes involved in isoprenoid and chlorophyll metabolism. *Plant Mol. Biol.* **2003**, *51*, 925–948.

84 Facchini, P.J., Chappell, J. Gene family for an elicitor-induced sesquiterpene synthase in tobacco. *Proc. Natl. Acad. Sci. USA* **1992**, *89*, 11088–11092.

85 Colby, S.M., Alonso, W.R., Katahira, E.J., McGarvey, D.J., Croteau, R. 4-*S*-Limonene synthase from the oil glands of spearmint (*Mentha spicata*): cDNA isolation, characterization and bacterial expression of the catalytically active monoterpene cyclase, *J. Biol. Chem.* **1993**, *268*, 23016–23024.

86 Mau, C.J.D., West, C.A. Cloning of casbene synthase cDNA: evidence for conserved structural features among terpenoid cyclases in plants. *Proc. Natl. Acad. Sci. USA* **1994**, *91*, 8479–8501.

87 Back, K., Chappell, J. Cloning and bacterial expression of a sesquiterpene cyclase from *Hyoscyamus muticus* and its molecular comparison to related terpene cyclases. *J. Biol. Chem.* **1995**, *270*, 7375–7381.

88 Chappell, J. The biochemistry and molecular biology of isoprenoid metabolism. *Plant Physiol.* **1995**, *107*, 1–6.

89 Chappell, J. Biochemistry and molecular biology of the isoprenoid biosynthetic pathway in plants. *Annu. Rev. Plant Physiol. Plant Mol. Biol.* **1995**, *46*, 521–547.

90 Bohlmann, J., Meyer-Gauen, G., Croteau, R. Plant terpenoid synthases: molecular biology and phylogenetic analysis. *Proc. Natl. Acad. Sci. USA* **1998**, *95*, 4126–4133.

91 Trapp, S.C., Croteau, R.B. Genomic organization of plant terpene synthases and molecular evolutionary implications. *Genetics* **2001**, *158*, 811–832.

92 Aubourg, S., Lecharny, A., Bohlmann, J. Genomic analysis of the terpenoid synthase (*AtTPS*) gene family of *Arabidopsis thaliana*. *Mol. Genet. Genomics* **2002**, *267*, 730–745.

93 Bohlmann, J., Martin, D., Oldham, N.J., Gershenzon, J. Terpenoid secondary metabolism in *Arabidopsis thaliana*: cDNA cloning, characterization, and functional expression of a myrcene/(*E*)-β-ocimene synthase. *Arch. Biochem. Biophys.* **2000**, *375*, 261–269.

94 Van Poeke, R.M.P., Posthumus, M.A., Dicke, M. Herbivore-induced volatile production by *Arabidopsis thaliana* leads to attraction of the parasitoid *Cotesia rubecula*: chemical, behavioral and gene-expression analysis. *J. Chem. Ecol.* **2001**, *27*, 1911–1928.

95 ** Chen, F., Tholl, D., D'Auria, J.C., Farooq, A., Pichersky, E., Gershenzon, J. Biosynthesis and emission of terpenoid volatiles from *Arabidopsis* flowers. *Plant Cell* **2003**, *15*, 481–494. [This paper was one of the first to show that *Arabidopsis* flowers emit monoterpenes and sesqui-terpenes as potential olfactory cues for pollinating insects.]

96 * Aharoni, A., Giri, A.P., Deuerlein, S., Griepink, F., de Kogel, W.-D.,

Verstappen, F.W.A., Verheoven, H.A., Jongsma, M.A., Schwab, W., Bouwmeester, H.J. Terpenoid metabolism in wild-type and transgenic *Arabidopsis* plants. *Plant Cell* **2003**, *15*, 2866–2884. [This paper was one of the first to report the occurrence of terpenoid metabolism in *Arabidopsis*. The work demonstrates the value of *Arabidopsis* to study the biosynthesis and ecological role of terpenoids in wild-type and transgenic plants.]

97 D'Auria, J.C., Gershenzon, J. The secondary metabolism of *Arabidopsis thaliana*: growing like a weed. *Curr. Opin. Plant Biol.* **2005**, *8*, 308–316.

98 Steeghs, M., Bais, H.P., de Gouw, J., Goldan, P., Kuster, W., Northway, M., Fall, R., Vivanco, J.M. Proton-transfer-reaction mass spectrometry as a new tool for real time analysis of root-secreted volatile organic compounds in *Arabidopsis*. *Plant Physiol.* **2004**, *135*, 47–58.

99 Chen, F., Ro, D.-K., Petri, J., Gershenzon, J., Bohlmann, J., Pichersky, E., Tholl, D. Characterization of a root-specific *Arabidopsis* terpene synthase responsible for the formation of the volatile monoterpene 1, 8-cineole. *Plant Physiol.* **2004**, *135*, 1956–1966.

100 Tholl, D., Chen, F., Petri, J., Gershenzon, J., Pichersky, E. Two sesquiterpene synthases are responsible for the complex mixture of sesquiterpenes emitted from *Arabidopsis* flowers. *Plant J.* **2005**, *42*, 757–771.

101 Fazio, G.C., Xu, R., Matsuda, S.P. Genome mining to identify new plant triterpenoids. *J. Am. Chem. Soc.* **2004**, *126*, 5678–5679.

102 Herrera, J.B., Bartel, B., Wilson, W.K., Matsuda, S.P. Cloning and characterization of the *Arabidopsis thaliana* lupeol synthase gene. *Phytochemistry* **1998**, *49*, 1905–1911.

103 Segura, M.J., Meyer, M.M., Matsuda, S.P. *Arabidopsis thaliana* LUP1 converts oxidosqualene to multiple triterpene alcohols and a triterpene diol. *Organic Lett.* **2000**, *2*, 2257–2259.

104 Kushiro, T., Shibuya, M., Masuda, K., Ebizuka, Y. A novel multifunctional triterpene synthase from *Arabidopsis*

thaliana. *Tetrahedron Lett.* **2000**, *41*, 7705–7710.

105 * Yu, J., Hu, S., Wang, J., Wong, G.K.-S., Li, S., et al. A draft sequence of the rice genome (*Oryza sativa* L. spp. *Indica*). *Science* **2002**, *296*, 79–92. [This paper reports the first extensive analysis of the rice genome sequence.]

106 McGarvey, D., Croteau, R. Terpenoid metabolism. *Plant Cell* **2005**, *7*, 1015–1026.

107 Xu, M., Hillwig, M.L., Prisic, S., Coates, R.M., Peters, R.J. Functional identification of rice *syn*-copalyl diphosphate synthase and its role in initiating biosynthesis of diterpenoid phytoalexin/allelopathic natural products. *Plant J.* **2004**, *39*, 309–318.

108 Otomo, K., Kenmoku, H., Oikawa, H., Konig, W.A., Toshima, H., Mitsuhashi, W., Yamane, H., Sassa, T., Toyomasu, T. Biological functions of *ent*- and *syn*-copalyl diphosphate synthases in rice: key enzymes for the branch point of gibberellin and phytoalexin biosynthesis. *Plant J.* **2004**, *39*, 886–893.

109 Prisic, S., Xu, M., Wilderman, P.R., Peters, R.J. Rice contains two disparate *ent*-copalyl diphosphate synthases with distinct metabolic functions. *Plant Physiol.* **2004**, *136*, 4228–4236.

110 Gai, X.W., Lal, S., Xing, L.Q., Brendel, V., Walbot, V. Gene discovery using the maize genome database ZmDB. *Nucleic Acids Res.* **2000**, *28*, 94–96.

111 Köllner, T.G., Schnee, C., Gershenzon, J., Degenhardt, J. The variability of sesquiterpenes emitted from two *Zea mays* cultivars is controlled by allelic variation of two terpene synthase genes encoding stereoselective multiple product enzymes. *Plant Cell* **2004**, *16*, 1115–1131.

112 Achnine, L., Huhman, D.V., Farag, M.A., Sumner, F.W., Blount, J.W., Dixon, R.A. Genomics-based selection and functional characterization of triterpene glycosyl-transferases from the model legume *Medicago truncatula*. *Plant J.* **2005**, *41*, 875–887.

113 Ro, D.-K., Arimura, G.-I., Lau, S.Y.W., Piers, E., Bohlmann, J. Loblolly pine abietadienol/abietadienal oxidase PtAO (CYP720B1) is a multifunctional, multisubstrate cytochrome P450 mono-oxygenase. *Proc. Natl. Acad. Sci. USA* **2005**, *102*, 8060–8065.

114 Martin, D., Fäldt, J., Bohlmann, J. Functional characterization of nine Norway spruce *TPS* genes and evolution of gymnosperm terpene synthases of the *TPS-d* subfamily. *Plant Physiol.* **2004**, *135*, 1908–1927.

115 * Gershenzon, J., McCaskill, D., Rajaonarivony, J.I.M., Mihaliak, C., Karp, F., Croteau, R. Isolation of secretory cells from plant glandular trichomes and their use in biosynthetic studies of mono-terpenes and other gland products. *Anal. Biochem.* **1992**, *200*, 130–138. [This paper describes an efficient method for the isolation of mint glandular trichomes, which allowed the establishment of cell type-specific cDNA libraries rich in terpenoid biosynthetic genes.]

116 * Lange, B.M., Wildung, M.R., Stauber, E.J., Sanchez, C., Pouchnik, D., Croteau, R. Probing essential oil biosynthesis and secretion by functional evaluation of expressed sequence tags from mint glandular trichomes. *Proc. Natl. Acad. Sci. USA* **2000**, *97*, 2934–2939. [This paper describes the first characterization of a cell type-specific EST database from mint glandular trichomes.]

117 Aziz, N., Paiva, N.L., May, G.D., Dixon, R.A. Transcriptome analysis of alfalfa glandular trichomes. *Planta* **2005**, *221*, 28–38.

118 McCaskill, D., Gershenzon, J., Croteau, R. Morphology and monoterpene bio-synthetic capabilities of secretory cell clusters isolated from glandular trichomes of peppermint *Mentha* × *piperita* L. *Planta* **1992**, *187*, 445–454.

119 Alonso, W.R., Rajaonarivony, J.I.M., Gershenzon, J., Croteau, R. Purification of 4S-limonene synthase, a monoterpene cyclase from the glandular trichomes of peppermint (*Mentha* × *piperita*) and spearmint (*Mentha spicata*). *J. Biol. Chem.* **1992**, *267*, 7582–7587.

120 Bertea, C.M., Schalk, M., Karp, F., Maffei, M., Croteau, R. Demonstration that menthofuran synthase of mint (*Mentha*) is a cytochrome P450 mono-oxygenase: cloning, functional expres-sion, and characterization of the

responsible gene. *Arch. Biochem. Biophys.* **2001**, *390*, 279–286.

121 Ringer, K.L., McConkey, M.E., Davis, E.M., Rushing, G.W. Croteau, R. Mono-terpene double-bond reductases of the (–)-menthol biosynthetic pathway: isola-tion and characterization of cDNAs encoding (–)-isopiperitenone reductase and (+)-pulgeone reductase of pepper-mint. *Arch. Biochem. Biophys.* **2003**, *418*, 80–92.

122 Ringer, K.L., Davis, E.M., Croteau, R. Monoterpene metabolism. Cloning, expression and characterization of (–)-isopiperitenol/(–)-carveol dehydrogenase of peppermint and spearmint. *Plant Physiol.* **2005**, *137*, 863–872.

123 Davis, E.M., Ringer, K.L., McConkey, M.E., Croteau, R. Monoterpene metabolism. Cloning, expression, and characterization of menthone reductases from peppermint. *Plant Physiol.* **2005**, *137*, 873–881.

124 Iijima, Y., Gang, D.R., Fridman, E., Lewinsohn, E., Pichersky, E. Character-ization of geraniol synthase from the peltate glands of sweet basil. *Plant Physiol.* **2004**, *134*, 370–379.

125 Iijima, Y., Davidovich-Rikanati, R., Fridman, E., Gang, D.R., Bar, E., Lewinsohn, E., Pichersky, E. The biochemical and molecular basis for the divergent patterns in the biosynthesis of terpenes and phenylpropenes in the peltate glands of three cultivars of basil. *Plant Physiol.* **2004**, *136*, 3724–3736.

126 Lücker, J., Bowen, P., Bohlmann, J. *Vitis vinifera* terpenoid cyclases: functional identification of two sesquiterpene synthase cDNAs encoding (+)-valencene synthase and (–)-germacrene D synthase and expression of mono- and sesqui-terpene synthases in grapevine flowers and berries. *Phytochemistry* **2004**, *65*, 2649–2659.

127 Driesel, A.J., Lommele, A., Drescher, D., Topfer, R., Bell, M., Cartharius, I., Cheutin, N., Huck, J.-F., Kubiak, J., Regnard, P., Steinmetz, A. Towards the transcriptome of grapevine (*Vitis vinifera* L.) *Acta Hort.* **2003**, *603*, 239–249.

128 Jung, J.D., Park, H.-W., Hahn, Y., Jur, C.-G., In, D.S., Chung, H.-J., Liu, J.R., Choi, D.-W. Discovery of genes for

ginsenoside biosynthesis by analysis of ginseng expressed sequence tags. *Plant Cell Rep.* **2003**, *22*, 224–230.

129 Jennewein, S., Wildung, M.R., Chau, M., Walker, K., Croteau, R. Random sequencing of an induced *Taxus* cell cDNA library for identification of clones involved in Taxol biosynthesis. *Proc. Natl. Acad. Sci. USA* **2004**, *101*, 9149–9154.

130 Kingston, D.G.I., Jagtap, P.G., Yuan, H., Samala, L. The chemistry of Taxol and related taxoids. *Prog. Chem. Organic Nat. Prod.* **2002**, *84*, 53–225.

131 Ko, J.-H., Chow, K.-S., Han, K.-H. Transcriptome analysis reveals novel features of the molecular events occurring in the laticifers of *Hevea brasiliensis* (para rubber tree). *Plant Mol. Biol.* **2003**, *53*, 479–492.

132 Davis, W. The rubber industry's biological nightmare. *Fortune* **1997**, *4*, 86–95.

133 Dennis, M.S., Light, D.R. Rubber elongation factor from *Hevea brasiliensis*. Identification, characterization, and role in rubber biosynthesis. *J. Biol. Chem.* **1989**, *264*, 18608–18617.

134 Kant, M.R., Ament, K., Sabelis, M.W., Haring, M.A., Schuurink, R.C. Differential timing of spider mite-induced direct and indirect defenses in tomato plants. *Plant Physiol.* **2004**, *135*, 483–485.

135 Merke, P., Kappers, I.F., Verstappen, F.W.A., Vorst, O., Dicke, M., Bouwmeester, H.J. Combined transcript and metabolite analysis reveals genes involved in spider mite induced volatile formation in cucumber plants. *Plant Physiol.* **2004**, *135*, 2012–2024.

136 Dixon, R.A., Achnine, L., Kota, P., Liu, C.-J., Reddy, M.S.S., Wang, L. The phenylpropanoid pathway and plant defense – a genomics perspective. *Mol. Plant Pathol.* **2002**, *3*, 371–390.

137 Raguso, R.A., Pichersky, E. Floral volatiles from *Clarkia breweri* and *C. concinna* (Onagraceae): recent evolution of floral scent and moth pollination. *Plant System. Evol.* **1995**, *194*, 55–67.

138 Dudareva, N., Pichersky, E., Gershenzon, J. Biochemistry of plant volatiles. *Plant Physiol.* **2004**, *135*, 1893–1902.

139 Gang, D.R., Beuerle, T., Ullmann, P., Werk-Reichhart, D., Pichersky, E. Differential production of *meta* hydroxylated phenylpropanoids in sweet basil peltate glandular trichomes and leaves is controlled by the activities of specific acyltransferases and hydroxylases. *Plant Physiol.* **2002**, *130*, 1536–1544.

140 Wildermuth, M.C., Dewdney, J., Wu, G., Ausubel, F.M. Isochorismate synthase is required to synthesize salicylic acid for plant defence. *Nature* **2001**, *414*, 562–565.

141 Costa, M.A., Collins, R.E., Anterola, A.M., Cochrane, F.C., Davin, L.B., Lewis, N.G. An *in silico* assessment of gene function and organization of the phenylpropanoid pathway metabolic networks in *Arabidopsis thaliana* and limitations thereof. *Phytochemistry* **2003**, *64*, 1097–1112.

142 Costa, M.A., Bedgar, D.L., Moinuddin, S.G.A., Kim, K.-W., Cardenas, C.L., Cochrane, F.C., Shockey, J.M., Helms, G.L., Amakura, Y., Takahashi, H., Milhollan, J.K., Davin, L.B., Browse, J., Lewis, N.G. Characterization *in vitro* and *in vivo* of the putative multigene 4-coumarate:CoA ligase network in *Arabidopsis*: syringyl lignin and sinapate/sinapyl alcohol derivative formation. *Phytochemistry* **2005**, *66*, 2070–2091.

143 Hostel, W. *The Biochemistry of Plants*, Volume 7. Academic Press, New York, **1981**.

144 Whetten, R.W., MacKay, J.J., Sederoff, R.R. Recent advances in understanding lignin biosynthesis. *Annu. Rev. Plant Physiol. Plant Mol. Biol.* **1998**, *49*, 585–609.

145 Lim, E.-K., Li, Y., Parr, A., Jackson, R., Ashford, D.A., Bowles, D.J. Identification of glucosyltransferase genes involved in sinapate metabolism and lignin synthesis in *Arabidopsis*. *J. Biol. Chem.* **2001**, *276*, 4344–4349.

146 Jones, P., Messner, B., Nakajima, J.-I., Schäffner, A.R., Saito, K. UGT73C6 and UGT78D1, glycosyltransferases involved in flavonol glycoside biosynthesis in *Arabidopsis thaliana*. *J. Biol. Chem.* **2003**, *278*, 43910–43918.

147 Gachon, C.M.M., Langlois-Meurinne, M., Saindrenan, P. Plant secondary metabolism glycosyltransferases: the emerging functional analysis. *Trends Plant Sci.* **2005**, *10*, 542–549.

148 Kirst, M., Johnson, A.F., Baucom, C., Ulrich, E., Hubbard, K., Staggs, R., Paule, C., Retzel, E., Whetten, R., Sederoff, R. Apparent homology of expressed genes from wood-forming tissues of loblolly pine (*Pinus taeda* L.) with *Arabidopsis thaliana*. *Proc. Natl. Acad. Sci. USA* **2003**, *100*, 7383–7388.

149 Ehlting, J., Mattheus, N., Aischliman, D.S., Li, E., Hamberger, B., Cullis, I.F., Zhuang, J., Kaneda, M., Mansfield, S.D., Samuels, L., Ritland, K., Ellis, B.E., Bohlmann, J., Douglas, C.J. Global transcript profiling of primary stems from *Arabidopsis thaliana* identifies candidate genes for missing links in lignin biosynthesis and transcriptional regulators of fiber differentiation. *Plant J.* **2005**, *42*, 618–640.

150 Allona, I., Quinn, M., Shoop, E., Swope, K., St. Cyr, S., Carlis, J., Riedl, J., Retzel, E., Campbell, M.M., Sederoff, R., Whetten, R.W. Analysis of xylem formation in pine by cDNA sequencing. *Proc. Natl. Acad. Sci. USA* **1998**, *95*, 9693–9698.

151 Sate, Y., Wuli, B., Sederoff, R., Whetten, R. Molecular cloning and expression of eight laccase cDNAs in loblolly pine (*Pinus taeda*). *J. Plant Res.* **2001**, *114*, 147–155.

152 Ward, G., Hadar, Y., Bilkis, I., Konstantinovsky, L., Dosoretz, C.G. Initial steps of ferulic acid polymerization by lignin peroxidase. *J. Biol. Chem.* **2001**, *276*, 18734–18741.

153 McCaig, B.C., Meagher, R.B., Dean, J.F. Gene structure and molecular analysis of the laccase-like multicopper oxidase (*LMCO*) gene family in *Arabidopsis thaliana*. *Planta* **2005**, *221*, 619–636.

154 Egerstdotter, U., van Zyl, L.M., Mackay, J., Peter, G., Kirst, M., Clark, C., Whetten, R., Sederoff, R. Gene expression during formation of earlywood and latewood in loblolly pine: expression profiles of 350 genes. *Plant Biol.* **2004**, *6*, 654–663.

155 Kirst, M., Myburg, A.A., de León, J.P.G., Kirst, M.E., Scott, J., Sederoff, R.

Coordinated genetic regulation of growth and lignin revealed by quantitative trait locus analysis of cDNA microarray data in an interspecific backcross of *Eucalyptus*. *Plant Physiol.* **2004**, *135*, 2368–2378.

156 Kirst, M., Basten, C.J., Myburg, A.A., Zeng, Z.-B., Sederoff, R.R. Genetic architecture of transcript-level variation in differentiating xylem of a *Eucalyptus* hybrid. *Genetics* **2005**, *169*, 2295–2303.

157 Stasolla, C., van Zyl, L., Egertsdotter, U., Craig, D., Liu, W., Sederoff, R.R. Transcript profiles of stress-related genes in developing white spruce (*Picea glauca*) somatic embryos cultured with poly-ethylene glycol. *Plant Sci.* **2003**, *165*, 719–729.

158 Dixon, R.A., Sumner, L.W. Legume natural products: understanding and manipulating complex pathways for human and animal health. *Plant Physiol.* **2003**, *131*, 878–885.

159 Udvardi, M.K., Tabata, S., Parniske, M., Stougaard, J. *Lotus japonicus*: legume research in the fast lane. *Trends Plant Sci.* **2005**, *10*, 222–228.

160 Koes, R., Verweij, W., Quattrocchio, F. Flavonoids: a colorful model for the regulation and evolution of biochemical pathways. *Trends Plant Sci.* **2005**, *10*, 236–242.

161 Dixon, R.A., Ferreira, D. Molecules of interest: genistein. *Phytochemistry* **2002**, *60*, 205–211.

162 Bell, C.J., Dixon, R.A., Farmer, A.D., Flores, R., Inman, J., Gonzales, R.A., Harrison, M.J., Paiva, N.L., Scott, A.D., Weller, J.W., May, G.D. The *Medicago* Genome Initiative: a model legume database. *Nucleic Acids Res.* **2001**, *29*, 114–117.

163 Steele, C.L., Gijzen, M., Qutob, D., Dixon, R.A. Molecular characterization of the enzyme catalyzing the aryl migration reaction of isoflavonoid biosynthesis in soybean. *Arch. Biochem. Biophys.* **1999**, *367*, 146–150.

164 Jung, W., Yu, O., Lau, S.-M.C., O'Keefe, D.P., Odell, J., Fader, G., McGonigle, B. Identification and expression of isoflavone synthase, the key enzyme for biosynthesis of isoflavones in legumes. *Nature Biotechnol.* **2000**, *18*, 208–212.

165 Oldroyd, G.E., Geurts, R. *Medicago truncatula*, going where no plant has gone before. *Trends Plant Sci.* **2001**, *6*, 552–554.

166 Liu, C.-J., Huhman, D., Sumner, L.W., Dixon, R.A. Regiospecific hydroxylation of isoflavones by cytochrome P450 81E enzymes from *Medicago truncatula*. *Plant J.* **2003**, *36*, 471–484.

167 * Xie, D.-Y., Sharma, S.B., Paiva, N.L., Ferreira, D., Dixon, R.A. Role of anthocyanidin reductase, encoded by *BANYULS* in plant flavonoid biosynthesis. *Science* **2003**, *299*, 396–399. [This paper uses an EST approach to identify the BANYULS genes from *Arabidopsis* and *Medicago truncatula* that encode anthocyanidin reductase, which converts anthocyanidins to their corresponding 2, 3-*cis*-flavan-3-ols.]

168 Lawrence, B.M. Chemical components of labiate oils and their exploitation. In: Harley, R.M., Reynolds, T. (Eds.), *Advances in Labiate Science*. Royal Botanical Gardens, Kew, UK, **1992**, pp. 399–436.

169 Obeng-Ofori, D., Reichmuth, C. Bioactivity of eugenol, a major component of essential oil of *Ocimum suave* (Wild.) against four species of stored-product *Coleoptera*. *Int. J. Pest Management* **1997**, *43*, 89–94.

170 Adams, S., Weidenborner, M. Mycelial deformations of *Cladosporium herbarum* due to the application of eugenol or carvacrol. *J. Essential Oil Res.* **1996**, *8*, 535–540.

171 Manitto, P., Monti, D., Gramatica, P. Biosynthesis of phenylpropanoid compounds. I. Biosynthesis of eugenol in *Ocimum basilicum* L. *J. Chem. Soc. Perkin Trans. I* **1974**, *14*, 1727–1731.

172 Manitto, P., Gramatica, P., Monti, D. Biosynthesis of phenylpropanoid compounds. II. Incorporation of specifically labeled cinnamic acids into eugenol in basil (*Ocimum basilicum*). *J. Chem. Soc. Perkin Trans. I* **1975**, *16*, 1549–1551.

173 Klischies, M., Stockigt, J., Zenk, M.H. Biosynthesis of the allylphenols eugenol and methyleugenol in *Ocimum basilicum* L. *Chem Commun.* **1975**, *21*, 879–880.

174 Senanayake, U.M., Wills, R.H.B., Lee, T.H. Biosynthesis of eugenol and cinnamic aldehyde in *Cinnamomum zeylanicum*. *Phytochemistry* **1977**, *16*, 2032–2033.

175 Gang, D.R., Lavid, N., Zubieta, C., Chen, F., Beuelre, T., Lewinsohn, E., Noel, J.P., Pichersky, E. Characterization of phenylpropene *O*-methyltransferases from sweet basil: facile change of substrate specificity and convergent evolution within a plant *O*-methyltransferase family. *Plant Cell* **2002**, *14*, 505–519.

176 Facchini, P.J. Alkaloid biosynthesis in plants: biochemistry, cell biology, molecular regulation, and metabolic engineering applications. *Annu. Rev. Plant Physiol. Plant Mol. Biol.* **2001**, *52*, 29–66.

177 De Luca, V., Laflamme, P. The expanding universe of alkaloid biosynthesis. *Curr. Opin. Plant Biol.* **2001**, *4*, 225–233.

178 Hashimoto, T., Yamada, Y. New genes in alkaloid metabolism and transport. *Curr. Opin. Biotechnol.* **2003**, *14*, 163–168.

179 Morishige, T., Dubouzet, E., Choi, K.-B., Yazaki, K., Sato, F. Molecular cloning of columbamine *O*-methyltransferase from cultured *Coptis japonica* cells. *Eur. J. Biochem.* **2002**, *269*, 5659–5667.

180 Ziegler, J., Diaz-Chávez, M.L., Kramell, R., Ammer, C., Kutchan, T.M. Comparative macroarray analysis of morphine containing *Papaver somniferum* and eight morphine free *Papaver* species identifies an *O*-methyltransferase involved in benzylisoquinoline biosynthesis. *Planta* **2006**, *222*, 458–471.

181 Liscombe, D.K., MacLeod, B.P., Loukanina, N., Nandi, O.I., Facchini, P.J. Evidence for the monophyletic evolution of benzylisoquinoline alkaloid biosynthesis in angiosperms. *Phytochemistry* **2005**, *66*, 1374–1393.

182 Pilatzke-Wunderlich, I., Nessler, C.L. Expression and activity of cell-wall-degrading enzymes in the latex of opium poppy, *Papaver somniferum* L. *Plant Mol. Biol.* **2001**, *45*, 567–576.

183 Fahn, A. *Plant Anatomy*, 4th edition. Pergamon Press, Oxford, UK, **1990**.

184 Siminszky, B., Gavilano, L., Bowen, S.W., Dewey, R.E. Conversion of nicotine to nornicotine in *Nicotiana tabacum* is mediated by CYP82E4, a cytochrome P450 monooxygenase. *Proc. Natl. Acad. Sci. USA* **2005**, *102*, 14919–14924.

185 ** The *Arabidopsis* Genome Initiative. Analysis of the genome sequence of the flowering plant *Arabidopsis thaliana*. *Nature* **2000**, *408*, 796–815. [This paper describes the annotation and analysis of the first complete plant genome sequence from *Arabidopsis*.]

186 Devoto, A., Ellis, C., Magusin, A., Chang, H.-S., Chilcott, C., Zhu, T., Turner, J.G. Expression profiling reveals *CO11* to be a key regulator of genes involved in wound- and methyl jasmonate-induced secondary metabolism, defense, and hormone interactions. *Plant Mol. Biol.* **2005**, *58*, 497–513.

187 Cheong, Y.H., Chang, H.S., Gupta, R., Wang, X., Zhu, T., Luan, S. Transcriptional profiling reveals novel interactions between wounding, pathogen, abiotic stress, and hormonal responses in *Arabidopsis*. *Plant Physiol.* **2002**, *129*, 661–677.

188 Li, L., He, Z., Pandey, G.K., Tsuchiya, T., Luan, S. Functional cloning and characterization of a plant efflux carrier for multidrug and heavy metal detoxification. *J. Biol. Chem.* **2002**, *277*, 5360–5368.

189 Tsuji, J., Jackson, E.P., Gage, D.A., Hammerschmidt, R., Somerville, S.C. Phytoalexin accumulation in *Arabidopsis thaliana* during the hypersensitive reaction to *Pseudomonas syringae* pv *syringae*. *Plant Physiol.* **1992**, *129*, 1304–1309.

190 Carter, C.J., Thornburg, R.W. Tobacco nectarin V is a flavin-containing berberine bridge enzyme-like protein with glucose oxidase activity. *Plant Physiol.* **2004**, *134*, 460–469.

191 Kutchan, T.M., Dittrich, H. Characterization and mechanism of the berberine bridge enzyme, a covalently flavinylated oxidase of benzophen-anthridine alkaloid biosynthesis in higher plants. *J. Biol. Chem.* **1995**, *270*, 24475–24481.

192 Fabbri, M., Delp, G., Schmidt, O., Theopold, U. Animal and plant members of a gene family with similarity to alkaloid-synthesizing enzymes. *Biochem. Biophys. Res. Commun.* **2000**, *271*, 191–196.

193 Theopold, U., Samakovlis, C., Erdjument-Bromage, H., Dillon, N., Axelsson, B., Schmidt, O., Tempst, P., Hultmark, D. *Helix pomatia* lectin, an inducer of *Drosophila* immune response binds to hemomucin, a novel surface mucin. *J. Biol. Chem.* **1996**, *271*, 12708–12715.

194 Theopold, U., Li, D., Fabbri, M., Scherfer, C., Schmidt, O. The coagulation of insect hemolymph. *Cell. Mol. Life Sci.* **2002**, *59*, 363–372.

195 Theopold, U., Schmidt, O. *Helix pomatia* lectin and annexin V, two molecular probes for insect microparticles: possible involvement in hemolymph coagulation. *J. Insect Physiol.* **1997**, *43*, 667–674.

196 Morita, Y., Kodama, K., Shiota, S., Mine, T., Kataoka, A., Mizushima, T., Tsuchiya, T. NorM, a putative, multidrug efflux protein of *Vibrio parahaemolyticus* and its homologue in *Escherichia coli*. *Antimicrob. Agents Chemother.* **1998**, *42*, 1778–1782.

197 Shitan, N., Bazin, I., Dan, K., Obata, K., Kigawa, K., Ueda, K., Sato, F., Forestier, C., Yazaki, K. Involvement of CjMDR1, a plant multidrug-resistant-type ATP-binding cassette protein, in alkaloid transport in *Coptis japonica*. *Proc. Natl. Acad. Sci. USA* **2003**, *100*, 751–756.

198 Gould, S.J., Vrba, E.S. Exaptation: a missing term in the science of form. *Paleobiology* **1982**, *8*, 4–15.

11
Biotechnology of Solanaceae Alkaloids:
A Model or an Industrial Perspective?

Birgit Dräger

This chapter focuses on the plant family Solanaceae which, as parent plants for cell and tissue cultures, are noteworthy for three main reasons:

1. Solanaceae produce several alkaloids of medical and biotechnological importance, namely the tropane alkaloids hyoscyamine and scopolamine, steroid alkaloids, and nicotine (Figs. 11.1 and 11.2). Among these alkaloids, scopolamine is medicinally the most important, with consumption several-fold higher than that of hyoscyamine [1], mainly because it is used as the starting material for the semi-synthesis of several important drugs. Both, scopolamine and the structurally related hyoscyamine possess strong parasympatholytic activity, blocking the

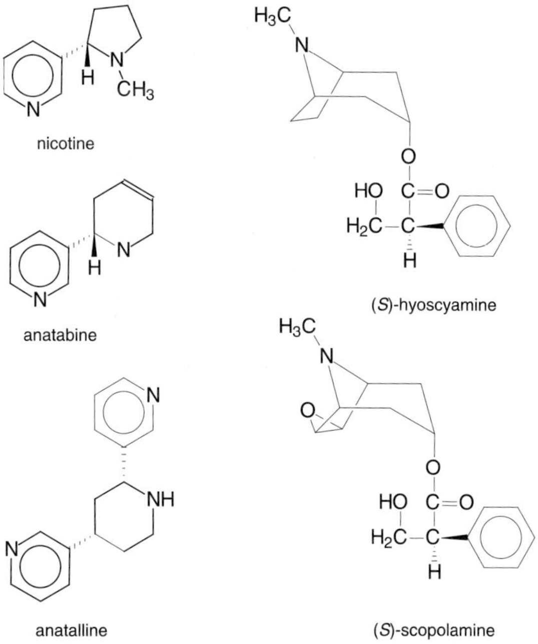

Fig. 11.1 Typical Solanaceae alkaloids.

Medicinal Plant Biotechnology. From Basic Research to Industrial Applications
Edited by Oliver Kayser and Wim J. Quax

Fig. 11.2 Sesquiterpenoid and glycoalkaloid phytoalexins from Solanaceae.

capsidiol

rishitin

α-solanine

solanidine

D-galactose

D-glucose   L-rhamnose

α-tomatine

tomatidine

D-galactose

D-glucose

D-xylose   D-xylose

parasympathicus action by binding to the muscarinic acetylcholine receptors in synapses, without exerting any intrinsic activity.

2. Many other Solanaceae are important crop plants grown worldwide (e.g., potato and tomato). Regulatory mechanisms concerning growth and biomass accumulation are of major interest.

3. Solanaceae tissue cultures often grow vigorously and regenerate more easily than do those of many other medicinal plants. Almost all tissue culture systems known for plants have been realized, and some have been developed, with Solanaceae; these include root cultures, shoot cultures, and de-differentiated cells as callus or cell suspension cultures.

11.1
Culture Systems

11.1.1
The Early Days of Solanaceae Culture

Solanaceae root cultures as experimental systems for the study of alkaloid metabolism were first prepared more than 50 years ago, initially by cutting roots from aseptically grown *Datura ferox* seedlings, followed by cultivation in liquid nutrient medium [2]. Root cultures of *Datura metel* also formed tropane alkaloids and were used to investigate the biosynthesis of the tropane ring structure [2–5]. Comparable root cultures of *Nicotiana* species in White's nutrient medium [6] produced nicotine; they showed differential preferences for nitrogen and carbon sources in the nutrient medium for optimal growth and proliferation [7]. Solanaceae root cultures could also be started from dedifferentiated cell suspensions of *Hyoscyamus niger* by transfer of single cells or small cell aggregates to Linsmaier–Skoog's medium [8] devoid of auxin and with no or low concentrations of cytokinin (e.g., 10^{-8} M benzyladenine) [9]. Root cultures from *Duboisia* species were obtained by cultivation of spontaneously formed root primordia on callus in liquid medium with 10^{-5} M indolebutyric acid as auxin supply [10]. Root cultures of *Hyoscyamus* species were started by the excision of roots from sterile seedlings proliferated even in the absence of phytohormones, and formed more than 1% dry mass (dm) tropane alkaloids [11]. Similarly, roots excised from sterile *Datura stramonium* seedlings showed doubling times of 6 to 19 days, and produced hyoscyamine (0.2–0.6% dm) and scopolamine (0.1–0.3% dm) [12].

In contrast, alkaloid formation by cell suspension cultures of Solanaceae is poor, with *H. niger* cell suspensions showing only traces of alkaloids [9,13,14]. Likewise, neither callus cultures of *D. stramonium* [15], *Duboisia leichhardtii* [16] nor several callus cultures of *Hyoscyamus* species [11] showed any accumulation of tropane alkaloids. Calcium alginate immobilization of *Datura innoxia* cells increased alkaloid production tenfold, but the final yield remained low (20 µg g^{-1} dm hyoscyamine and 8 µg g^{-1} dm scopolamine) [17]. Cell cultures of *Hyoscyamus muticus* produced only trace amounts of alkaloids, but were used for the selection of metabolic variants. Cells resistant to the amino acid analogue *p*-fluorophenylalanine accumulated high levels of cinnamoylputrescine. Root cultures regenerated from these cells produced more hyoscyamine than control root cultures [18]. Early grafting experiments [19,20] and an analysis of xylem sap [21] from Solanaceae had revealed that nicotine and tropane alkaloids are formed in the roots and transported into the aerial parts of the plants. A comparison of cultured roots with callus cultures of *H. muticus* after application of MeJas, an elicitor of alkaloid formation (see Section 11.2.2.4), showed that early enzymes of tropane alkaloid formation were induced only in the roots, and not in the callus [22]. This result confirmed the restriction of alkaloid formation to roots and, in addition, claimed that differentiated root tissue is a prerequisite for alkaloid accumulation of more than trace quantities. The transition from root cultures to de-differentiated cell suspension was induced by transfer to medium containing auxin (9×10^{-7} M 2,4-dichlorophenoxyacetic acid) and cytokinin (4×10^{-7} M 6-benzylaminopurine) [23], and was followed in *Datura stramonium* by using NMR after ^{15}N-nitrate and ^{15}N-ammonium application. A decrease in alkaloid production during de-differentiation was accompanied by an increase in production of the primary metabolites putrescine and γ-aminobutyric acid [24].

11.1.2
Hairy Roots

An important step towards exploitation of cultured Solanaceae roots for both, alkaloid production and study of biosynthesis and metabolism was the introduction of "hairy roots" generated by transformation with *Agrobacterium rhizogenes* [25–27]. Roots of tropane alkaloid-producing species were shown to grow rapidly in simple mineral nutrient media such as Gamborgs B5 [28] and, as with most other *Agrobacterium rhizogenes*-transformed root cultures, growth was independent of exogenous phytohormone addition. These roots produced high levels of alkaloids [29–32], for example 0.3% dm alkaloids for *D. stramonium* roots [29] (Table 11.1). When root cultures were induced by different *Agrobacterium* strains, substantial variation in alkaloid formation and growth characteristics as well as somaclonal variation occurred repeatedly [33–35]. The *rol* ABC genes of the Ri plasmid of *A. rhizogenes* were sufficient to sustain strong growth and high alkaloid production (8 mg g^{-1} dm), with scopolamine concentrations rising to 2.5-fold those of hyoscyamine [36]. Once established, root cultures have proven to be more stable in metabolism during repeated subcultures than comparable cell suspension cultures [37]. The frequency of chromosome alterations was low in root cultures of *D. stramonium*. Karyotypes of roots transformed with *A. rhizogenes* were even more stable than those cultured after excision from seedlings [38]. Root cultures of many other medicinal plants obtained by transformation with *A. rhizogenes* were examined as potential sources of high-value pharmaceuticals (for a summary, see [39]). For practical purposes, long-term storage of tissue cultures is advantageous and proved possible for *H. niger*, where root cultures successfully cryopreserved using a vitrification method were subsequently found to have a high regeneration rate of 93.3% [40].

Shoots emerge spontaneously from root cultures, whether transformed by *A. rhizogenes* or cut from sterile seedlings, thus proving the regeneration potency of Solanaceae cultured tissues [55]. Potato plants obtained from shoot-forming root cultures after *A. rhizogenes* transformation displayed typical phenotypic alterations such as wrinkled leaves and abundant root growth, with reduced geotropism. These alterations are attributed to the *rol* genes of the *A. rhizogenes* plasmid that is transferred to the plant tissue during transformation [56], causing alterations in phytohormones and polyamine metabolism in the plant tissues [57,58]. *H. muticus* root cultures obtained by transformation with *A. rhizogenes*-regenerated plants that

Table 11.1 Hyoscyamine and scopolamine accumulation in cell and tissue cultures.

Species	Hyoscyamine [% dm]	Hyoscyamine [mg L⁻¹]	Scopolamine [% dm]	Scopolamine [mg L⁻¹]	Details	Ref.
Callus and cell suspensions						
Atropa belladonna cell suspension	Trace amounts	–	Trace amounts	–	Shake flasks	41
Datura stramonium callus	Not detected	–	Not detected	–		42
Datura innoxia cell suspension	0.00025	–	0.00025	–	Shake flasks	17
D. innoxia cell suspension	0.0025	–	0.001	–	Addition of Ca²⁺ and alginate	17
Duboisia leichhardtii callus	Not detected	–	Not detected	–		16
Hyoscyamus niger cell suspension	0.015	0.84	0.001	0.067	Shake flasks	14
H. niger callus	0.01	0.694	0.002	0.227		14
H. niger cell suspension	0.045	5.6	–	–	Shake flasks	13
H. niger callus	0.03	–	0.004	–		13
H. niger cell suspension	0.019	–	0.003	–	Shake flasks	43
Root cultures						
A. belladonna	0.37	–	0.024	–	Transformed with A. rhizogenes	32
A. belladonna	0.14	–	0.013	–	Excised roots from seedlings	41
A. belladonna	0.95	–	0.09	–	Transformed with A. rhizogenes	31
A. belladonna	0.3	40.3	Traces	–	Transformed with A. rhizogenes	44
A. belladonna	0.23	–	0.57	–	Transformed with A. rhizogenes rol genes	36
Datura candida hybrid	0.11	–	0.57	–	Transformed with A. rhizogenes	45
Datura quercifolia	1.24	–	–	–	Transformed with A. rhizogenes	46
Datura stramonium	0.3	–	Traces	–	Transformed with A. rhizogenes	29
D. stramonium	0.62	6.2	0.33	3.5	Excised roots from seedlings	12
D. stramonium	0.45	6.2	–	–	Transformed with A. rhizogenes	47
D. stramonium	–	–	0.56	–	Transformed with A. rhizogenes	48
Duboisia leichhardtii	0.53	24.8	1.16	35.7	Root formation on callus	10
D. leichhardtii	–	–	1.8	78.0	Transformed with A. rhizogenes, selected for high scopolamine production	49
Duboisia myoporoides × D. leichhardtii	0.36	9.7	1.3	62.4	Transformed with A. rhizogenes	50
D. myoporoides	–	–	3.2	70.4	Transformed with A. rhizogenes, selected for high scopolamine production	51
Hyoscyamus albus	1.1	–	0.2	–	Excised roots from seedlings	11
Hyoscyamus muticus	1.1	180	0.1	17	Transformed with A. rhizogenes overexpressing H6H	52
H. niger	0.01	1.1	0.15	6.6	Root formation on callus	9
H. niger	0.2	20	4	411	Root culture overexpressing PMT and H6H	53
Scopolia japonica	1.3	–	0.5	–	Transformed with A. rhizogenes, selected for high alkaloid production	54

accumulated tropane alkaloids in the same concentration range as the non-regen-erated, non-transformed plants [59]. Similarly, root cultures of a *Duboisia* hybrid with a high scopolamine content formed shoots spontaneously. The regenerated plants showed various degrees of *Agrobacterium rol* gene effects [60] and a wide range of alkaloid content [50]. Concordant with the concept of alkaloid formation bound to root tissue, isolated shoot cultures of *Duboisia myoporoides* were poor in tropane alkaloids [61,62]. The rapid shoot-forming capacity of non-transformed So-lanaceae callus and root cultures is useful for the micropropagation of rare or en-dangered species of Solanaceae [63], and also for the *in-vitro* multiplication of cul-tivars with desired horticultural traits [64,65].

11.1.3
Further Culture Systems

Many diverse cell culture concepts such as photoautotrophic cell cultures were first realized with Solanaceae (e.g., for tomato and tobacco [66]), and consequently the metabolism of secondary products – and alkaloids in particular – was investigated. Tomato cell culture lines after elicitation with *Fusarium oxysporum* f. sp. *lycopersici* increased the incorporation of phenolics into the cell walls, though phytoalexins such as rishitin and capsidiol or the glycoalkaloid tomatine were not increased in photoautotrophic, photomixotrophic, or heterotrophic cell lines [67]. Tobacco cells, in contrast, accumulated the alkaloid nicotine in large quantities in heterotrophic cultures and in moderate concentrations in photomixotrophic cultures, but not under photoautotrophic conditions (Table 11.2). Treatment with a fungal elicitor from *Phytophtora megasperma* f. sp. *glycinea* led to an accumulation of the phytoa-lexin capsidiol in heterotrophic and photomixotrophic tobacco cells, whereas pho-toautotrophic cultures formed only low levels. Nicotine levels were not affected by elicitation with a fungal elicitor [68].

Glycoalkaloids such as solanine or tomatine, besides tropane alkaloids and nico-tine, form another interesting group of Solanaceae alkaloids, because they may

Table 11.2 Tobacco alkaloid production in cell suspensions and root cultures.

Species	Alkaloid	% dm	mg L⁻¹	Details	Ref.
Cell suspension					
N. tabacum	Nicotine	0.032	14.4	Heterotrophic cell suspension	68
N. tabacum	Nicotine	0.003	0.64	Photoautotrophic cell suspension	68
N. tabacum cv. By-2	Anatalline	0.054	–	Cell suspension, MeJas-induced	69
	Anatabine	0.48	–	Cell suspension, MeJas-induced	69
Root culture					
N. tabacum	Nicotine	1.0	–	Transformed with A. rhizogenes	70
N. tabacum	Nicotine	1.1	–	Transformed with rol genes of A. rhizogenes	71

serve as a potential alternative to diosgenin as source material for commercial steroid drug synthesis (e.g., progesterone and cortisone). Glycoalkaloids occur mainly in the genus *Solanum* and in the closely related *Lycopersicon*, which recently has been grouped into *Solanum* [72]. Most glycoalkaloid terpenoid skeletons possess similar steroid structures (Fig. 11.2) and may readily be converted to 16-dehydropregnenolone acetate, a key intermediate in the synthesis of steroid drugs. Glycoalkaloids are found predominantly in Solanaceae shoots and leaves; in fact, they are formed *de novo* upon the greening of tissues. A prominent example is the greening of potato tubers in daylight, with concomitant accumulation of bitter and toxic glycoalkaloids. Among dark-grown Solanaceae cell culture systems, root cultures of *Solanum aviculare* proved superior to cell suspensions and callus cultures for glycoalkaloid accumulation, and showed growth-dependent alkaloid accumulation. The maximal alkaloid content was 0.6% dm in roots grown in Murashige Skoog mineral salt mixture [73] with 8% sucrose in the medium [74]. Hairy root cultures of *Solanum mauritianum* grew rapidly but contained only 0.013% dm solasodine [75]. Although the optimization of media composition provided better production, higher levels of solasodine were toxic towards the productive root cells. Exogenous solasodine added to the medium inhibited growth, with steroidal alkaloid production declining to negative values when >10 mg L^{-1} solasodine was added, indicating that the cells degraded or converted solasodine when present in high concentrations. *In-situ* removal of the alkaloid was recommended for solasodine-producing cell cultures systems [76]. Attempts were made to create immobilization systems; for example, *Solanum xanthocarpum* cells were immobilized in calcium alginate gel beads, but the solasodine content was only ca. twofold that in free cells, and maximal in cells from stationary phase cultures. After elicitation of cells of *Solanum eleagnifolium* (using a fungal elicitor obtained from *Alternaria* species), solasodine production was increased from 0.9 to 1.5 mg g^{-1} dm in suspension cultures, and from 0.75 to 1.4 mg g^{-1} dm in immobilized cells [77]. However, these contents were insufficient for economic production by cell cultures. Green shoot cultures of *Solanum dulcamara* were obtained by transformation with *Agrobacterium tumefaciens* and examined following the observation of glycoalkaloid accumulation, predominantly in green photosynthetic tissues. Different glycosylation patterns on the steroid alkaloid basic molecule were stated in *in-vitro* tissues, which would basically not limit the use of steroid structures. The total glycoalkaloid content was only ca. 0.8% g^{-1} dm shoot tissue [78], but the growth rate of the *in-vitro* shoots was sufficiently rapid to be economically competitive.

In summary, Solanaceae are amenable to many tissue culture systems, with properties of good growth and easy handling. Secondary compounds – and alkaloids in particular – are found in most of these cultures, in variable concentrations. Following the optimization of culture conditions or their precursors, the exploitation of bioactive compounds for industrial production appears possible.

11.2
Alkaloid Production

11.2.1
Choice of the Best Culture System

Tropane alkaloid-producing Solanaceae were among the first cell and tissue cultures to be examined for the feasibility of economic production of alkaloids in fermenters. Tropane alkaloid concentrations in cell suspension or callus cultures were low and, as a first attempt, high-producing cells were selected repeatedly in order to exploit genetic variation among individual cell clones. Despite this, high-producing and stable cell lines were not obtained [14]. Tropane alkaloid contents in cell suspension cultures did not rise considerably by medium nutrient variation or by elicitation, with the highest hyoscyamine content being only 0.045% in *H. niger* suspension cultures (Table 11.1) [13].

Root cultures were evaluated for alkaloid productivity, and both, roots cultured with growth-promoting phytohormones and those obtained after *A. rhizogenes* transformation proved more promising. Large variations are evident when comparing root cultures of different origin for alkaloid productivity (e.g., see Table 11.1). The method of root culture initiation, for example by excision from seedlings or by transformation with *A. rhizogenes*, appears not to be decisive for alkaloid production. Two factors, however, were repeatedly described as influential:

- Rapid growth and biomass accumulation of cultured root is contrary to high alkaloid accumulation *per* biomass. *Hyoscyamus* root cultures, for example, grew faster with more auxin in the nutrient medium, but alkaloid biosynthesis decreased [11]. *Duboisia myoporoides* root cultures selected for high alkaloid production revealed those lines as highest producers that showed comparatively slow growth [51].

- The parent plant has a major influence on the alkaloid productivity of the root strain. Root cultures of *Duboisia* species repeatedly proved to be those with the highest total alkaloid contents. In particular, they produce large amounts of scopolamine, which is of high market value. This is in concordance with *Duboisia* hybrid plants also being the solanaceous species with high alkaloid contents (3–7% tropane alkaloids of dry leaf mass). The major fraction in most *Duboisia* plants is scopolamine [79,80].

11.2.2
Optimization Strategies

Consequently, yield optimization strategies were performed not only with *Duboisia* root cultures but also with other species such as *Datura* sp. or *Atropa belladonna*. The approaches were the same as routinely chosen for cell suspension cultures, and mainly included:

- optimization of light, gas and culture vessels;
- feeding of alkaloid precursors;
- nutrient medium variation; and
- elicitation

The following sections provide examples of the success of these optimization methods.

11.2.2.1 Large-Scale Culture, Light, and Aeration

Roots of *D. stramonium* were successfully cultivated in a modified 14-L-stirred-tank reactor. Both, batch or and continuous fermentation modes were performed, and productivity was optimized by variation of incubation temperature and nutrient supply. Nutrient consumption was dependent on the plant species and on growth rates [81]. The highest production rate of 8.2 mg L^{-1}·day hyoscyamine was obtained with half-strength B5 medium and for roots growth at 30 °C [82]. Alkaloid release into the culture medium was temperature-dependent [83]. Using an airlift bioreactor (4 L) with an additional vertical mesh inserted to increase root anchorage, treatment with Tween 20 encouraged both growth and alkaloid productivity of cultured roots of *Datura metel*. Scopolamine was produced at 0.84 mg L^{-1} · day, and 70% was excreted. The scopolamine released into the culture medium was separated with an Amberlite XAD-2 column located in the media exit [84]. Transformed root cultures of *Atropa belladonna* were cultivated in 3-L and 30-L modified stirred bioreactors. After one month, 1.5 g tropane alkaloids were produced by the 30-L reactor, comprising 5.4 mg atropine, 0.9 mg scopolamine, 1.6 mg 6-OH-hyoscyamine, and 2.0 mg littorine g^{-1} dm, respectively [85]. A hairy root clone of *Duboisia leichhardtii* was cultured in a bioreactor connected to an Amberlite XAD-2 column for entrapment of scopolamine. Nutrient medium was continuously exchanged. After 11 weeks, 0.5 g L^{-1} scopolamine was obtained in the XAD-2 column. Polyurethane foam or stainless-steel mesh to support the hairy roots in the bioreactor increased scopolamine recovery. In a two-stage culture (initially in medium optimized for hairy root growth and subsequently in medium for scopolamine release), 1.3 g L^{-1} scopolamine was recovered during 11 weeks of culture [86]. Similar extractive fermentation techniques were examined for the production of other of plant alkaloids [87].

The illumination of root cultures of *D. stramonium*, *Hyoscyamus albus*, and *A. belladonna* had variable effects. Within four weeks of light exposure, *A. belladonna* roots turned green, and chlorophyll-containing plastids became visible microscopically. The roots retained their typical anatomy, but alkaloid production decreased [88]. Root cultures of *H. albus* after illumination showed slightly higher concentrations of alkaloids, and in particular scopolamine was increased [89]. Root cultures of *Datura innoxia* developed photosynthetically active chloroplasts under illumination and, in addition, produced significantly more tropane alkaloids (1.2% dm in heterotrophic roots; 2.9% dm in photoautotrophic roots) than control root cultures in the dark [90].

Enhanced oxygen gas supplementation in root cultures of *D. myoporoides* shifted the alkaloid pattern in favor of more scopolamine production [91]. For *A. belladonna* roots, the specific growth rates, biomass yields and atropine levels were maximum at around 150% air saturation, demonstrating that roots cultivated in reactors with air sparging are oxygen-limited [92]. By better oxygen saturation, root growth to high density became possible and yielded more than $1 \, g \, L^{-1}$ scopolamine within three weeks of culture [93]. Various culture vessels and fermentation strategies have been designed and successfully applied for plant cell cultures, though most of them are not applicable to cultured roots [94]. *A. belladonna* roots were successfully cultivated in airlift reactors, but their alkaloid productivity was lower compared with shake flasks [95].

11.2.2.2 Feeding of Alkaloid Precursors

Precursors of hyoscyamine and scopolamine biosynthesis (Fig. 11.3) were added to the nutrient medium in order to overcome bottlenecks in metabolism and to enhance alkaloid accumulation. The results, however, were not overwhelming. Alkaloid accumulation was seen as variable, when *N*-methylputrescine, tropine, phenylalanine, and tropic acid were applied to root cultures of *H. niger* [13]. In roots of *D. innoxia*, addition of various precursors alone was ineffective in stimulating hyoscyamine production. A short treatment with Tween 20 combined with phenylalanine or phenyllactic acid application increased the level of hyoscyamine by 40–60% [96]. Polyamine feeding (e.g., with putrescine and spermidine) showed limited success for scopolamine content in *D. myoporoides* roots. The putrescine analogues diaminomethane, diaminoethane, diaminopropane, and cadaverine increased the scopolamine from 0.56% to 0.86%, 0.90%, 1.06%, and 1.04%, respectively [97]. The results should be interpreted rather as an unspecific elicitation effect (see Section 11.2.2.4) than as a specific precursor result. Diaminopropane and cadaverine are not precursors of the tropane ring system; rather, they inhibit the first specific enzyme of the pathway, putrescine *N*-methyltransferase (PMT) *in vitro* (see Section 11.3.2). The influences of precursors of the tropic acid moiety and the tropane bicyclic ring of hyoscyamine were tested separately in *D. stramonium* root cultures. Feeding precursors for the tropic acid moiety either had no influence or had a detrimental effect on hyoscyamine accumulation. Feeding putrescine, agmatine or tropine did not enhance alkaloid accumulation, but rather resulted in a lowering of hyoscyamine levels [98]. Tropinone application doubled hyoscyamine formation in root cultures of *D. stramonium*, but the absolute hyoscyamine level was rather low (0.3% dm) in these roots [99].

11.2.2.3 Nutrient Medium Variation

As fast growth and alkaloid production had repeatedly been observed as oppositional, a two-stage culture method, in which the first stage promotes lateral root induction and the second stage enhances root elongation, was developed for nontransformed root cultures of *D. myoporoides*. In the first culture stage (7 days), roots were cultivated in Nitsch and Nitsch medium supplemented with 3% sucrose and 10 µM indolebutyric acid as synthetic auxin. In the second stage (14 days), roots received 5% sucrose in Nitsch and Nitsch medium and no auxin. Production of scopolamine reached 2.5 g L^{-1}, when this two-stage culture was combined with a high-density culture method [100]. A similar increase in alkaloids after increased sucrose supply was reported for root cultures of *Datura stramonium*, *Datura quercifolia*, *Atropa baetica*, and *A. belladonna*. When sucrose in *D. stramonium* root culture medium (Gamborg's B5 [28]) was varied systematically from 1% to 6%, the total alkaloid content rose tenfold [101]. In root cultures of *D. quercifolia* accordingly, 5% sucrose in B5 medium was found optimal for hyoscyamine production (1.24% dm) [46]. Root cultures of *D. myoporoides* and *D. leichhardtii* grew optimally in MS medium with 7–10% sucrose, and showed best tropane alkaloid production under these conditions [102]. In other root cultures, an increase of sucrose alone was not as efficient as concomitant variation of carbohydrate and mineral supply in the medium. Root cultures of *A. baetica* in half-strength Gamborg's B5-medium with the usual 3% sucrose grew slower than in full-strength medium, but contained ca. 50% more hyoscyamine and threefold more scopolamine [103]. For *A. belladonna* roots, a decrease in mineral supply (half-strength Gamborg's B5) was combined with enhanced sucrose supply (5%) and increased both, hyoscyamine and calystegines, which are nortropane alkaloids derived from the tropane alkaloid pathway (Fig. 11.3) [44]. A detailed study of the sugar effect revealed contrasting results for several monosaccharides, for example, mannose, glucose, fructose, and sorbitol on growth and alkaloid formation. It was concluded that carbohydrates not only serve as carbon sources but also show differential effects as signal compounds [44, 88]. The correlation of high carbohydrate and high alkaloids, however, may not be generalized. Root cultures of a hybrid of *Datura candida* × *D. aurea* in half-strength Gamborg's B5 medium supplemented with 5% sucrose were the best for root growth, whereas full-strength B5 medium was optimal for hyoscyamine (0.36% dm) and scopolamine (0.17% dm) accumulation [104].

11.2.2.4 **Elicitation**

The reaction of plant cells to stress of various nature by enhanced formation of secondary product is termed "elicitation". Elicitors are thus chemicals of biotic or abiotic origin that provoke a plant-typical stress response [105]. Jasmonic acid and the ester methyljasmonate (MeJas) were detected as signal compounds (Fig. 11.4) following wounding or microbial infection of plant tissue [106–108]. They induce the formation of many secondary product biosynthetic pathways, including several alkaloids. Nicotine formation was enhanced after wounding of *Nicotiana sylvestris*, and MeJas was the signal molecule [109]. The key enzyme of nicotine biosynthesis, PMT (see Fig. 11.3) was induced upon elicitation with MeJas. Transcripts of the tobacco PMT gene were detected only in the root, and enhanced transcription was antagonized by ethylene [110,111]. In cell cultures of *Nicotiana tabacum*, mRNA for ornithine decarboxylase, *S*-adenosylmethionine synthase and PMT were induced simultaneously by MeJas. These enzymes provide precursors for nic-

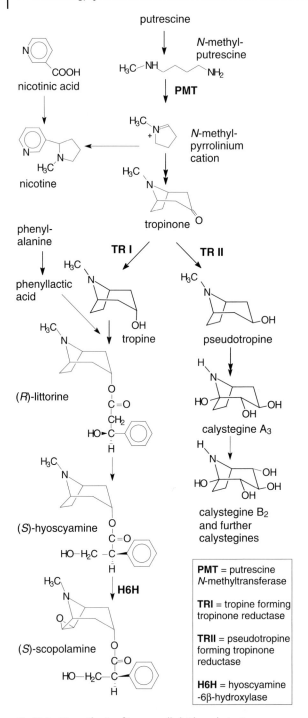

Fig. 11.3 Biosynthesis of tropane alkaloids and nicotine.

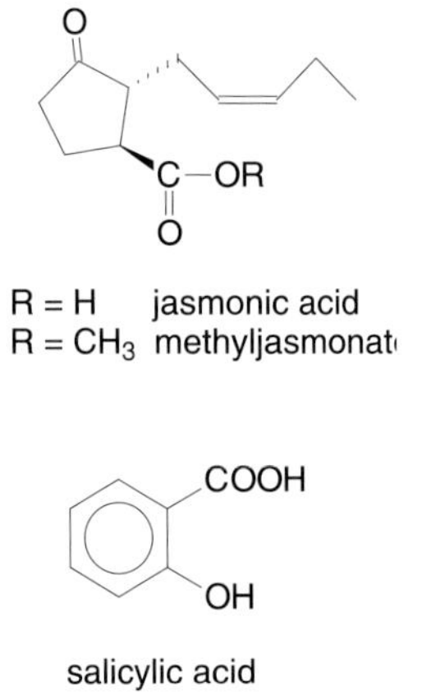

Fig. 11.4   Signal molecules in stress response.

R = H   jasmonic acid
R = CH$_3$ methyljasmonate

salicylic acid

otine biosynthesis, and nicotine levels were enhanced subsequently. Auxin addition to the cell culture medium significantly reduced the accumulation of mRNA for nicotine biosynthetic enzymes [112]. Other alkaloids deriving from nicotinic acid, such as anatalline (4.8 mg g^{-1} dm) and anatabine (0.54 mg g^{-1} dm) (see Fig. 11.1) were enhanced after MeJas treatment of *Nicotiana tabacum* cv. By-2 cell cultures [69]. MeJas, however, does not only affect genes coding for tobacco alkaloid biosynthesis. Transcript profiling of elicited tobacco cells revealed an extensive MeJas-mediated genetic reprogramming of metabolism, which correlated with shifts in the biosynthesis of the metabolites investigated [113].

Tropane alkaloids, which share the same key enzyme PMT at the start of the pathway, respond to elicitation in some cases only, and by a less drastic total alkaloid increase. Copper and cadmium salts (1 mM), identified as abiotic elicitors in other plant cells, induced rapid accumulation of sesquiterpenoid defense compounds, for example, lubimin and 3-hydroxylubimin in root cultures of *D. stramonium*. The sesquiterpenoids were undetectable in non-elicited cultures. However, no change was seen in the total tropane alkaloid content. A considerable and rapid release of alkaloid into the nutrient medium was observed (50–75% of total alkaloids within 40–60 h) upon copper and cadmium elicitation. In cultures treated with copper ions, the alkaloids were reabsorbed [114]. MeJas increased alkaloid formation in hairy root cultures of *D. stramonium* as well as a cell wall preparation from baker's yeast and oligogalacturonides. In all cases, this was associated with an increase in tropine but a decline in phenyllactate concentrations. Increased tropane alkaloid synthesis was due to the differential enhancement of tropine biosynthesis [115]. MeJas, 100 nM in the root culture medium, was found to be the most effective concentration [101]. Root cultures of *Hyoscyamus muticus* responded to elicitation with chitosan; 50–500 µg mL^{-1} culture enhanced the production of hyoscyamine fivefold. The response varied and was dependent upon the chitosan concentration and the time course of elicitation. Chitosan, like copper ions in *D. stramonium* roots, affected the permeability of the transformed roots, releasing hyoscyamine into the medium [116]. When examining the responses to MeJas treatment in detail, large increases in *N*-methylputrescine levels in normal and hairy roots of *H. muticus* were detected. Levels of free putrescine and perchloric acid-soluble conjugated putrescine, spermidine and spermine also increased dramatically. Although treatment of root cultures with MeJas enhanced the precursors putrescine and *N*-methylputrescine, it provoked only modest increases in tropane alkaloid tissue levels [117]. It appears that every individual tropane alkaloid-producing root culture reacts in a specific way to elicitation. In *Brugmansia candida* root cultures, several biotic and abiotic elicitors were tested. Salicylic acid (Fig. 11.4) increased the release of hyoscyamine and scopolamine and enhanced their production. $AgNO_3$ similarly increased scopolamine release (threefold) and total alkaloid accumulation (five- to eightfold) in the roots; the inhibitory effects of $AgNO_3$ and salicylic acid on ethylene could partly explain these responses. Yeast extract enhanced the intracellular content of both alkaloids (ca. threefold), and increased the release of scopolamine (sevenfold). $CaCl_2$ had little effect on the accumulation or release of either alkaloid, while $CdCl_2$ released both alkaloids (three- to 24-fold), but was highly detrimental to growth [118]. Similar effects for salicylic acid and MeJas were reported for root cultures of *Scopolia parviflora* [119]. In root cultures, 200 µM MeJas increased the alkaloid accumulation 25-fold up to a level of 1 mg g^{-1} fresh mass, while the flavonoid quercetin enhanced alkaloid production tenfold to 0.4 mg g^{-1} fresh mass within 24 h. In contrast, 100 µM salicylic acid decreased alkaloids to a level of 1 µg g^{-1} fresh mass [120]. Bacterial elicitors from *Pseudomonas aeruginosa*, *Bacillus cereus*, and *Staphylococcus aureus* were also investigated and altered the ratio between scopolamine and hyoscyamine. Absolute alkaloid concentrations were enhanced only slightly [121]. Considering the induction of tropane alkaloids by MeJas in root cultures, it was tempting to reinvestigate callus cultures, which are closer to dedifferentiated cell suspensions suited for large-scale production in fermenters than to roots. Putrescine and *N*-methylputrescine increased upon treatment with MeJas in root cultures of *H. muticus*, whereas in callus cultures *N*-methylputrescine levels were not affected. Arginine decarboxylase, ornithine decarboxylase and diamine oxidase activities in root cultures were strongly stimulated by treatment with MeJas, but were inhibited in callus cultures. Exposure to MeJas also enhanced PMT activity in root cultures more than in callus cultures [22]. These results match to the finding that the gene coding for PMT is transcribed specifically in the pericycle of young roots of *Atropa belladonna*. Treatment of these *A. belladonna* roots with MeJas did not up-regulate the expression of β-glucuronidase or the endogenous PMT genes [122]. Accordingly, a root culture of *A. belladonna* did not respond with increased total alkaloid concentration upon elicitation. *A. belladonna*, like many Solanaceae, produces calystegines in addition to the tropane ester alkaloid hyoscyamine. Calystegines were monitored upon chitosan and MeJas treatment within one to six days, and they decreased in concentration [44].

In summary, the contents of tropane alkaloids in Solanaceae root culture systems were optimized to impressive yields. Nonetheless, the production of tropane alkaloids from root cultures remains uneconomic in comparison with very good plant sources available [79,80]. The impact of Solanaceae cell cultures lies in providing material for the elucidation of alkaloid biosynthetic steps and their regulation.

11.3
Alkaloid Biosynthesis

11.3.1
Tropane Alkaloid Metabolites, Enzymes and Products

Despite the biosynthetic pathway of tropane alkaloids being depicted almost universally in textbooks of plant natural products, only two enzymes specific to the biosynthesis of hyoscyamine have been isolated and characterized (Fig. 11.3): PMT (EC 2.1.1.53) and the tropine-forming tropinone reductase (EC 1.1.1.206). Further, hyoscyamine-6-hydroxylase (EC 1.14.11.11), which catalyzes the formation of scopolamine from hyoscyamine, has been cloned and characterized in detail. Although most metabolic steps in tropane alkaloid formation have been elucidated using radioactive precursors and subsequent metabolite analysis, this method is prone to errors if not interpreted with caution, since assumed precursor molecules may be metabolized without participating in the plant's endogenous biosynthetic pathway (e.g., see [123]).

It is only 15 years ago that hydroxylated nortropane alkaloids, the calystegines, were structurally elucidated in extracts from roots of *Calystegia sepium*, Convolvulaceae [124]. Based on the nortropane structure, it was soon hypothesized that calystegines are formed via the tropane alkaloid pathway, and consequently tropane alkaloid-containing Solanaceae were investigated for calystegines [125,126]. Specific extraction and purification schemes were developed for the hydrophilic alkaloids, together with modified chromatographic procedures [127,128]. Calystegines were measured in many Solanaceae tissues; they also accumulate in root cultures but not in dedifferentiated cell suspension (Table 11.3). Calystegines resemble monosaccharides in structure, and were shown (somewhat unsurprisingly) to be potent inhibitors of glycosidase activity [129]. They are known to be widespread among the Solanaceae, not only in the species forming the tropane ester alkaloids but also in members of the large genus *Solanum* (e.g., potato) [130]. Erythroxyla-

Table 11.3 Calystegine accumulation in root cultures.

Species/root culture	Total calystegines		Details	Reference
	% dm	mg L⁻¹		
Atropa belladonna	0.2	17.1	Transformed with *A. rhizogenes*	44
Hyoscyamus albus	0.16	–	Excised roots from seedlings	126
Hyoscyamus aureus	0.08	–	Excised roots from seedlings	126
Hyoscyamus muticus	0.11	–	Excised roots from seedlings	126
Hyoscyamus niger	0.11	–	Excised roots from seedlings	126
Hyoscyamus pusillus	0.04	–	Excised roots from seedlings	126
For comparison: *Calystegia sepium* Convolvulaceae	016	22.8	Transformed with *A. rhizogenes*	133

ceae also contain calystegines, but they are taxonomically remote from Solanaceae [131]. *Erythroxylum* species, however, are renowned for their cocaine content; the cocaine molecule also contains a tropane bicyclus. Biosynthesis of calystegines and their tropane nucleus in *Erythroxylum* species awaits elucidation. Current knowledge on calystegine biosynthesis and biochemistry has been summarized [132].

11.3.2
Putrescine *N*-methyltransferase (PMT)

Formation of the tropane bicyclus in Solanaceae begins by methylation of the ubiquitous diamine putrescine (see Fig. 11.3). This reaction is common to both, tropane and nicotine biosynthesis, and the enzyme PMT was initially extracted and measured from tobacco plant roots [134–137] and callus cultures [138]. Root cultures of *D. stramonium* [136,139,140], of *Hyoscyamus albus* [141,142], and of *H. niger* [143] contain PMT with catalytic properties similar to those of the tobacco enzyme. The cDNA of the *pmt* gene was cloned from tobacco (EMBL Accession No. D28506) [144] and from *Nicotiana sylvestris* (EMBL Accession Nos. AB004322, AB004323, AB004324) [110].

The *pmt* gene was shown to be expressed exclusively in the root pericycle of *A. belladonna* [122] and in the endodermis, xylem and outer cortex cells of *Nicotiana sylvestris* roots [110]. Expression was monitored by fusion of the 5′-flanking regions of the PMT genes to the β-glucuronidase reporter gene. The results offer an explanation for the inability of dedifferentiated cell suspension to synthesize tropane alkaloids: specific pericycle cells are required for *pmt* expression in tropane alkaloid-containing plants. Tobacco *pmt*, in comparison, is expressed in several cell types, among them root cortex parenchyma cells, to which dedifferentiated suspension cells may be similar. In tobacco, *pmt* is stress-responsive and inducible by MeJas [109–112]. In contrast to the tobacco *pmt* promoter [145], no MeJas responsive element was identified in the promoter region of *A. belladonna pmt* [122]. A detailed deletion analysis of the *N. tabacum pmt* promoter showed that as little as 111 bp upstream of the transcriptional start site were sufficient to confer MeJas-responsiveness. Deletion of a conserved G-box element (GCACGTTG) at −103 to −96 bp completely abolished MeJA-responsiveness. Further mutagenesis studies revealed that, in addition to a functional G-box, MeJA-responsiveness of the PMT promoter also required a TA-rich region and a GCC-motif (TGCGCCC) located at −80 to −69 bp and −62 to −56 bp relative to the start site, respectively, indicating multiple intersecting signal transduction pathways and different transcriptional regulatory factors involved in MeJas-response of PMT expression in tobacco [146]. Some *pmt* expression was also observed in tobacco leaves after mechanical wounding. This expression was highly localized around the wound site and proved to be transient, with levels being maximal immediately after wounding but diminishing after 2–4 h [147]. PMT expression is also regulated by auxin; in cell suspensions and root cultures, auxin was repeatedly observed to decrease nicotine and tropane alkaloid accumulation (see Sections 11.2.1 and 11.2.2.3). In fact, *pmt* gene transcription is suppressed upon auxin application [88, 112, 148].

The PMT protein resembles spermidine synthase (SPDS; EC 2.5.1.16), a ubiquitous enzyme which is considered to be the evolutionary ancestor of PMT in tobacco [149]. As calystegines are assumed to be formed via the tropane biosynthetic pathway, *Solanum tuberosum* and calystegine-forming Convolvulaceae should also contain PMT activity in order to build up the tropane alkaloid skeleton. Alternatively, it may be hypothesized that PMT is not required, because the bicyclic nortropane skeleton arises directly from putrescine without methylation via oxidation and condensation with three carbons. The alternative pathway for calystegines does not contain tropinone reductases and PMT [150]. In order to answer this question for potato, the cloning of a putative PMT gene was undertaken and, after expression, the corresponding cDNA yielded an enzyme with catalytic properties similar to those of tobacco and *D. stramonium* PMTs [151]. cDNA sequences homologous to *pmt* were also isolated from *C. sepium*, the cDNA is currently expressed (M. Teuber and B. Dräger, unpublished results). The results suggest a similar biosynthetic sequence for the formation of the tropane bicyclus in all these species with PMT as key enzyme.

The subsequent steps from *N*-methylputrescine to tropinone have been deduced mainly from precursor feeding. A diamine oxidase is thought to be involved in *N*-methylputrescine oxidation in tropane alkaloid, as well as in nicotine formation [152], but a specific enzyme was only partially purified from *H. niger* [153]. Later, a methylputrescine oxidase which differed from diamine oxidase was purified from tobacco roots [154]. The reactions from 4-aminobutanal, the *N*-methylpyrrolinium cation, and the formation of the tropane bicyclus are somewhat hypothetical [155].

11.3.3
Tropinone Reductases

For a long time – mainly because of their strong pharmacological effects – the tropine esters hyoscyamine and scopolamine were taken as the major tropane alkaloids in Solanaceae, on the basis of both quantity and activity. It was thought accordingly that during the course of its biosynthesis, tropinone was reduced stereospecifically to tropine (3α-tropanol) (Fig. 11.3) and not to the isomeric pseudotropine (3β-tropanol). Measurement of tropinone-reducing enzyme activities in *D. stramonium* protein extracts confirmed this view: tropine only was found as reduction product, and pseudotropine was not formed [156]. The first tropinone reductase purification from *H. niger*, however, yielded an enzyme specific for pseudotropine formation [157]. In addition, pseudotropine was proved not to be isomerized into tropine in plant tissues [158]. Consequently, as that enzyme was not responsible for tropine formation, the existence of another tropinone reductase forming tropine was postulated. Two separate tropinone reductases were purified from *H. niger* root cultures [159], and also from *D. stramonium* root cultures [160]. *A. belladonna* root cultures also contained two specific enzymes [161]; the tropine-forming enzyme was termed TRI (EC 1.1.1.206), and the pseudotropine-forming enzyme TRII (EC 1.1.1.236). TRII activity was found to be strong in many Solanaceae tissues; for example, shortly after the application of tropinone, pseudotropine accumulated faster than tropine [161]. Esters of pseudotropine (e.g., of acetic acid or tiglic acid) were identified only as minor alkaloids in those plants, and the metabolic role of TRII and the destination of pseudotropine formation were enigmatic, until calystegines were brought into the context of tropane alkaloid biosynthesis. The structure of calystegines contains an equatorial hydroxyl group in position 3, the typical feature of pseudotropine (Fig. 11.3). Accordingly, a typical TRII was isolated and characterized from potato tubers that contain calystegines, but not the tropane esters hyoscyamine and scopolamine [162]. It is now accepted that the pseudotropine-forming TRII is responsible for calystegine biosynthesis, while TRI is required for the formation of tropine, which is integrated into hyoscyamine and scopolamine.

A comparison of TRI and TRII enzymes from *D. stramonium* [160,163] and from *H. niger* [159] revealed proteins with similar properties, but with different catalytic and kinetic behavior. The molecular mass of the protein subunits was determined to be between 28 000 and 30 000 Da in each case. Sequencing of cDNA coding for *D. stramonium* TRs confirmed the protein subunits to consist of 273 (TRI) and 260 (TRII) amino acids with molecular masses of 29 615 Da and 28 310 Da, respectively [164]. Amino acid sequence homology (167 identical amino acid residues, 64%) and comparison of conserved amino acid motifs grouped both tropinone reductases into the family of short-chain dehydrogenases with their typical amino acid motifs [165,166]. All TRs require NADPH as reducing co-substrate; the enzymes are somewhat permissive for the ketone substrates, but strictly specific for the positioning of the resulting alcohol group.

The similarity in protein and catalysis of both reductases, but the apparent differences in reaction stereospecificity, were intriguing. Differential tropinone acceptance and fixation were suspected to be responsible for the selective formation of tropine and pseudotropine [163]. Reaction velocity, substrate affinity, and pH optima for TRI and TRII are conspicuously different (summarized in [167]). After heterologous expression in *E. coli*, sufficient enzyme protein was available for crystallization and protein structure elucidation of TRI and TRII [168–170]. Modeling of the tropinone binding site of TRI and TRII proved two different methods of substrate fixation. Positioning of the substrate in an optimal angle for hydride transfer from NADPH is considered an important prerequisite for efficient catalysis, and the enzyme proteins must be able to adjust to each transition state of the reaction [171].

11.3.4
Hyoscyamine-6β-Hydroxylase

Subsequent esterification of tropine was shown to occur with phenyllactic acid, the first esterified alkaloid being littorine, which also accumulates in some Solanaceae and in root cultures of the respective plants [172–177]. Rearrangement of the phenyllactic acid moiety of littorine to yield hyoscyamine was demonstrated by labeled precursors and NMR [178–180], but not elucidated on an enzymatic level. There are various suggestions for the rearrangement mechanism, for example, an oxida-

tive reaction [181–183] or, alternatively, a radical mechanism with *S*-adenosylme-
thionine as source of a 5'-deoxyadenosyl radical, which initiates the rearrangement
[184,185].

Hyoscyamine is oxidized to form scopolamine by an oxoglutarate-dependent
dioxygenase, the hyoscyamine-6β-hydroxylase (H6H) [186]. The enzyme, which
was purified and characterized from a *H. niger* root culture [186,187], performs a
two-step reaction, first hydroxylating hyoscyamine in 6-position and subsequently
forming the epoxy group of scopolamine [188,189]. Antibodies against the purified
enzyme enabled localization of the H6H protein in the pericycle of root diameters
of *H. niger* [190]. GUS-fusions to the promoter region of the *h6h* gene from *A. belladonna* and immunohistochemistry confirmed pericycle expression of *h6h* [191].
This finding similar to specific localization of PMT enforces the conclusion of differentiated root tissue being necessary for tropane alkaloid biosynthesis. Cloning
of the *h6h* gene [192] and transformation of *A. belladonna* with *h6h* cDNA yielded
plants with a drastically increased scopolamine production [193].

11.4
Solanaceae Model Systems for Transformation and Overexpression

H6H overexpression in *A. belladonna* was the first proof of functional heterologous
expression of a tropane alkaloid pathway enzyme and, thereby, directed alteration
of the alkaloid yield and pattern. The success triggered many further efforts to use
the cloned genes of tropane alkaloid formation for overexpression. Transformation
protocols applied either *A. rhizogenes* transformation or particle bombardment by
a biolistic device. Efficient regeneration of fertile plants after transformation, however – even with Solanaceae – was a major obstacle, and remains difficult. A regeneration protocol after transformation of *H. muticus* with particle bombardment was
optimized [194].

PMT overexpression in transgenic plants of *Nicotiana sylvestris* increased nicotine content; suppression of endogenous PMT activity severely decreased the nicotine content and induced abnormal morphologies. In contrast, PMT-overexpress-
ing transformants of *A. belladonna* were phenotypically normal and had hyoscyamine levels (1.3–2.4 mg g^{-1} dm) comparable to those for the wild-type and vector
controls. The only difference was some accumulation of *N*-methylputrescine [195].
Transgenic hairy root clones had a fivefold increase of the PMT transcript, but scopolamine, hyoscyamine, tropine, pseudotropine, tropinone, and calystegines were
found unaltered or somewhat decreased. Auxin addition reduced tropane alkaloids
in control roots as seen before, while in *pmt*-overexpressing roots, all alkaloids remained unaltered [148]. The unchanged alkaloid profiles after augmented PMT expression indicated that enforcing this enzyme alone was not sufficient to increase
tropane alkaloid synthesis in *A. belladonna* plants and hairy roots. After *pmt* overexpression in hairy root cultures of *D. metel*, both hyoscyamine and scopolamine
production were improved, whereas in *H. muticus* only hyoscyamine contents were
increased by *pmt* gene overexpression. The results indicate that the same biosyn-

thetic pathway in related plant species is regulated differently, and overexpression of a given gene does not necessarily lead to a similar accumulation pattern of secondary metabolites [196].

For nicotine production, PMT appears to be largely regulating the whole subsequent biosynthetic pathway. Suppression of *pmt* transcripts by virus-induced gene silencing [197], by *pmt* antisensing, or by *pmt* inverted-repeats yielded plants with decreased nicotine contents that were preferred by herbivorous insects [198,199]. *pmt* antisensing did not reduce the transcript levels of other genes encoding enzymes of nicotine biosynthesis (e.g., quinolinate phosphoribosyltransferase regulating the synthesis of nicotinic acid). The pyridine ring is used for both, nicotine and anatabine synthesis. Elevated anatabine levels in antisense-PMT roots were observed and may be a direct consequence of a relative oversupply of nicotinic acid [200].

The activities of tropinone-reducing enzymes are usually high and considered as not limiting for tropine formation [159]; consequently, these enzymes were not primary targets for overexpression. The feasibility of TRI overexpression was shown in tobacco plants that, after tropinone application, formed tropine and acetyltropine as esterification product [201]. Overexpression of tropine-forming tropinone reductases became interesting with the understanding that tropinone reduction forms a branch point in tropane alkaloid metabolism (Fig. 11.3). Overexpression of either tropinone reductase was attempted in species that accumulate both tropine-derived alkaloids and calystegines. Transformation of *A. belladonna* with cDNA of tropinone reductases altered the ratio of tropine-derived alkaloids *versus* pseudotropine-derived alkaloids [202].

Root cultures of *A. belladonna* overexpressing *h6h* contained 0.4% dm total alkaloids, with 0.3% dm scopolamine. In control roots, hyoscyamine (0.3% dm) is the major ester alkaloid [203]. Tobacco plants overexpressing *h6h* were equally able to convert externally applied hyoscyamine into scopolamine [204]. Leaves of tobacco plants simultaneously transformed with *pmt* and *h6h* (T_1 progeny) were fed with tropinone and hyoscyamine. They converted externally applied hyoscyamine into scopolamine and, besides the expected TRI reaction product tropine, acetyltropine was generated. In addition, leaves of the transgenic plants showed three- to 13-fold higher nicotine content and nicotine-related compounds such as anatabine, nornicotine, bipyridine, anabasine, and myosmine [201]. *h6h* was also introduced into *H. muticus*, and root cultures were obtained. The best root clone produced 17 mg L^{-1} scopolamine – approximately 100 times more than the control clones [52]. Simultaneous introduction and overexpression of PMT and H6H in transgenic *H. niger* hairy root cultures yielded higher levels of scopolamine than wild-type and transgenic lines harboring a single gene (*pmt* or *h6h*). The best root line produced 411 mg L^{-1} scopolamine, ninefold the corresponding wild-type with 43 mg L^{-1} [53]. This productivity is comparable with the production of paclitaxel in cell suspension culture of *Taxus* species, which presents the latest example of industrially applied secondary compound production in plant cell cultures [205].

The decision as to whether tropane alkaloid production in root cultures now crosses the line between a model system and an economically feasible alternative to field cultivation remains hampered by good plant sources and thereby comparable cheap prices for these alkaloids. For other alkaloids (e.g., taxanes), production in fermenters appeared highly attractive, when a similar productivity was achieved [206].

11.5
Conclusion and Future Aspects

11.5.1
More Genes and Enzymes

For successful metabolic engineering of tropane alkaloid formation, irrespective of whether this is intended for tissue cultures or intact plants, all enzymatic steps of the biosynthesis must be characterized. The genes encoding the enzymes and the corresponding regulatory gene sequences also await characterization. Alkaloid biosynthetic pathways in plants other than Solanaceae are better known at present, with more enzymes and genes having been isolated, sequenced, and characterized. A prominent example is the recent metabolic engineering of benzoquinoline alkaloid biosynthesis based on the particular knowledge of pathway enzymes [207], where transgenic plants of opium poppy (*Papaver somniferum*) have been blocked for morphine production [208, 209]. Another example of a secondary compound pathway, which has been largely elucidated during the past years, is taxane formation in *Taxus* species [210]. Natural enzymes will not remain the only instruments for overexpression, as many secondary pathway enzymes show some substrate flexibility and accept synthetic analogues of their natural ligands [211, 212]. Directed evolution of enzymes (for a review, see [213]) and alteration of enzymatic specificities – also termed "combinatorial biochemistry" (for a summary, see [214]) – followed by the introduction of the corresponding genes into plant tissue will open the way to variant and novel alkaloids and other natural products. In flavonoid biosynthesis, where many enzymes and their genes have been investigated in detail, impressive examples of targeted manipulation in the biosynthetic pathway have been published (summarized in [215, 216]).

The complex flavonoid biosynthesis was also the first secondary product pathway, where individual transcription factors for gene activity were described [217, 218]. The elucidation of more regulatory mechanisms – for alkaloid biosynthesis in particular – will become mandatory for successful metabolic engineering. The first example for transcription factors regulating alkaloid biosynthesis are ORCAs (octadecanoid responsive *Catharanthus* AP2-domain proteins), which are involved in jasmonate signaling inducing strictosidine synthase for monoterpene alkaloid formation [219,220]. Comparable regulatory genes at hand, induction and fine-tuning of tropane alkaloid biosynthesis will become less like trial and error. Other gene elements have also been found to regulate plant transgene expression, for example the cis-acting amplification-promoting sequences (*aps*) from the non-transcribed spacer region of tobacco ribosomal RNA. Analysis of transgenic tobacco plants showed that *aps* increased the copy number and transcription of the adjacent heterologous genes and both, increased transgene copy number and enhanced expression were stably inherited [221]. This and other regulatory elements will be helpful tools for designing plants or tissue cultures with desired metabolic traits.

The question remains as to which approaches are useful for straightforward identification of genes involved in alkaloid biosynthesis? With high-throughput hybridization and sequencing systems at hand, transcript profiling of jasmonate-elicited tobacco cells was combined with alkaloid metabolite analysis. Upon MeJas treatment, extensive genetic reprogramming of metabolism was observed [113]. Careful selection of promising target sequences and in-depth analysis is prone to detect genes regulating alkaloid biosynthesis in tobacco in the future. A similar concept was successfully applied for the detection of genes of morphine biosynthesis in *Papaver somniferum*, with macroarrays with cDNA having been prepared for differentiating expression between morphine-containing *P. somniferum* plants and eight other *Papaver* species that accumulate other benzylisoquinolines instead of morphine. Among three cDNAs showing increased expression in *P. somniferum* compared to all other *Papaver* species, one encoded a novel *O*-methyltransferase, thus proving the potency of the concept [222]. In the future, potent genomics tools will be combined with metabolic profiling to identify key genes that serve for engineering secondary product pathways [223]

11.5.2
Compartmentation, Transport, and Excretion

For large-scale production in fermenters, the excretion of alkaloids into the medium is desirable. When the *h6h* gene from *H. niger* was overexpressed in *Nicotiana tabacum*, externally applied hyoscyamine was taken up, and up to 85% of the scopolamine formed was released to the culture medium [224]. The mechanism of secretion is unknown in this case, but membrane transport may be hypothesized. When ATP-binding cassette (ABC) transporter genes from yeast were expressed in cultured tobacco cells, exogenously applied nicotine, hyoscyamine and scopolamine were less toxic than in control cells. It was assumed that the engineered tobacco cells were able to excrete the alkaloids [225].

The toxicity of alkaloids to the cells that produce them must be considered as a major constraint of high-level production, and efficient compartmentation of the toxic products is necessary. Storage in the vacuole is often assumed, but this demands transport of the metabolites through the tonoplast as well as secretion through the plasmalemma. A cDNA encoding an ABC-transporter was cloned from a berberine-producing *Coptis japonica* cell culture [226], whereupon functional analysis showed that the encoded protein transported berberine and was localized in the plasma membrane of *C. japonica* cells. In the plant, the protein is localized in the xylem of the rhizome and is held responsible for berberine transport from the root to the rhizome [227, 228].

Further examples of membrane transport being fairly specific and highly regulated for each secondary metabolite also exist. Genes for such transporters will be important for systematic metabolic engineering aimed at increasing the productivity of valuable secondary metabolites in plants [227]. In addition, consecutive enzymes of secondary product pathways may be exclusively localized to individual specialized cell types and demand transport of metabolites between those cells. Examples are essential oil biosynthesis in glandular trichomes [229], formation of terpenoid indole alkaloids in several cells types of *Catharanthus* leaf ([230]), or morphine biosynthesis in vascular bundles and lactifers of opium poppy [231, 232]. Highly ordered translocation processes that involve transport proteins must be elucidated when attempting to understand how plants regulate the formation and accumulation of natural product profiles, and which are the limitations of biosynthesis and accumulation [233].

Acknowledgment

The author thanks Stefan Biastoff of the Department of Pharmaceutical Biology, Martin-Luther University for critically reading and correcting the manuscript.

References

1 Bruhn, J.G. *Acta Pharmaceutica Nordica* **1989**, *1*, 117–130.

2 Romeike, A. *Naturwissenschaften* **1959**, *46*, 492–493.

3 Liebisch, H.W., Ramin, H., Schoeffinius, I., Schuette, H.R. *Z. Naturforschung, Teil B: Anorganische Chemie, Organische Chemie, Biochemie, Biophysik, Biologie* **1965**, *20*, 1183–1185.

4 Liebisch, H.W., Peisker, K., Radwan, A.S., Schuette, H.R. *Z. Pflanzenphysiologie* **1972**, *67*, 1–9.

5 Gross, D., Schuette, H.R. *Arch. Pharm.* **1963**, *296*, 1–6.

6 Singh, M., Krikorian, A.D. *Ann. Botany* **1981**, *47*, 133–139.

7 Alcalde, M.B. *Farmacognosia* **1956**, *16*, 283–455.

8 Linsmaier, E.M., Skoog, F. *Physiol. Plant.* **1965**, *18*, 100–127.

9 Hashimoto, T., Yamada, Y. *Planta Med.* **1983**, *47*, 195–199.

10 Endo, T., Yamada, Y. *Phytochemistry* **1985**, *24*, 1233–1236.

11 Hashimoto, T., Yukimune, Y., Yamada, Y. *J. Plant Physiol.* **1986**, *124*, 61–75.

12 Maldonado, M., Ayora-Talavera, I.T.-D.R., Loyola, V. *In Vitro Cell. Dev. Biol. Plant* **1992**, *28P*, 67–72.

13 Hashimoto, T., Yamada, Y. *Agric. Biol. Chem.* **1987**, *51*, 2769–2774.

14 Yamada, Y., Hashimoto, T. *Plant Cell Rep.* **1982**, *1*, 101–103

15 Hiraoka, N., Tabata, M., Konoshima, M. *Phytochemistry* **1973**, *12*, 795–799.

16 Yamada, Y., Endo, T. *Plant Cell Rep.* **1984**, *3*, 166-168.

17 Gontier, E., Sangwan, B.S., Barbotin, J.N. *Plant Cell Rep.* **1994**, *13*, 533–536.

18 Medina, B.F., Flores, H.E. *Plant Physiol.* **1995**, *108*, 1553–1560.

19 Mothes, K. *Angew. Chem.* **1952**, *64*, 254–258.

20 Hills, K.L., Trautner, E.M., Rodwell, C.N. *Aust. J. Sci.* **1946**, *9*, 24–25.

21 Dawson, R.F. *Science* **1941**, *94*, 396–397.

22 Biondi, S., Scaramagli, S., Oksman-Caldentey, K.-M., Poli, F. *Plant Sci.* **2002**, *163*, 563–569.

23 Mesnard, F., Girard, S., Fliniaux, O., Bhogal, R.K., Gillet, F., Lebreton, J., Fliniaux, M.A., Robins, R.J. *Plant Sci.* **2001**, *161*, 1011–1018.

24 Fliniaux, O., Mesnard, F., Raynaud-Le Grandic, S., Baltora-Rosset, S., Bieneme, C., Robins, R.J., Fliniaux, M.-A. *J. Exp. Bot.* **2004**, *55*, 1–8.

25 White, F.F., Ghidossi, G., Gordon, M.P., Nester, E.W. *Proc. Natl. Acad. Sci. USA* **1982**, *79*, 3193–3197.

26 Cardarelli, M., Spano, L., De-Paolis, A., Mauro, M.L., Vitali, G., Costantino, P. *Plant Mol. Biol.* **1985**, *5*, 385–392.

27 Chilton, M.D., Tepfer, D.A., Petit, A., David, C., Casse-Delbart, F., Tempe, J. *Nature* **1982**, *295*, 432–434.

28 Gamborg, O.L., Miller, R.A., Ojima, K. *Exp. Cell Res.* **1968**, *50*, 151–158.

29 Payne, J., Hamill, J.D., Robins, R.J., Rhodes, M.-J.C. *Planta Med.* **1987**, *53*, 474–478.

30 Deno, H., Yamagata, H., Emoto, T., Yoshioka, T., Yamada, Y., Fujita, Y. *J. Plant Physiol.* **1987**, *131*, 315–324.

31 Jung, G., Tepfer, D. *Plant Sci.* **1987**, *50*, 145–152.

32 Kamada, H., Okamura, N., Satake, M., Harada, H., Shimomura, K. *Plant Cell Rep.* **1986**, *5*, 239–242.

33 Sevon, N., Hiltunen, R., Oksman-Caldentey, K.M. *Planta Med.* **1998**, *64*, 37–41.

34 Aoki, T., Matsumoto, H., Asako, Y., Matsunaga, Y., Shimomura, K. *Plant Cell Rep.* **1997**, *16*, 282–286.

35 Giulietti, A.M., Parr, A.J., Rhodes, M.-J.C. *Planta Med.* **1993**, *59*, 428–431.

36 Bonhomme, V., Laurain, M.D., Lacoux, J., Fliniaux, M., Jacquin, D.A. *J. Biotechnol.* **2000**, *81*, 151–158.

37 Toivonen, L. *Biotechnol. Progress* **1993**, *9*, 12–20.

38 Baiza, A.M., Quiroz, M.A., Ruiz, J.A., Loyola, V. *Plant Cell Tiss. Organ Cult.* **1999**, *59*, 9–17.

39 Sevon, N., Oksman-Caldentey, K.M. *Planta Med.* **2002**, *68*, 859–868.

40 Jung, D.-W., Sung, C.K., Touno, K., Yoshimatsu, K., Shimomura, K. *J. Plant Physiol.* **2001**, *158*, 801–805.

41 Hartmann, T., Witte, L., Oprach, F., Toppel, G. *Planta Med.* **1986**, *52*, 390–395.

42 Staba, E.J., Jindra, A. *J. Pharm. Sci.* **1968**, *57*, 701–704.

43 Oksman-Caldentey, K.M., Vuorela, H., Strauss, A., Hiltunen, R. *Planta Med.* **1987**, *53*, 349–354.

44 Rothe, G., Garske, U., Dräger, B. *Plant Sci.* **2001**, *160*, 1043–1053.

45 Christen, P., Roberts, M.F., Phillipson, J.D., Evans, W.C. *Plant Cell Rep.* **1989**, *8*, 75–77.

46 Dupraz, J.M., Christen, P., Kapetanidis, I. *Planta Med.* **1993**, *60*, 158–162.

47 Pinol, M.T., Palazon, J., Cusido, R., Serrano, M. *Botanica Acta* **1996**, *109*, 133–138.

48 Jaziri, M., Legros, M., Homes, J., Vanhaelen, M. *Phytochemistry* **1988**, *27*, 419–420.

49 Mano, Y., Ohkawa, H., Yamada, Y. *Plant Sci.* **1989**, *59*, 191–202.

50 Celma, C.R., Palazon, J., Cusido, R.M., Pinol, M.T., Keil, M. *Planta Med.* **2001**, *67*, 249–253.

51 Yukimune, Y., Hara, Y., Yamada, Y. *Biosci. Biotechnol. Biochem.* **1994**, *58*, 1443–1446.

52 Jouhikainen, K., Lindgren, L., Jokelainen, T., Hiltunen, R., Teeri, T.H., Oksman-Caldentey, K.M. *Planta* **1999**, *208*, 545–551.

53 Zhang, L., Ding, R., Chai, Y., Bonfill, M., Moyano, E., Oksman-Caldentey, K.-M., Xu, T., Pi, Y., Wang, Z., Zhang, H., Kai, G., Liao, Z., Xiaofen Sun, X., Tang, K. *Proc. Natl. Acad. Sci. USA* **2004**, *101*, 6786–6791.

54 Mano, Y., Nabeshima, S., Matsui, C., Ohkawa, H. *Agric. Biol. Chem.* **1986**, *50*, 2715–2722.

55 Zelcer, A., Soferman, O., Izhar, S., Volcani, C. *Plant Cell Rep.* **1983**, *2*, 252–254.

56 Ooms, G., Karp, A., Burrell, M.M., Twell, D., Roberts, J. *Theoret. Appl. Genetics* **1985**, *70*, 440–446.

57 Martin, T.J., Tepfer, D., Burtin, D. *Plant Sci.* **1991**, *80*, 131–144.

58 Martin, T.J., Corbineau, F., Burtin, D., Ben-Hayyim, G., Tepfer, D. *Plant Sci.* **1993**, *93*, 63–76.

59 Oksman-Caldentey, K.M., Kivela, O., Hiltunen, R. *Plant Sci.* **1991**, *78*, 129–136.

60 Dehio, C., Grossmann, K., Schell, J., Schmulling, T. *Plant Mol. Biol.* **1993**, *23*, 1199–1210.

61 Khanam, N., Khoo, C., Close, R., Khan, A.G. *Ann. Botany* **2000**, *86*, 745–752.

62 Khanam, N., Khoo, C., Close, R., Khan, A.G. *Phytochemistry* **2001**, *56*, 59–65.

63 Panaia, M., Senaratna, T., Brunn, E., Dixon, K.W., Sivasithamparam, K. *Plant Cell Tiss. Organ Cult.* **2000**, *63*, 23–19.

64 Kang, Y.M., Min, J.Y., Moon, H.S., Karigar, C.S., Prasad, D.T., Lee, C.H., Choi, M.S. *Plant Cell Rep.* **2004**, *23*, 128–133.

65 Hiraoka, N., Shinohara, C., Ogata, H., Chang, J.I., Bhatt, I.D. *Natural Med.* **2004**, *58*, 98–103.

66 McHale, N.A. *Plant Physiol.* **1985**, *77*, 204–242.

67 Beimen, A., Witte, L., Barz, W. *Botanica Acta* **1992**, *105*, 152–160.

68 Ikemeyer, D., Barz, W. *Plant Cell Rep.* **1989**, *8*, 479–482.

69 Haekkinen, S.T., Rischer, H., Laakso, I., Maaheimo, H., Seppaenen-Laakso, T., Oksman-Caldentey, K.M. *Planta Med.* **2004**, *70*, 936–941.

70 Parr, A.J., Hamill, J.D. *Phytochemistry* **1987**, *26*, 3241–3246.

71 Palazon, J., Cusido, R.M., Roig, C., Pinol, M.T. *Plant Cell Rep.* **1998**, *17*, 384–390.

72 Olmstead, R.G., Palmer, J.D. *Systematic Bot.* **1997**, *22*, 19–29.

73 Murashige, T., Skoog, F. *Physiol. Plant.* **1962**, *15*, 473–497.

74 Kittipongpatana, N., Hock, R.S., Porter, J.R. *Plant Cell Tiss. Organ Cult.* **1998**, *52*, 133–143.

75 Drewes, F.E., Van-Staden, J. *Plant Growth Regul.* **1995**, *17*, 27–31.

76 Yu, S., Kwok, K.H., Doran, P.M. *Enzyme Microb. Technol.* **1996**, *18*, 238–243.

77 Quadri, L.E.N., Giulietti, A.M. *Enzyme Microb. Technol.* **1993**, *15*, 1074–1077.

78 Ehmke, A., Ohmstede, D., Eilert, U. *Plant Cell Tiss. Organ Cult.* **1995**, *43*, 191–197.

79 Lean, J.A., Ralph, C.S. *J. Proc. Royal Soc. New South Wales* **1944**, *77*, 96–98.

80 Hills, K.L., Bottomley, W., Mortimer, P.I. *Aust. J. Appl. Sci.* **1954**, *5*, 283–291.

81 Hilton, M.G., Wilson, P.-D.G. *Planta Med.* **1995**, *61*, 345–350.

82 Hilton, M.G., Rhodes, M.J. *Appl. Biochem. Biotechnol.* **1990**, *33*, 132–138.

83 Hilton, M.G., Rhodes, M.-J.C. *Planta Med.* **1993**, *59*, 340–344.

84 Cusido, R.M., Palazon, J., Pinol, M.T., Bonfill, M., Morales, C. *Planta Med.* **1999**, *65*, 144–148.

85 Lee, K.T., Suzuki, T., Yamakawa, T., Kodama, T., Igarashi, Y., Shimomura, K. *Plant Cell Rep.* **1999**, *18*, 567–571.

86 Muranaka, T., Ohkawa, H., Yamada, Y. *Appl. Biochem. Biotechnol.* **1993**, *140*, 219–223.

87 Chang, H.N., Sim, S.J. *Curr. Opin. Biotechnol.* **1995**, *6*, 209–212.

88 Rothe, G., Dräger, B. *Plant Sci.* **2002**, *163*, 979–985.

89 Sauerwein, M., Wink, M., Shimomura, K. *J. Plant Physiol.* **1992**, *140*, 147–152.

90 Flores, H.E., Dai, Y.R., Cuello, J.L., Maldonado, M., Loyola, I.V. *Plant Physiol.* **1993**, *101*, 363–371.

91 Yukimune, Y., Yamagata, H., Hara, Y., Yamada, Y. *Biosci. Biotechnol. Biochem.* **1994**, *58*, 1824–1827.

92 Williams-Gary, R.C., Doran, P.M. *Biotechnol. Bioeng.* **1999**, *64*, 729–740.

93 Yukimune, Y., Tabata, H., Hara, Y., Yamada, Y. *Biosci. Biotechnol. Biochem.* **1994**, *58*, 1447–1450.

94 Roberts, S.C., Shuler, M.L. *Curr. Opin. Biotechnol.* **1997**, *8*, 154–159.

95 Sharp, J.M., Doran, P.M. *J. Biotechnol.* **1990**, *16*, 171–186.

96 Boitel, C.M., Laberche, J.C., Lanoue, A., Ducrocq, C., Sangwan-Norreel, B.S. *Plant Cell Tiss. Organ Cult.* **2000**, *60*, 131–137.

97 Yoshioka, T., Yamagata, H., Ithoh, A., Deno, H., Fujita, Y., Yamada, Y. *Planta Med.* **1989**, *55*, 523–524.

98 Robins, R.J., Parr, A.J., Bent, E.G., Rhodes, M.-J.C. *Planta* **1991**, *183*, 185–195.

99 Dräger, B., Portsteffen, A., Schaal, A., MacCabe, P.H., Peerless, A.-C.J., Robins, R.J. *Planta* **1992**, *188*, 581–586.

100 Yukimune, Y., Tabata, H., Hara, Y., Yamada, Y. *Biosci. Biotechnol. Biochem.* **1994**, *58*, 1820–1823.

101 Saenz, C.L., Loyola, V. *Appl. Biochem. Biotechnol.* **1997**, *61*, 321–337.

102 Kitamura, Y., Sugimoto, Y., Samejima, T., Hayashida, K., Miura, H. *Chem. Pharm. Bull.* **1991**, *39*, 1263–1266.

103 Zarate, R. *Plant Cell Rep.* **1999**, *18*, 418–423.

104 Nussbaumer, P., Kapetanidis, I., Christen, P. *Plant Cell Rep.* **1998**, *17*, 405–409.

105 Poulev, A., O'Neal, J.M., Logendra, S., Pouleva, R.B., Timeva, V., Garvey, A.S., Gleba, D., Jenkins, I.S., Halpern, B.T., Kneer, R., Cragg, G.M., Raskin, I. *J. Med. Chem.* **2003**, *46*, 2542–2547.

106 Gundlach, H., Mueller, M.J., Kutchan, T.M., Zenk, M.H. *Proc. Natl. Acad. Sci. USA* **1992**, *89*, 2389–2393.

107 Creelman, R.A., Tierney, M.L., Mullet, J.E. *Proc. Natl. Acad. Sci. USA* **1992**, *89*, 4938–4941.

108 Blechert, S., Brodschelm, W., Hoelder, S., Kammerer, L., Kutchan, T.M., Mueller, M.J., Xia, Z.Q., Zenk, M.H. *Proc. Natl. Acad. Sci. USA* **1995**, *92*, 4099–4105.

109 Baldwin, I.T., Zhang, Z.P., Diab, N., Ohnmeiss, T.E., McCloud, E.S., Lynds, G.Y., Schmelz, E.A. *Planta* **1997**, *201*, 397–404.

110 Shoji, T., Yamada, Y., Hashimoto, T. *Plant Cell Physiol.* **2000**, *41*, 831–839.

111 Shoji, T., Nakajima, K., Hashimoto, T. *Plant Cell Physiol.* **2000**, *41*, 1072–1076.

112 Imanishi, S., Hashizume, K., Nakakita, M., Kojima, H., Matsubayashi, Y., Hashimoto, T., Sakagami, Y., Yamada, Y., Nakamura, K. *Plant Mol. Biol.* **1998**, *38*, 1101–1111.

113 Goossens, A., Haekkinen, S.T., Laakso, I., Seppaenen-Laakso, T., Biondi, S., De Sutter, V., Lammertyn, F., Nuutila, A.M., Soederlund, H., Zabeau, M., Inze, D., Oksman-Caldentey, K.M. *Proc. Natl. Acad. Sci. USA* **2003**, *100*, 8595–8600.

114 Furze, J.M., Rhodes, M.-J.C., Parr, A.J., Robins, R.J., Whitehead, I.M., Threlfall, D.R. *Plant Cell Rep.* **1991**, *10*, 111–114.

115 Zabetakis, I., Edwards, R., O'Hagan, D. *Phytochemistry* **1999**, *50*, 53–56.

116 Sevon, N., Hiltunen, R., Oksman-Caldentey, K.M. *Pharm. Pharmacol. Lett.* **1992**, *2*, 96–99.

117 Biondi, S., Fornale, S., Oksman-Caldentey, K.M., Eeva, M., Agostani, S., Bagni, N. *Plant Cell Rep.* **2000**, *19*, 691–697.

118 Pitta, A.S., Spollansky, I.T.C., Giulietti, A.M. *Enzyme Microb. Technol.* **2000**, *26*, 252–258.

119 Kang, S.M., Jung, H.Y., Kang, Y.M., Yun, D.J., Bahk, J.D., Yang, J.K., Choi, M.S. *Plant Sci.* **2004**, *166*, 745–751.

120 Zayed, R., Wink, M. *Z. Naturforschung, C: J. Biosci.* **2004**, *59*, 863–867.

121 Jung, H.Y., Kang, S.M., Kang, Y.M., Kang, M.J., Yun, D.J., Bahk, J.D., Yang, J.K., Choi, M.S. *Enzyme Microb. Technol.* **2003**, *33*, 987–990.

122 Suzuki, K., Yamada, Y., Hashimoto, T. *Plant Cell Physiol.* **1999**, *40*, 289–297.

123 Hashimoto, T., Yamada, Y., Leete, E. *J. Am. Chem. Soc.* **1989**, *111*, 1141–1142.

124 Goldmann, A., Milat, M.L., Ducrot, P.H., Lallemand, J.Y., Maille, M., Lepingle, A., Charpin, I., Tepfer, D. *Phytochemistry* **1990**, *29*, 2125–2128.

125 Dräger, B., van Almsick, A., Mrachatz, G. *Planta Med.* **1995**, *61*, 577–579.

126 Dräger, B., Funck, C., Hoehler, A., Mrachatz, G., Nahrstedt, A., Portsteffen, A., Schaal, A., Schmidt, R. *Plant Cell Tiss. Organ Cult.* **1994**, *38*, 235–240.

127 Dräger, B. *Phytochem. Analyt.* **1995**, *6*, 31–37.

128 Dräger, B. *J. Chromatogr. A* **2002**, *978*, 1–35.

129 Asano, N., Nash, R.J., Molyneux, R.J., Fleet, G.W. *Tetrahedron: Asymmetry* **2000**, *11*, 1645–1680.

130 Keiner, R., Dräger, B. *Plant Sci.* **2000**, *150*, 171–179.

131 Brock, A., Bieri, S., Christen, P., Dräger, B. *Phytochemistry* **2005**, *66*, 1231–1240.

132 Dräger, B. *Nat. Prod. Rep.* **2004**, *21*, 211–223.

133 Scholl, Y., Höke, D., Dräger, B. *Phytochemistry* **2001**, *58*, 883–889.

134 Mizusaki, S., Tanabe, Y., Noguchi, M., Tamaki, E. *Plant Cell Physiol.* **1973**, *14*, 103–110.

135 Mizusaki, S., Tanabe, Y., Noguchi, M., Tamaki, E. *Plant Cell Physiol.* **1971**, *12*, 633–640.

136 Feth, F., Arfmann, H.A., Wray, V., Wagner, K.G. *Phytochemistry* **1985**, *24*, 921–923.

137 Wagner, R., Feth, F., Wagner, K.G., *Physiol. Plant.* **1986**, *68*, 667–672.

138 Feth, F., Wagner, R., Wagner, K.G. *Planta* **1986**, *168*, 402–407.

139 Walton, N.J., Robins, R.J., Peerless, A.-C.J. *Planta* **1990**, *182*, 136–141.

140 Walton, N.J., Peerless, A.-C.J., Robins, R.J., Rhodes, M.-J.C., Boswell, H.D., Robins, D.J. *Planta* **1994**, *193*, 9–15.

141 Hashimoto, T., Yukimune, Y., Yamada, Y. *Planta* **1989**, *178*, 131–137.

142 Hashimoto, T., Yukimune, Y., Yamada, Y. *Planta* **1989**, *178*, 123–130.

143 Hibi, N., Fujita, T., Hatano, M., Hashimoto, T., Yamada, Y. *Plant Physiol.* **1992**, *100*, 826–835.

144 Hibi, N., Higashiguchi, S., Hashimoto, T., Yamada, Y. *Plant Cell* **1994**, *6*, 723–735.

145 Hashimoto, T., Shoji, T., Mihara, T., Oguri, H., Tamaki, K., Suzuki, K.i., Yamada, Y. *Plant Mol. Biol.* **1998**, *37*, 25–37.

146 Xu, B., Timko, M. *Plant Mol. Biol.* **2004**, *55*, 743–761.

147 Sachan, N., Falcone, D.L. *Phytochemistry* **2002**, *61*, 797–805.

148 Rothe, G., Hachiya, A., Yamada, Y., Hashimoto, T., Dräger, B. *J. Exp. Bot.* **2003**, *54*, 2065–2070.

149 Hashimoto, T., Tamaki, K., Suzuki, K., Yamada, Y. *Plant Cell Physiol.* **1998**, *39*, 73–79.

150 Asano, N., Kato, A., Miyauchi, M., Kizu, H., Tomimori, T., Matsui, K., Nash, R.J., Molyneux, R.J. *Eur. J. Biochem.* **1997**, *248*, 296–303.

151 Stenzel, O., Teuber, M., Dräger, B. *Planta* **2006**, *223*, 200–212.

152 Walton, N.J., McLauchlan, W.R. *Phytochemistry* **1990**, *29*, 1455–1457.

153 Hashimoto, T., Mitani, A., Yamada, Y. *Plant Physiol.* **1990**, *93*, 216–221.

154 McLauchlan, W.R., Mckee, R.A., Evans, D.M. *Planta* **1993**, *191*, 440–445.

155 Robins, R.J., Walton, N.J. *The Alkaloids* **1993**, *44*, 115–187.

156 Koelen, K.J., Gross, G.G. *Planta Med.* **1982**, *44*, 227–230.

157 Dräger, B., Hashimoto, T., Yamada, Y. *Agric. Biol. Chem.* **1988**, *52*, 2663–2667.

158 Yamada, Y., Hashimoto, T., Endo, T., Yukimune, Y., Kohno, J., Hamaguchi, N., Drager, B. *Proc. Phytochem. Soc. Europe* **1990**, *30*, 227–242.

159 Hashimoto, T., Nakajima, K., Ongena, G., Yamada, Y. *Plant Physiol.* **1992**, *100*, 836–845.

160 Portsteffen, A., Dräger, B., Nahrstedt, A. *Phytochemistry* **1992**, *31*, 1135–1138.

161 Dräger, B., Schaal, A. *Phytochemistry* **1994**, *35*, 1441–1447.

162 Keiner, R., Kaiser, H., Nakajima, K., Hashimoto, T., Dräger, B. *Plant Mol. Biol.* **2002**, *48*, 299–308.

163 Portsteffen, A., Dräger, B., Nahrstedt, A. *Phytochemistry* **1994**, *37*, 391–400.

164 Nakajima, K., Hashimoto, T., Yamada, Y. *Proc. Natl. Acad. Sci. USA* **1993**, *90*, 9591–9595.

165 Jornvall, H., Persson, B., Krook, M., Atrian, S., Gonzalez, D.R., Jeffery, J., Ghosh, D. *Biochemistry* **1995**, *34*, 6003–6013.

166 Joshi, C.P., Chiang, V.L. *Plant Mol. Biol.* **1998**, *37*, 663–674.

167 Dräger, B. *Phytochemistry* **2006**, *67*, 327–337.

168 Nakajima, K., Yamashita, A., Akama, H., Nakatsu, T., Kato, H., Hashimoto, T., Oda, J., Yamada, Y. *Proc. Natl. Acad. Sci. USA* **1998**, *95*, 4876–4881.

169 Yamashita, A., Nakajima, K., Kato, H., Hashimoto, T., Yamada, Y., Oda, J., *Acta Crystallogr., Section D: Biol. Crystallogr.* **1998**, *D54*, 1405–1407.

170 Yamashita, A., Kato, H., Wakatsuki, S., Tomizaki, T., Nakatsu, T., Nakajima, K., Hashimoto, T., Yamada, Y., Oda, J.i. *Biochemistry* **1999**, *38*, 7630–7637.

171 Yamashita, A., Endo, M., Higashi, T., Nakatsu, T., Yamada, Y., Oda, J., Kato, H. *Biochemistry* **2003**, *42*, 5566–5573.

172 Ansarin, M., Woolley, J.G. *J. Nat. Prod.* **1993**, *56*, 1211–1218.

173 Ansarin, M., Woolley, J.G. *Phytochemistry* **1993**, *32*, 1183–1187.

174 Ansarin, M., Woolley, J.G. *Phytochemistry* **1994**, *35*, 935–939.

175 Robins, R.J., Woollcy, J.G., Ansarin, M., Eagles, J., Goodfellow, B.J. *Planta* **1994**, *194*, 86–94.

176 Nakanishi, F., Sasaki, K., Shimomura, K. *Plant Cell Rep.* **1998**, *18*, 249–251.

177 Kitamura, Y., Nishimi, S., Miura, H., Kinoshita, T. *Phytochemistry* **1993**, *34*, 425–427.

178 Chesters, N.C.J.E., O'Hagan, D., Robins, R.J. *J. Chem. Soc., Perkin Trans. 1: Organic Bio-Org. Chem. (1972-1999)* **1994**, 1159–1162.

179 Chesters, N.C.J.E., O'Hagan, D., Robins, R.J. *J. Chem. Soc., Chem. Commun.* **1995**, 127–128.

180 Duran, P.R., O'Hagan, D., Hamilton-John, T.G., Wong, C.W. _Phytochemistry_ **2000**, _53_, 777–784.

181 Robins, R.J., Chesters, N.C.J.E., O'Hagan, D., Parr, A.J., Walton, N.J., Woolley, J.G. _J. Chem. Soc. Perkin Trans. 1_ **1995**, _4_, 481–485.

182 Chesters, N.C.J.E., Walker, K., O'Hagan, D., Floss, H.G. _J. Am. Chem. Soc._ **1996**, _118_, 925–926.

183 Zabetakis, I., Edwards, R., Hamilton, J.-T.G., O'Hagan, D. _Plant Cell Rep._ **1998**, _18_, 341–345.

184 Ollagnier, S., Kervio, E., Retey, J. _FEBS Lett._ **1998**, _437_, 309–312.

185 Patterson, S., O'Hagan, D. _Phytochemistry_ **2002**, _61_, 323–329.

186 Hashimoto, T., Yamada, Y. _Plant Physiol._ **1986**, _81_, 619–625.

187 Hashimoto, T., Yamada, Y. _Eur. J. Biochem._ **1987**, _164_, 277–285.

188 Hashimoto, T., Kohno, J., Yamada, Y. _Plant Physiol._ **1987**, _84_, 144–147.

189 Hashimoto, T., Matsuda, J., Yamada, Y. _FEBS Lett._ **1993**, _329_, 35–39.

190 Hashimoto, T., Hayashi, A., Amano, Y., Kohno, J., Iwanari, H., Usuda, S., Yamada, Y. _J. Biol. Chem._ **1991**, _266_, 4648–4653.

191 Suzuki, K., Yun, D.J., Chen, X.Y., Yamada, Y., Hashimoto, T. _Plant Mol. Biol._ **1999**, _40_, 141–152.

192 Matsuda, J., Okabe, S., Hashimoto, T., Yamada, Y. _J. Biol. Chem._ **1991**, _266_, 9460–9464.

193 Yun, D.J., Hashimoto, T., Yamada, Y. _Proc. Natl. Acad. Sci. USA_ **1992**, _89_, 11799–11803.

194 Zeef Leo, A.H., Christou, P., Leech, M.J. _Transgenic Res._ **2000**, _9_, 163–168.

195 Sato, F., Hashimoto, T., Hachiya, A., Tamura, K.I., Choi, K.B., Morishige, T., Fujimoto, H., Yamada, Y. _Proc. Natl. Acad. Sci. USA_ **2001**, _98_, 367–372.

196 Moyano, E., Jouhikainen, K., Tammela, P., Palazon, J., Cusido, R.M., Pinol, M.T., Teeri, T.H., Oksman-Caldentey, K.M. _J. Exp. Bot._ **2003**, _54_, 203–211.

197 Saedler, R., Baldwin, I.T. _J. Exp. Bot._ **2004**, _55_, 151–157.

198 Steppuhn, A., Gase, K., Krock, B., Halitschke, R., Baldwin, I.T. _PLoS Biology_ **2004**, _2_, 1074–1080.

199 Voelckel, C., Krugel, T., Gase, K., Heidrich, N., Van Dam, N.M., Winz, R., Baldwin, I.T. _Chemoecology_ **2001**, _11_, 121–126.

200 Chintapakorn, Y., Hamill, J.D. _Plant Mol. Biol._ **2003**, _53_, 87–105.

201 Rocha, P., Stenzel, O., Parr, A., Walton, N., Christou, P., Dräger, B., Leech, M.J. _Plant Sci._ **2002**, _162_, 905–913.

202 Richter, U., Rothe, G., Fabian, A.-K., Rahfeld, B., Dräger, B. _J. Exp. Bot._ **2005**, _56_, 645–652.

203 Hashimoto, T., Yun, D.J., Yamada, Y. _Phytochemistry_ **1993**, _32_, 713–718.

204 Yun, D.J., Hashimoto, T., Yamada, Y. _Biosci. Biotechnol. Biochem._ **1993**, _57_, 502–503.

205 Zhong, J.J. _J. Biosci. Bioeng._ **2002**, _94_, 591–599.

206 Jennewein, S., Croteau, R. _Appl. Microbiol. Biotechnol._ **2001**, _57_, 13–19.

207 Kutchan, T.M. _Alkaloids_ **1998**, _50_, 257–316.

208 Millgate, A.G., Pogson, B.J., Wilson, I.W., Kutchan, T.M., Zenk, M.H., Gerlach, W.L., Fist, A.J., Larkin, P.J. _Nature_ **2004**, _431_, 413–414.

209 Page, J.E. _Trends Biotechnol._ **2005**, _23_, 331–333.

210 Ketchum, R.E.B., Rithner, C.D., Qiu, D., Kim, Y.S., Williams, R.M., Croteau, R.B. _Phytochemistry_ **2003**, _62_, 901–909.

211 Boswell, H.D., Dräger, B., McLauchlan, W.R., Portsteffen, A., Robins, D.J., Robins, R.J., Walton, N.J. _Phytochemistry_ **1999**, _52_, 871–878.

212 Boswell, H.D., Dräger, B., Eagles, J., McClintock, C., Parr, A., Portsteffen, A., Robins, D.J., Robins, R.J., Walton, N.J., Wong, C. _Phytochemistry_ **1999**, _52_, 855–869.

213 Sterner, R., Hoecker, B. _Chemical Rev._ **2005**, _105_, 4038–4055.

214 Oksman-Caldentey, K.M., Inze, D. _Trends Plant Sci._ **2004**, _9_, 433–440.

215 Dixon, R.A. _Curr. Opin. Plant Biol._ **2005**, _8_, 329–336.

216 Dixon, R.A. _Nature_ **2001**, _411_, 843–847.

217 Memelink, J., Kijne, J.W., vom Endt, D. _Phytochemistry_ **2002**, _61_, 107–114.

218 Memelink, J. _Curr. Opin. Plant Biol._ **2005**, _8_, 230–235.

219 Menke-Frank, L.H., Champion, A., Kijne, J.W., Memelink, J. *EMBO J.* **1999**, *18*, 4455–4463.

220 Memelink, J., Verpoorte, R., Kijne, J.W. *Trends Plant Sci.* **2001**, *6*, 212–219.

221 Borisjuk, N., Borisjuk, L., Komarnytsky, S., Timeva, S., Hemleben, V., Gleba, Y., Raskin, I. *Nature Biotechnol.* **2000**, *18*, 1303–1306.

222 Ziegler, J., Diaz-Cha'vez, M.L., Kramell, R., Ammer, C., Kutchan, T.M., *Planta* **2005**, *222*, 458–471.

223 Oksman-Caldentey, K.M., Saito, K. *Curr. Opin. Biotechnol.* **2005**, *16*, 174–179.

224 Haekkinen, S.T., Moyano, E., Cusido, R.M., Palazon, J., Pinol, M.T., Oksman-Caldentey, K.M. *J. Exp. Bot.* **2005**, *56*, 2611–2618.

225 Goossens, A., Hakkinen, S.T., Laakso, I., Oksman-Caldentey, K.M., Inze, D. *Plant Physiol.* **2003**, *131*, 1161–1164.

226 Yazaki, K., Shitan, N., Takamatsu, H., Ueda, K., Sato, F. *J. Exp. Bot.* **2001**, *52*, 877–879.

227 Yazaki, K. *Curr. Opin. Plant Biol.* **2005**, *8*, 301–307.

228 Shitan, N., Bazin, I., Dan, K., Obata, K., Kigawa, K., Ueda, K., Sato, F., Forestier, C., Yazaki, K. *Proc. Natl. Acad. Sci. USA* **2003**, *100*, 751–756.

229 Turner, G.W., Croteau, R. *Plant Physiol.* **2004**, *136*, 4215–4227.

230 Burlat, V., Oudin, A., Courtois, M., Rideau, M., St. Pierre B. *Plant J.* **2004**, *38*, 131–141.

231 Bird, D.A., Franceschi, V.R., Facchini, P.J. *Plant Cell* **2003**, *15*, 2626–2635.

232 Weid, M., Ziegler, J., Kutchan, T.M. *Proc. Natl. Acad. Sci. USA* **2004**, *101*, 13957–13962.

233 Kutchan, T.M. *Curr. Opin. Plant Biol.* **2005**, *8*, 292–300.